Viral Strategies and Cellular Countermeasures that Regulate mRNA Access to the Translation Apparatus

Viral Strategies and Cellular Countermeasures that Regulate mRNA Access to the Translation Apparatus

Guest Editor

Christopher U. T. Hellen

MDPI

Basel • Beijing • Wuhan • Barcelona • Belgrade • Novi Sad • Cluj • Manchester

Guest Editor
Christopher U. T. Hellen
Department of Cell Biology
State University of New York
Downstate Health Sciences
University
Brooklyn, NY
USA

Editorial Office
MDPI AG
Grosspeteranlage 5
4052 Basel, Switzerland

This is a reprint of the Special Issue, published open access by the journal *Viruses* (ISSN 1999-4915), freely accessible at: https://www.mdpi.com/journal/viruses/special_issues/6H1D0JZ83S.

For citation purposes, cite each article independently as indicated on the article page online and as indicated below:

Lastname, A.A.; Lastname, B.B. Article Title. *Journal Name* **Year**, *Volume Number*, Page Range.

ISBN 978-3-7258-4969-7 (Hbk)
ISBN 978-3-7258-4970-3 (PDF)
https://doi.org/10.3390/books978-3-7258-4970-3

Contents

About the Editor

Christopher U. T. Hellen

Christopher U. T. Hellen DPhil (Oxon) is Professor in the Department of Cell Biology, State University of New York Downstate Health Sciences University. His scientific research interests include the molecular mechanisms of translation and associated processes in eukaryotes and their subversion by viruses.

Preface

This Special Issue focuses on the evolving measures and countermeasures employed by viruses and their hosts to either enable or to restrict utilization by viral and cellular mRNAs of the cytoplasmic translation apparatus. These competing processes variously promote the synthesis of viral proteins, thereby allowing viral genome replication and encapsidation to occur, suppress viral translation as part of the cellular innate immune response, or prevent the translation of innate immune effectors, thereby disarming cellular antiviral defenses. These translation-associated processes are consequently important determinants of pathogenesis. Appreciation of their diversity and understanding of their mechanisms are advancing rapidly, and now is therefore an opportune moment to survey the state of art in the field.

This volume brings together reviews and research papers from leading experts, covering topics that include the mode of action of innate immune effectors such as Interferon-induced proteins with tetratricopeptide repeats (IFITs), protein kinase R (PKR), the PKR-like endoplasmic reticulum kinase (PERK),the zinc finger antiviral protein ZAP and shiftless (SFL), as well as mechanisms by which coronaviruses, alphaviruses and poxviruses evade their activities to sustain viral translation. Numerous viruses abrogate 5′-terminal cap-mediated initiation of translation of cellular mRNAs, while themselves exploiting non-canonical initiation mechanisms that are independent of the 5′-terminal cap and the canonical cap-binding initiation factor eIF4E. Several of these mechanisms are described here, including initiation mediated by binding of an alternative cap-binding complex to a hypermethylated 5′-terminal cap, and initiation mediated by structurally and mechanistically diverse picornavirus, calicivirus and dicistrovirus internal ribosomal entry sites (IRESs). Papers in this volume also address evolutionary adaptations of viral mRNAs to maximize exploitation of their coding capacity of viral genomes, by translation of overlapping reading frames by IRES-mediated initiation at alternative initiation codons or by programmed ribosomal frame-shifting. This volume will therefore be of interest to virologists and to scientists researching mechanisms and regulation of gene expression.

Finally, I thank all of the authors for their informative and thought-provoking contributions, Corrina Zhai for her professionalism and enthusiastic assistance, and other members of the editorial and production staff of *Viruses* for their contributions at all stages in the production of this Special Issue.

Christopher U. T. Hellen
Guest Editor

viruses

MDPI

Commentary

Viral Strategies and Cellular Countermeasures That Regulate mRNA Access to the Translation Apparatus

Christopher U. T. Hellen

Department of Cell Biology, SUNY Downstate Health Sciences University, Brooklyn, NY 11203, USA; christopher.hellen@downstate.edu

Abstract: The papers introduced in the Commentary present new insights and review aspects of current knowledge concerning the competition between viruses and their hosts for the cellular translation apparatus. Viruses depend on this apparatus and utilize diverse mechanisms to usurp it for the translation of viral mRNAs and to suppress synthesis of cellular proteins. Virus-induced modification of translation factors, selective abrogation of mRNA binding to ribosomes and degradation of cellular mRNAs all impair elements of the innate immune response, thereby undermining host defenses against infection. Various cellular mechanisms prevent translation of viral mRNAs, by modifying components of the translation apparatus to effect a generalized shut-off of translation or by binding of host proteins to viral mRNAs to induce their degradation or to prevent their engagement with the translation apparatus. Viruses have in turn evolved countermeasures to evade these defenses, for example by encoding proteins that impair the activity of host factors or via alterations in the sequence and structure of viral mRNAs. Such changes enable viral mRNAs to avoid recognition by host factors or to support translation initiation by specialized mechanisms that involve only a subset of the factors that are required by cellular mRNAs.

Keywords: translational control; mRNA translation; IRES; interferon-stimulated gene; Shiftless; ITAF; picornavirus; calicivirus; coronavirus; poxvirus

Viruses depend on the translation apparatus of host cells for the synthesis of viral proteins. Cells employ various countermeasures to restrict the translation of viral mRNAs, and viruses have responded to these cellular defenses by evolving mechanisms to evade them. The outcome of viral infection is determined by these competing processes. In this Special Issue, leading experts present and review data regarding new developments in the understanding of mechanisms by which viruses exploit the cellular translation apparatus and counter or evade cellular mechanisms that restrict viruses.

1. Mechanism of mRNA Translation in Eukaryotes

Translation consists of four stages: initiation, elongation, termination, and ribosome recycling. During the canonical initiation process [1], an mRNA with a 5′-terminal m^7G cap and a 3′-terminal poly(A) tail is selected by eukaryotic initiation factor (eIF) 4F, a heterotrimer that consists of eIF4E (the cap-binding subunit), eIF4A (an RNA helicase), and eIF4G (which binds to eIF4E, eIF4A, eIF3, and the poly(A)-binding protein (PABP)). Initiator tRNA (Met-$tRNA_i^{Met}$) forms a ternary complex with eIF2•GTP, and it binds with eIF1, eIF1A, and eIF3 to a ribosomal 40S subunit to form a 43S preinitiation complex. This complex is recruited to the capped 5′-end of the mRNA by eIF4F and scans downstream, and when it encounters an initiation codon, eIF2-bound GTP is hydrolyzed and initiator tRNA is released into the ribosomal peptidyl (P) site to form a 48S complex. eIF5B promotes the joining of a 60S ribosomal subunit to this complex, yielding an 80S ribosome that can begin elongation. eIF2B recycles eIF2•GDP to eIF2•GTP for the next round of initiation. Elongation involves cycles of delivery of cognate aminoacyl-tRNA to the ribosomal aminoacyl (A) site by eukaryotic elongation factor (eEF) 1A, peptide bond formation, and translocation promoted by eEF2 of the base-paired mRNA and peptidyl-tRNA from A to P sites. The entry of a stop codon into the A site signals the termination of translation, which involves the binding of eRF1/eRF3•GTP to the A site, the recognition of the stop codon by eRF1, GTP hydrolysis, the release of eRF3•GDP, the cleavage of peptidyl-tRNA, and the release of the completed polypeptide [2]. The resulting post-termination complex, still bound to eRF1, is disassembled by ABCE1, freeing the 60S subunit. mRNA and deacylated tRNA are released from the 40S subunit by redundant pathways involving eIF1, eIF1A, and eIF3 or specific recycling factors (MCTS1/DENR or eIF2D). The 40S and 60S subunits can then participate in the next translation cycle.

2. Regulation of Translation by the Innate Immune Response

Host cells employ Toll-like, retinoic acid-inducible gene-I (RIG-I)-like, and other pattern-recognition receptors to detect viral nucleic acids and proteins, and to activate immune signaling pathways to induce the expression of interferon (IFN)-stimulated genes (ISGs). Specific elements of the innate immune response target different stages in the translation process to abrogate the translation of viral mRNAs, and in some instances, they also impair the translation of cellular mRNAs. Viruses have in turn developed diverse strategies to counteract or evade the activity of innate immune effectors.

Smart et al. [3] reviewed the activities of ISGs that target mRNAs, leading to the virus-specific abrogation of translation, or that target components of the translation apparatus, resulting in either the repression of viral translation or the activation of the translation of IFN-inducible antiviral proteins. Interferons induce the phosphorylation and consequent deactivation of the eIF4E-binding protein (4E-BP) by activating signaling pathways downstream of mammalian target of rapamycin (mTOR), leading to the release of eIF4E from 4E-BP, thereby promoting the translation of antiviral proteins. ISGs inhibit major stages in initiation, including the recruitment of mRNA and initiator tRNA to ribosomes. The interferon-induced proteins with tetratricopeptide repeats (IFITs) are abundantly expressed early in the antiviral response, and they have been reported to bind to eIF3 but also directly to non-self RNAs to inhibit their translation [4]. IFIT1 binds to the 5′-terminal region of mRNAs lacking methylation on the first cap-proximal nucleotide ('cap0'), and its association with IFIT2-IFIT3 yields a high-affinity heterotrimeric complex that competitively inhibits the binding of eIF4F. Protein Kinase R (PKR) phosphorylates eIF2α, causing it to sequester eIF2B, thereby limiting the recycling of eIF2•GDP to eIF2•GTP and globally repressing translation. PKR is one of several components of the integrated stress response that converges on the phosphorylation of eIF2α to repress translation. The synthesis of viral proteins in infected cells commonly overwhelms the endoplasmic reticulum (ER), activating the PKR-like ER kinase (PERK), which also phosphorylates eIF2α and downregulates viral translation. The phosphorylation of eIF2α is reversed by protein phosphatase 1, which is directed to p-eIF2α by the Constitutive Repressor of eIF2α Phosphorylation (CReP) and by the stress-inducible Growth Arrest and DNA Damage-inducible 34 (GADD34). Zinc finger antiviral protein (ZAP) binds to viral mRNAs, with a preference for CpG-rich regions, and to eIF4A, impairing its interaction with eIF4G, preventing the formation of eIF4F, and impairing translation in a virus-specific manner. ZAP also promotes the degradation of some viral mRNAs and binds to coronavirus frameshift elements, impairing the −1 programmed ribosomal frameshifting that is required for the expression of replicase proteins. Shiftless (SHFL) impairs frameshifting on human immunodeficiency virus 1 (HIV-1) mRNA, and has other less-well-characterized inhibitory activities. Other ISGs directly impair the elongation process. Thus, Schlafen 11 is a tRNA endonuclease that cleaves leucine, serine, and other class II tRNAs, counteracting changes in the cellular tRNA repertoire induced by viral infection, thereby repressing the translation of specific viral mRNAs in a codon-usage-dependent manner. Other ISGs target viral mRNAs for degradation or modification, either directly or indirectly. These include ISG20, an RNAse, and adenosine deaminase acting on RNA 1 (ADAR1), which deamidates adenosines to inosines, leading to the miscoding of modified viral mRNAs. The 2′-5′ oligoadenylate synthase activates RNAse L, leading to the non-specific degradation of viral and cellular mRNAs.

3. Viral Mechanisms for Evasion of ISG-Mediated Repression of Translation

Different virus families have evolved strategies to evade aspects of ISG-mediated repression of translation by inhibiting the activity of individual ISGs or by exploiting alternative mechanisms for translation initiation. These strategies may involve the expression of viral proteins that inhibit or sequester ISGs or the acquisition of structural elements in viral mRNAs that enable them to exploit non-canonical initiation mechanisms. In either case, the lack of compatibility in different cell types may restrict viral growth, thus influencing tissue specificity and host range.

Alphaviruses such as Sindbis virus (SV) and Western equine encephalitis virus (WEEV) infect vertebrate hosts and arthropod (mosquito) vectors, causing moderate-to-high shut-off of host translation in mammalian cells but not in mosquitoes. Ventoso et al. [5] summarized the IFN-induced effectors that block host and viral translation in mammalian alphavirus-infected cells, principally PKR, ZAP, IFIT1, and IFIT3, and describe adaptive changes in alphavirus mRNAs that lead to the resistance to IFIT1 or ZAP. These changes include the presence of a stable 5′-terminal hairpin that impairs the binding of IFIT1, and the suppression of CpG content. Notably, such suppression is not evident in alphaviruses that infect only insects, which do not encode ZAP orthologs. Other structural features in alphaviruses confer resistance to the PKR-mediated phosphorylation of eIF2α by enabling an alternative, eIF2-independent mechanism of initiation.

PKR is an important factor in restricting viral infection [6], and its evasion is an important contributor to the tropism of poxviruses. They encode two proteins that suppress the activation of PKR: the vaccinia virus E3L protein and its orthologs bind dsRNA and prevent PKR dimerization and its consequent activation, whereas the K3L protein and its orthologs act as pseudo-substrates for PKR's eIF2α-binding site. Megawati et al. [7] described the results of experiments performed to determine the ability of Yaba monkey tumor virus and Tanapox virus K3L orthologs to inhibit PKR from different primate species, and to restore the replication of a vaccinia virus strain lacking both E3L and K3L in different primate-derived cell lines. Their results showed that K3L inhibited PKR in primate cells in a species-dependent manner, and that K3L moieties from different viruses inhibit PKR in specific cell lines with varying efficiency. These findings provide insights into the influence of viral PKR antagonists as determinants of poxvirus host range.

Recent studies have established that coronaviruses influence the integrated stress response in different ways, ranging from the suppression of eIF2α phosphorylation by Middle East respiratory syndrome (MERS) coronavirus to its cell type-dependent activation by severe acute respiratory syndrome coronavirus 2 (SARS-CoV-2) [8]. Sodium arsenite is a strong inducer of PERK-mediated eIF2α phosphorylation, but Dolliver et al. [9] reported that this activity and the activation of PKR by polyinosinic-polycytidylic acid (poly(I:C), a dsRNA mimic) are both inhibited by human coronavirus OC43 (OC43) infection. Although GADD34 expression was induced by OC43, this was not responsible for the suppression of eIF2α phosphorylation in cells infected by this virus. Therefore, the mechanism by which OC43 influences this process remains to be determined.

Shiftless (SHFL) was identified as an inhibitor of -1 programmed ribosomal frameshifting (PRF) during the synthesis of the HIV-1 Gag-Pol precursor [10], but this activity is not essential for its role in inhibiting HIV infectivity [11]. The implication that SHFL may inhibit more than one facet of the translation process is consistent with the observation of its antiviral activity against viruses that are not dependent on frameshifting. Moreover, although SHFL expression is IFN-induced, it is expressed at a low level in a constitutive manner, suggesting that it may have cellular functions in addition to a role as an innate immune effector. Kelly and Dinman [12] reported the results of their investigation of the potential constitutive roles of SHFL that may have been co-opted to promote antiviral immunity. They reported that SHFL influences +1 PRF as well as -1 PRF, confirmed that it suppresses stop codon readthrough and that it promotes reporter gene expression, potentially by suppressing spontaneous frameshifting. Overexpression and silencing experiments indicated that SHFL influenced the stability of mRNA, and gene silencing experiments linked SHFL to the ribosome-associated quality control (RQC) process.

4. Structure of Viral mRNAs

The structures of viral mRNAs and the organization of their coding regions are influenced (a) by genome size (limited by the encapsidation process, and for many RNA viruses, by the potentially lethal accumulation of mutations in larger genomes due to the low fidelity of replication and the lack of proofreading), (b) by the need to accommodate *cis*-acting elements required for replication and for initiation of translation in a single molecule, and (c) by evolutionary changes in the sequence, structure, and post-transcriptional modifications of mRNAs that enables them to evade the innate immune restriction of viral translation. The constraint on size and characteristics of the canonical translation initiation process in eukaryotes favor genome compression for most RNA viruses.

Khan and Fox [13] discussed a selection of the structural and functional classes of RNA elements present in viral mRNAs. Several of these elements enable viral genomes to maximize the exploitation of the limited coding capacity of viral genomes, including alternative initiation codons and frameshifting signals for the translation of overlapping reading frames and internal ribosomal entry sites (IRESs), termination–reinitiation signals, and the synthesis of subgenomic mRNAs to enable the translation of 3′-terminal open reading frames. TURBS (Termination Upstream Ribosome Binding Site) elements, which occur in mRNAs of influenza B virus (*Orthomyxoviridae*), members of the genus *Hapavirus* of *Rhabdoviridae*, and members of several genera of *Caliciviridae* [14,15], are dedicated termination–reinitiation elements that subvert ribosome recycling by tethering post-termination 40S ribosomal subunits to mRNA to promote the reinitiation of translation on bicistronic viral mRNAs in the vicinity of the open reading frame 1 (ORF1) stop codon.

Several viral mRNAs contain elements that promote non-canonical modes of initiation, independent of eIF2 and/or eIF4F, to evade the influence of innate immune effectors that limit canonical cap-dependent initiation. These elements include IRESs in the 5′-untranslated region (5′ UTR) of positive-sense RNA genomes and in the intercistronic regions of dicistronic RNA genomes (discussed in more detail below) and upstream ORFs that regulate the initiation of translation of downstream ORFs [16].

3′-Cap Independent Translation Enhancer (3′-CITE) elements occur in the 3′-terminal regions of the genomes of members of several genera of *Tombusviridae* [17]. They are structurally diverse and function by distinct mechanisms that include binding directly to subunits of eIF4F or to the 80S ribosome or its 60S subunit, and then engaging with sequences proximal to the 5′-terminus via long-range base pairing, thereby directing the translation apparatus to the 5′-end of the mRNA in a cap-independent manner. Many of the viral translational control elements discussed by Khan and Fox [13] are not known to have cellular homologues, but repressive elements related to the Gamma-interferon-Activated Inhibitor of Translation (GAIT) element that occurs in several inflammation-related human mRNAs have been recently identified in respiratory syncytial virus and two coronaviruses. Members of the subgenus *Sarbecovirus* of genus *Betacoronavirus* also contain a GAIT-related Sarbecoviral Pan-End Activating RNA (SPEAR) element that is required for the inducible translation of coronavirus subgenomic mRNAs rather than for suppressing translation.

5. IRES Structure and the Mechanism of IRES-Mediated Initiation of Translation

Viral IRESs, which are structurally and mechanistically diverse, have been assigned to six major groups [18]. This classification excludes several less-well-characterized groups of IRESs, e.g., those in retrovirus, iflavirus, and some pegivirus mRNAs; moreover, additional IRES classes likely remain to be discovered because several validated IRESs have sequences and structures that are unrelated to those of known classes. The single-stranded positive-sense RNA genomes of dicistroviruses contain two principal open reading frames, ORF1 and ORF2, both translated following the initiation mediated by an IRES. The intergenic region (IGR) IRES promotes the translation of ORF2 by completely bypassing the conventional initiation process, enabling translation to proceed even when eIF2 and eIF4F are inactivated. Some dicistrovirus genomes diverge from this paradigm. For example, the Bemisia-associated dicistrovirus 2 (BaDV-2) IGR IRES has an atypical structure, and the upstream ORF1 is split into the overlapping ORF1a and ORF1b, so that the translation of the complete non-structural polyprotein would require a −1 frameshift. Chen et al. [19] confirmed that such a frameshift does occur and identified sequence determinants of this process. They also determined that the IGR IRES is only 140 nt long and differs from conventional IGR IRESs that are 190–200 nt long in that it does not bind in a direct, factor-independent manner to ribosomes or to ribosomal subunits, initiates translation at an AUG codon, and requires at least a subset of initiation factors. These findings emphasize both the diversity of mechanisms used by IGR IRESs and the diversity of IRES structures.

Type 1 IRESs occur in the genomes of enteroviruses such as poliovirus and enterovirus A71. They are over 450 nt long and are located downstream of a 5′-terminal cloverleaf that is a *cis*-acting replication element required for negative-strand synthesis. Infection by enteroviruses induces a profound shut-off of cellular translation caused by the cleavage of eIF4G by a viral protease into an N-terminal fragment that binds eIF4E and PABP and a C-terminal fragment that binds eIF4A and eIF3, and by the dephosphorylation and consequent activation of the translation repressor 4E-BP. Type 1 IRESs bind directly to the C-terminal cleavage product of eIF4G, which is thought to recruit other factors (e.g., eIF4A) and the ribosomal 43S preinitiation complex. Ribosomal entry may be followed by initiation at the 3′-border of the IRES (yielding an accessory protein) or by scanning downstream to the initiation codon for the viral polyprotein. A distinctive characteristic of type 1 IRESs is their dependence on IRES *trans*-acting factors (ITAFs), which are cellular RNA-binding proteins that activate IRES function, likely by inducing conformational changes in the IRES, but that are not involved in cellular translation. Differences in the level of their expression may be sufficient to restrict viral translation in some cell types and may therefore influence tissue specificity and host range. Abedeera et al. [20] surveyed cellular proteins that bind to these IRESs and influence their function, including a panel of 'negative ITAFs' that bind to these IRESs and inhibit their activity, likely by impairing their functional interactions with positive ITAFs or with canonical components of the translation apparatus. Significantly, ITAFs are commonly nuclear proteins that are redistributed to the cytoplasm as a consequence of infection, which is promoted by the cleavage of components of the nuclear pore complex by viral proteases. Therefore, viral infection shapes the intracellular environment to favor viral translation.

Type 2 IRESs are ~450 nt long and were initially identified in picornaviruses of the genera *Cardiovirus* (e.g., encephalomyocarditis virus) and *Aphthovirus* (e.g., foot and mouth disease virus (FMDV)). Type 2 IRESs were recently also identified in the 5′UTRs of some caliciviruses, having likely been acquired from picornaviruses by recombination. Arhab et al. [21] characterized the mechanism of initiation on representative type 2 IRESs from grey teal calicivirus (GTCV) and *Caliciviridae* sp. isolate yc-13 (also known as red-crowned crane calicivirus (RaCV1)) by the in vitro reconstitution of the initiation process. The canonical components of the translation apparatus required for initiation on these IRESs are analogous to those required for initiation on picornavirus type 2 IRESs, but their ITAF requirements showed interesting differences: the activity of both IRESs was enhanced by Ebp1 (also known as $ITAF_{45}$, and initially identified as an FMDV-specific ITAF), whereas the GTCV IRES was further stimulated by the pyrimidine tract-binding protein (PTB), and the initiation mediated by the RaCV1 IRES was enhanced by Ebp1 and poly(rC)-binding protein 2 (PCBP2), previously thought to be specific to type 1 IRESs. These observations concerning overlapping ITAF requirements suggest that the mechanisms of initiation on type 1 and type 2 IRESs may be even more closely related than previously assumed. The analysis of calicivirus genomes revealed that many of those that contain type 2 IRESs also contained a 197–317-codon-long open reading frame (ORF1*) that overlapped with ORF1, the principal open reading frame. The use of overlapping reading frames is therefore another strategy employed by caliciviruses to maximize the exploitation of the coding capacity of their genomes, in addition to the synthesis of a polyprotein that is proteolytically processed to yield mature non-structural proteins, the synthesis of subgenomic mRNAs encoding capsid proteins, and termination–reinitiation to enable the synthesis of a minor capsid protein.

6. eIF4E-Independent Initiation Mediated by an Alternative Cap-Binding Complex

The primary human immunodeficiency virus 1 (HIV-1) transcript is capped, polyadenylated, and extensively spliced to yield over fifty mRNAs that encode fifteen proteins [22]. Infection leads to the transient activation of PKR followed by the repression of its activity, and the inhibition of the canonical initiation process as a result of HIV protease-mediated cleavage of eIF4G1, eIF3d and PABP, and the promotion of 4E-BP dephosphorylation by the HIV-1 Vpr protein, leading to the sequestration of eIF4E. Boris-Lawrie et al. [23] reviewed the alternate eIF4E-independent mechanisms of translation initiation employed by HIV-1, including IRES-mediated initiation, but they focused on the role of an alternative cap-binding complex that binds preferentially to a hypermethylated cap structure. The guanosine of the cap structure of unspliced HIV-1 transcripts is hypermethylated to trimethylguanosine (TMG) in a reaction that is promoted by RNA helicase A (also known as DHX9), and leads to the stable binding of an alternative cap-binding complex that consists of CBP80 and NCBP3 subunits. This complex recruits the 43S complex possibly via interactions with eIF3, but many details of this mechanism remain to be elucidated.

7. Conclusions

Many of the mechanisms by which host cells restrict the translation of viral mRNAs and by which viruses evade these restrictions and subvert the translation apparatus have been identified, but several of them remain incompletely characterized. The papers in this Special Issue contribute new insights into and review progress in understanding these processes. The various reports describe the mechanisms of action of different restriction factors, the mechanisms of viral translation, the role of viral proteins in counteracting the restriction of viral translation by cellular factors, and the role of viral RNA sequences and structures in evading innate immune antiviral responses and in enabling viral mRNAs to usurp the translation apparatus. These papers also highlight unresolved questions, emphasizing that much remains to be learnt on this topic and that further relevant insights can be anticipated.

Funding: This research was funded by National Institutes of Health (NIH) awards 5 R01 GM097014-09 and 1 R21 AI188505-01 to C.H. and by award 1 R21 AI166944-02 to T.V. Pestova and C.H.

Conflicts of Interest: The author declares no conflicts of interest.

References

1. Jackson, R.J.; Hellen, C.U.T.; Pestova, T.V. The mechanism of eukaryotic translation initiation and principles of its regulation. *Nat. Rev. Mol. Cell. Biol.* **2010**, *11*, 113–127. [CrossRef] [PubMed]
2. Hellen, C.U.T. Translation termination and ribosome recycling in eukaryotes. *Cold Spring Harb. Perspect. Biol.* **2018**, *10*, a032656. [CrossRef] [PubMed]
3. Smart, A.; Gilmer, O.; Caliskan, N. Translation inhibition mediated by interferon-stimulated genes during viral infections. *Viruses* **2024**, *16*, 1097. [CrossRef] [PubMed]
4. Mears, H.V.; Sweeney, T.R. Better together: The role of IFIT protein-protein interactions in the antiviral response. *J. Gen. Virol.* **2018**, *99*, 1463–1477. [CrossRef]
5. Ventoso, I.; Berlanga, J.J.; Toribio, R.; Díaz-López, I. Translational control of alphavirus-host interactions: Implications in viral evolution, tropism and antiviral response. *Viruses* **2024**, *16*, 205. [CrossRef]
6. Cesaro, T.; Michiels, T. Inhibition of PKR by viruses. *Front. Microbiol.* **2021**, *12*, 757238. [CrossRef]
7. Megawati, D.; Stroup, J.N.; Park, C.; Clarkson, T.; Tazi, L.; Brennan, G.; Rothenburg, S. Tanapox virus and Yaba monkey tumor virus K3 orthologs inhibit primate protein kinase R in a species-specific fashion. *Viruses* **2024**, *16*, 1095. [CrossRef]
8. Li, Y.; Renner, D.M.; Comar, C.E.; Whelan, J.N.; Reyes, H.M.; Cardenas-Diaz, F.L.; Truitt, R.; Tan, L.H.; Dong, B.; Alysandratos, K.D.; et al. SARS-CoV-2 induces double-stranded RNA-mediated innate immune responses in respiratory epithelial-derived cells and cardiomyocytes. *Proc. Natl. Acad. Sci. USA* **2021**, *118*, e2022643118. [CrossRef]
9. Dolliver, S.M.; Galbraith, C.; Khaperskyy, D.A. Human betacoronavirus OC43 interferes with the integrated stress response pathway in infected cells. *Viruses* **2024**, *16*, 212. [CrossRef]
10. Wang, X.; Xuan, Y.; Han, Y.; Ding, X.; Ye, K.; Yang, F.; Gao, P.; Goff, S.P.; Gao, G. Regulation of HIV-1 Gag-Pol expression by Shiftless, an inhibitor of programmed-1 ribosomal frameshifting. *Cell* **2019**, *176*, 625–635.e14. [CrossRef]
11. Jäger, N.; Ayyub, S.A.; Peske, F.; Liedtke, D.; Bohne, J.; Hoffmann, M.; Rodnina, M.V.; Pöhlmann, S. The inhibition of Gag-Pol expression by the restriction factor Shiftless is dispensable for the restriction of HIV-1 infection. *Viruses* **2024**, *16*, 583. [CrossRef] [PubMed]
12. Kelly, J.A.; Dinman, J.D. Shiftless Is a novel member of the ribosome stress surveillance machinery that has evolved to play a role in innate immunity and cancer surveillance. *Viruses* **2023**, *15*, 2296. [CrossRef] [PubMed]
13. Khan, D.; Fox, P.L. Host-like RNA elements regulate virus translation. *Viruses* **2024**, *16*, 468. [CrossRef]
14. Zinoviev, A.; Hellen, C.U.T.; Pestova, T.V. Multiple mechanisms of reinitiation on bicistronic calicivirus mRNAs. *Mol. Cell* **2015**, *57*, 1059–1073. [CrossRef]
15. Sherlock, M.E.; Langeberg, C.J.; Segar, K.E.; Kieft, J.S. A conserved class of viral RNA structures regulates translation reinitiation through dynamic ribosome interactions. *Cell Rep.* **2025**, *44*, 115236. [CrossRef]

16. Jan, E.; Mohr, I.; Walsh, D. A cap-to-tail guide to mRNA translation strategies in virus-infected cells. *Annu. Rev. Virol.* **2016**, *3*, 283–307. [CrossRef]
17. Truniger, V.; Miras, M.; Aranda, M.A. Structural and functional diversity of plant virus 3′-cap-independent translation enhancers (3′-CITEs). *Front. Plant Sci.* **2017**, *8*, 2047. [CrossRef]
18. Arhab, Y.; Bulakhov, A.G.; Pestova, T.V.; Hellen, C.U.T. Dissemination of internal ribosomal entry sites (IRES) between viruses by horizontal gene transfer. *Viruses* **2020**, *12*, 612. [CrossRef] [PubMed]
19. Chen, Y.; Chapagain, S.; Chien, J.; Pereira, H.S.; Patel, T.R.; Inoue-Nagata, A.K.; Jan, E. Factor-dependent internal ribosome entry site and -1 programmed frameshifting signal in the Bemisia-associated dicistrovirus 2. *Viruses* **2024**, *16*, 695. [CrossRef]
20. Abedeera, S.M.; Davila-Calderon, J.; Haddad, C.; Henry, B.; King, J.; Penumutchu, S.; Tolbert, B.S. The repurposing of cellular proteins during Enterovirus A71 infection. *Viruses* **2023**, *16*, 75. [CrossRef]
21. Arhab, Y.; Pestova, T.V.; Hellen, C.U.T. Translation of overlapping open reading frames promoted by type 2 IRESs in avian calicivirus genomes. *Viruses* **2024**, *16*, 1413. [CrossRef] [PubMed]
22. de Breyne, S.; Ohlmann, T. Focus on translation initiation of the HIV-1 mRNAs. *Int. J. Mol. Sci.* **2018**, *20*, 101. [CrossRef] [PubMed]
23. Boris-Lawrie, K.; Liebau, J.; Hayir, A.; Heng, X. Emerging roles of m7G-Cap hypermethylation and nuclear cap-binding proteins in bypassing suppression of eIF4E-dependent translation. *Viruses* **2025**, *17*, 372. [CrossRef] [PubMed]

viruses

MDPI

Review

Translation Inhibition Mediated by Interferon-Stimulated Genes during Viral Infections

Alexandria Smart [1,†], Orian Gilmer [1,†] and Neva Caliskan [1,2,*]

1 Helmholtz Institute for RNA-Based Infection Research, Helmholtz Centre for Infection Research (HIRI-HZI), Josef-Schneider-Strasse 2, 97080 Würzburg, Germany; alexandria.smart@helmholtz-hiri.de (A.S.); orian.gilmer@helmholtz-hiri.de (O.G.)

2 Regensburg Center for Biochemistry (RCB), University of Regensburg, 93053 Regensburg, Germany

* Correspondence: neva.caliskan@ur.de

† These authors contributed equally to this work.

Abstract: Viruses often pose a significant threat to the host through the exploitation of cellular machineries for their own benefit. In the context of immune responses, myriad host factors are deployed to target viral RNAs and inhibit viral protein translation, ultimately hampering viral replication. Understanding how "non-self" RNAs interact with the host translation machinery and trigger immune responses would help in the development of treatment strategies for viral infections. In this review, we explore how interferon-stimulated gene products interact with viral RNA and the translation machinery in order to induce either global or targeted translation inhibition.

Keywords: interferon-stimulated genes; viral RNA; translation; antiviral response; recoding

1. Introduction

Viruses are obligatory intracellular parasites that manipulate and hijack various cellular machineries for their own benefit, frequently posing a significant threat to their host [1]. From entry to egress, they evolved mechanisms to exploit host cellular machinery at every stage of their life cycle. Therefore, virus–host interactions are paramount to the progress of viral replication cycles, especially those involved in the translation of viral RNA (vRNA) into functional viral proteins [2]. Due to the reliance of viruses on host translational mechanisms, host cells can impede viral infection through the activation of the innate immune response by the recognition of pathogen-associated molecular patterns (PAMPs). PAMPs, which usually originate from viral nucleic acids [3] or proteins [4,5], are sensed by pathogen recognition receptors (PRRs). Upon recognition, PPRs induce a signaling cascade, ultimately leading to the production of various cytokines to orchestrate an antiviral immune response (Figure 1).

Among the cytokines secreted upon infection, those from the interferon (IFN) family have been shown to modulate numerous host processes and are well known for their ability to provide a strong first line of defense against invading pathogens [6]. IFNs induce an antiviral state in infected and neighboring cells through autocrine and paracrine signaling pathways [7,8]. One way this occurs is by IFNs binding to cell surface receptors, which initiates a signaling cascade involving the downstream activation of the JAK-STAT pathway. This ultimately leads to the transcriptional upregulation of IFN-stimulated genes (ISGs) [9,10]. However, viruses can also repress the activation of IFN pathways in host cells, thus increasing their chances of evading an antiviral response [11–15].

Upon infection, IFN-stimulated host restriction factors that slow viral biogenesis can target every phase of the viral replication process [9]. Driven by genetic conflicts with frequently evolving viral counterparts, restriction factors also tend to exhibit signs of rapid evolution [16]. This dynamic is often analogized to the "Red Queen's interaction", due to the imperative nature of continual adaptation for both parties. Thus, the interactions of viruses and host factors exist in a perpetual state of adversarial co-evolution [17]. In this review, first, we will examine the direct mechanisms of vRNA translation inhibition mediated by ISG cascades. Second, we will delve into the indirect regulation resulting from the degradation of vRNA (Table 1, Figure 2).

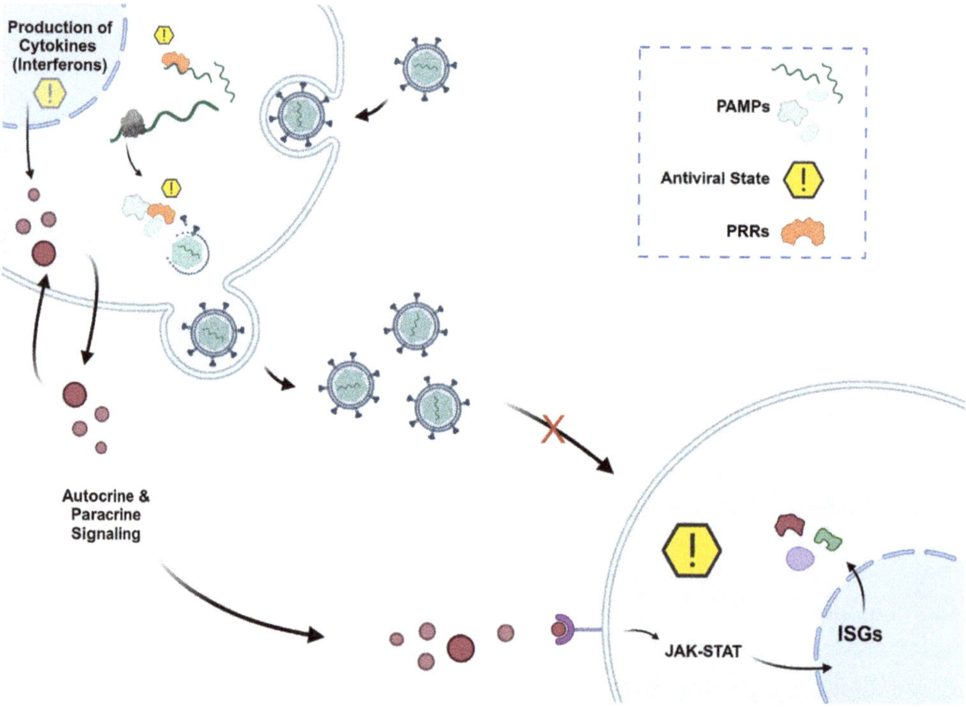

Figure 1. Interferon response triggered by viral infection. Upon viral infection, cells detect specific pathogen-associated molecular patterns (PAMPs) via pattern recognition receptors (PRRs), triggering the initiation of an antiviral response in the infected cell. This process begins with the production of cytokines, notably interferons (IFNs). Released IFNs not only increase the antiviral response in the infected cells but also induce one in neighboring cells through both autocrine and paracrine signaling mechanisms, respectively. IFNs bind to their respective cell surface receptors, initiating signaling cascades primarily via the JAK-STAT pathway or its variants. This activation leads to the engagement of transcription factors capable of promoting the expression of genes governed by interferon-stimulated response elements (ISREs) or gamma-activated sequences. Consequently, ISGs are synthesized, actively contributing to the establishment of the antiviral state within the cell, even preemptively prior to viral infiltration.

Table 1. Host restriction factors and their mechanisms.

ISG	Target (RNA/TM)	Specific Target	Mechanism of Action	References
Protein Kinase R	Translation machinery	eIF2α	Phosphorylation	[18,19]
2′-5′-Oligoadenylate Synthetase (OAS)	RNA (both viral and host)	Activation of RNase L	RNA degradation (including rRNA—translation inhibition)	[20,21]

Table 1. *Cont.*

ISG	Target (RNA/TM)	Specific Target	Mechanism of Action	References
Zinc finger antiviral protein (ZAP) (PARP13, ZC3HAV1)	Both	eIF4A, mRNA	Initiation inhibition, mRNA decay, −1PRF modulator	[22–30]
TRIM25	Both	vRNA and ZAP	Ubiquitination of eIFs ZAP	[28,30–33]
Shiftless Antiviral Inhibitor of Ribosomal Frameshifting (SHFL) (RyDEN, IRAV, C19ORF66)	Both	mRNA, ribosome	−1PRF modulator, antiviral factor	[28,34–38]
IFN-induced protein with tetratricopeptide repeats 1 (IFIT1) (P56 or ISG56) IFIT5 is paralog	Translation machinery	eIF3e, ternary complex, eEF1A, PKR	Initiation inhibition of vRNA with non-2′O-methylated 5′ cap	[39–47]
IFN-induced protein with tetratricopeptide repeats 2 (IFIT2) (P54 or ISG54)	Translation machinery	eIF3c, eIF3e IFIT1	Destabilizing ternary complex, inhibit 48S pre-initiation complex	[47–49]
IFIT3	Both	PPP RNA, IFIT1	Binds PPP-RNA, associates with IFIT1	[43,44,47]
ISG20	RNA (Upregulation of IFIT1)	vRNA, IFIT1	3′-5′ exonuclease, distinguishes between self and non-self RNA, upregulates other ISGs, IFIT1	[50–55]
Eukaryotic Initiation Factor 4E-Binding Protein (4E-BP)	Translation machinery	eIF4E	Initiation inhibition	[56–62]
Schlafen 11 (SLFN11)	Translation machinery	tRNA	tRNA cleavage by codon-bias discrimination	[63–65]
Indoleamine-2,3-dioxygenase (IDO)	Translation machinery	Trp metabolism	Amino acid starvation	[66]
PARP12	Both	Initiation factors ADP-ribosylation of Ago2	Inhibition of viral translation, immunomodulation	[28,67–69]
Adenosine Deaminase Acting on RNA 1 (ADAR1)	RNA	miRNA, vRNA	dsRNA recognition, vRNA modification, modulation through miRNA pathway	[70–78]
MOV10	RNA		Modulation through miRNA pathway, RNP granules relocalization RNA structure unfolding	[79–87]

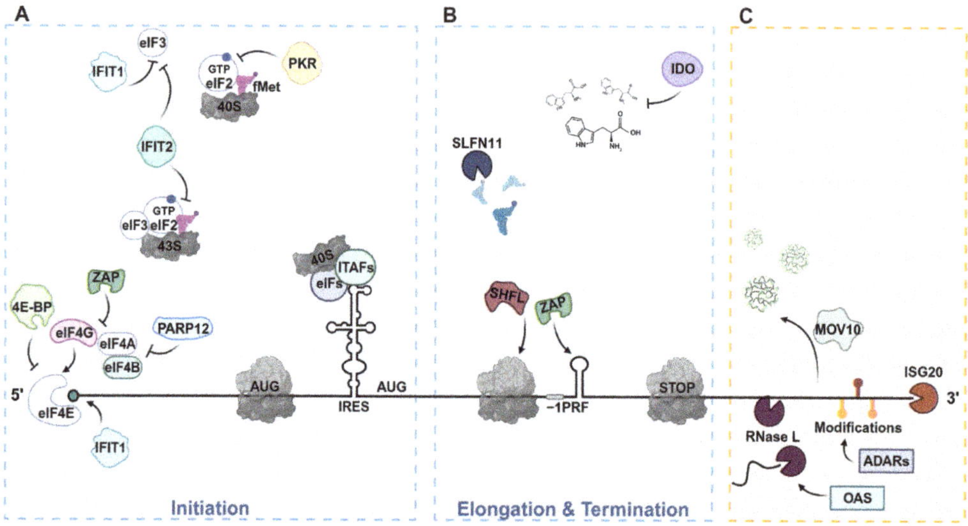

Figure 2. The myriad host interferon responses. Host restriction factors can either inhibit the translation of vRNA or remove vRNA from the host translation machinery. Due to the complex interplay and rapid evolution of virus–host mechanisms, there are multiple ways to restrict viral protein translation during ribosomal initiation (**A**), elongation, and termination (**B**) as well as degradation or localization of vRNAs with exonucleases (**C**). Specific RNA elements (e.g., IRES or −1PRF signal) act as sites for protein binding to regulate individual steps of translation.

2. Overview of Host Translation

Translation is a fundamental biological process conserved across all kingdoms of life. In eukaryotic cells, the translation process involves initiation, elongation, termination, and recycling with initiation occurring through either 5′ cap-dependent or cap-independent mechanisms (Figure 2A) [88–91]. In cap-dependent processes, the eukaryotic initiation factor (eIF) 4F complex—comprising eIF4A (an RNA helicase), eIF4E (a cap-binding protein), and eIF4G (a scaffolding protein)—recognizes and binds to the 5′ m7G cap of the mRNA [92]. Simultaneously, polyA-binding proteins (PABP) interact with both the polyA tail at the 3′ end and the eIF4G protein, bringing the two ends of the mRNA closer together. This interaction enhances the efficiency of translation and increases the affinity of the eIF4F complex for the 5′ cap [93]. Translation initiation begins with the recruitment of the 43S pre-initiation complex (43S), which includes eIF1, eIF1A, eIF3, eIF5, and a ternary complex consisting of eIF2, guanosine 5′-triphosphate, and methionine initiator transfer RNA (Met-tRNA$_i^{Met}$), to the 40S small ribosomal subunit via the eIF4F complex, thus forming the 48S initiation complex (48S) (Figure 2A) [94,95]. Canonically, the 48S complex scans the 5′ end of the mRNA, unwinding secondary and tertiary structures aided by the eIF4A RNA helicase. Upon recognition of an initiation codon by the Met-tRNA$_i^{Met}$ in the ribosomal P site, the 60S ribosomal subunit is recruited, and eIFs are released.

Though initiation in eukaryotes generally occurs in this way, for many viral and some cellular mRNAs, a cap-independent mechanism occurs through an mRNA structure known as the internal ribosome entry site (IRES) (Figure 2A) [96,97]. In high-stress conditions, endogenous IRES initiation can be more prevalent, indicating that IRES-mediated initiation within a cell is likely compensating for cap-dependent initiation especially when cap-dependent initiation is actively being inhibited in response to infection [97]. Though IRESs are diverse in sequence and structure, initiation at an IRES occurs by *cis*-regulatory elements on the mRNA such as RNA binding motifs or modifications as well as by *trans*-regulation such as initiation factors or IRES-transacting factors (ITAFs) [96]. IRES types are classified into four categories based on the regulation required for initiation as well as the type of secondary and tertiary structure elements present [91]. Type I IRESs, such as those in Poliovirus (PV) [98], require all initiation factors except eIF4E and recruit the ribosome for standard 5′ to 3′ scanning for the start codon. Conversely, type II IRESs, such as those found in Encephalomyocarditis virus (EMCV) [99], recruit translation machinery directly to a start site without a scanning step. Type III IRESs, such as the one in the Hepatitis C Virus (HCV) [100], require only eIF2, eIF3, and eIF5 to load a ribosome directly at a start codon but are generally composed of more complex RNA structures such as pseudoknots. Also including pseudoknot and complex higher order structured elements are type IV IRESs, which can be found in Cricket paralysis virus (CrPV) [101]. These IRESs do not require translation initiation factors and can promote translation initiation without a genuine start codon or initiator tRNA and are not limited to 5′UTR regions [91].

Once initiation occurs, elongation involves the 80S ribosome translocating along the mRNA, advancing three nucleotides at a time to synthesize the polypeptide chain using aminoacylated tRNAs in a codon-anticodon dependent manner. Upon encountering a stop codon, eukaryotic release factors (eRFs) facilitate the liberation of the newly synthesized peptide. Finally, the ribosomes disassemble into 40S and 60S subunits, poised for a new round of translation.

During viral infection, both the virus and host require host translation machinery to proliferate. Therefore, mechanisms are employed by the pathogens and the host to target two main components: mRNAs and the ribosome. Viruses have evolved numerous host shutoff strategies to interfere with cellular mRNA translation, such as the interruption of processes related to mRNA biogenesis, the degradation of cellular mRNA, or the inhibition of translation [102]. Conversely, cells utilize comparable strategies to impede viral infection, often through the action of ISGs.

3. Targeting Translation

Viral replication relies on the host's translation machinery, making it a prime target for efficient antiviral defense. Here, we explore the intricate landscape of interferon-stimulated factors (ISFs) that either globally inhibit translation (e.g., PKR, IFIT1/P56, IFIT2/P54 2, 4E-BP, and PARP12) or selectively hinder viral translation (e.g., SHFL, SLFN11, IDO, and ZAP). Understanding these mechanisms provides valuable insights into potential therapeutic interventions and the delicate balance between combating viral infections while preserving essential cellular functions.

Interferon-stimulated factors swiftly induce translational arrest by disrupting ribosome initiation (Figure 2A). Chief among them, Protein Kinase R (PKR) phosphorylates eukaryotic initiation factor 2-alpha (eIF2α), leading to a global translational arrest. This widespread impact affects both viral and cellular mRNAs, highlighting the evolutionary importance of a swift response to viral infection [18,19]. Similarly, interferon-induced proteins with tetratricopeptide repeats (IFITs) IFIT1 and IFIT2 act as inhibitors of initiation by binding to eukaryotic initiation factor 3 (eIF3) [103]. This interaction prevents eIF3 from associating with the 40s ribosomal subunit and impairs the recruitment of the mRNA and ternary complex [39–42,104]. In the context of IRES initiation, IFIT1 has been shown to directly suppress ribosome-vRNA complex formation both in vivo and in vitro [41]. IFITs are able to act as both a sensor and effector to inhibit translation through binding of the 5′ end of non-self RNA. This binding is sequence non-specific, through the phosphate backbone, and targets mainly 5′-triphosphate ends or 5′ capped ends lacking 2′-O-methylation [105]. The recognition of the 5′ end induces a competition between IFITs and the eIF4F complex for the cap binding. Finally, the IFIT antiviral activity seems to be enhanced when IFITs act as a heterodimer (IFIT1:IFIT2) or trimer (IFIT1:IFIT2:IFIT3) [106].

Importantly, the interferon-stimulated gene 20 (ISG20), discussed in more detail later in this manuscript, has been shown to detect foreign RNA and stimulate IFIT1 upregulation [52]. This adds yet another layer of regulation and demonstrates the complex interplay of signaling in interferon response pathways. A related factor, IFIT2/P54, binds to both eIF3e and eIF3c subunits to destabilize the ternary complex, further impeding the formation of the 48S [48].

Another key player in translational regulation during an interferon response is the eIF4E-Binding Proteins (4E-BPs), which directly competes with eIF4G to sequester eIF4E activity [107]. Mammalian target of rapamycin complex 1 (mTORC1)- and downstream target p70 S6 kinase-mediated phosphorylation of 4E-BPs releases the 4E-BP from eIF4E, resulting in the recruitment of eIF4G to the 5′ cap, thereby allowing the translation initiation to proceed [56–58,108]. However, hypophosphorylated 4E-BPs, or more specifically the 4E-BP1 isoform, binds tightly to eIF4E to inhibit cap-dependent translation (Figure 2A) [56–62]. Furthermore, 4E-BP1-negative cell and mouse models are more sensitive to the antiviral effects of IFN treatment upon infection than their 4E-BP1 positive counterparts [60]. Altogether, interferon-activated mediation of 4E-BP allows for immediate regulation of cap-dependent mRNA translation.

Upon viral infection, Poly-(ADP-Ribose) Polymerase 12 (PARP12) can be recruited to initiation factors to modulate the translation machinery and inhibit translation globally. Co-immunoprecipitation studies have shown that PARP12 directly interacts with eIF4B and eIF4A1, while the long isoform (PARP12L) can also associate with ribosomes [67]. The recruitment of PARP12 is dependent on Zinc Finger (ZnF) domains, which allows it to act as a conformational molecular switch further emphasizing the intricate nature of PARP12's influence on mRNA translation [69]. Though the global shutdown of translation by interfering with initiation is a swift way to combat viral replication, it can have detrimental effects on the cell. For this reason, hosts have evolved more selective inhibitory mechanisms that target the translation of viral mRNAs specifically.

4. A Selective Approach: Targeting Viral-Specific Translation

In contrast to global inhibition of translation, some ISGs have evolved to target viral genomes more selectively with limited effect on the cellular translation machinery (Figure 2B). Since virus and host mRNAs can have significant differences in nucleotide composition, one example of selective translation inhibition exploits the codon-usage bias in viral replication [109]. Upon dephosphorylation, Schlafen 11 (SLFN11) selectively binds tRNAs that are stoichiometrically affected upon infection. Acting as a tRNA endonuclease, SLFN11 dynamically adjusts the tRNA pool, offering a nuanced approach to hinder viral protein synthesis [63–65,110]. Another example of altering tRNA pools through an interferon response is by Indoleamine-2,3-dioxygenase (IDO). IDO is involved in the catabolism of tryptophan, an essential amino acid for viral replication. By depleting tryptophan, IDO creates a hostile environment for viral protein translation [66]. Though limiting specific amino acids and tRNAs affect viral protein synthesis and thus slows viral replication, host translation is not entirely unaffected. Accordingly, some host factors have also evolved to interact with specific RNA elements to impact viral translation to achieve even more specialized antiviral responses.

One such host factor is the multifaceted approach of zinc finger antiviral protein (ZAP, also known as PARP13), which represses viral translation at target mRNAs. Though there are four isoforms of ZAP, (ZAP-S, ZAP-M, ZAP-L, and ZAP-XL), ZAP-S and ZAP-L have been studied in most detail [111]. Antiviral and translation inhibitory effects and IFN responsiveness of ZAP variants differ, suggesting these isoforms have evolved to mediate distinct antiviral and cellular functions [26,111]. Initially, researchers thought ZAP was slowing infections by contributing to vRNA decay specifically interacting with G-C-rich regions [22–24,27,112]. It was later clarified ZAP can also inhibit the translation of vRNA through blocking initiation [25,113]. ZAP-S can associate with 5′ and 3′ UTRs of viral mRNAs to interfere with the interaction between eIF4A and eIF4G, which is required for translational initiation (Figure 2A) [25]. Beyond interference with initiation, ZAP-S has recently been shown to also modulate −1 programmed ribosomal frameshifting (−1PRF) by interacting with specific SARS coronaviruses SARS-CoV-1 and SARS-CoV-2 RNA elements, while not affecting the frameshifting in host mRNAs [29]. Since −1PRF is essential for the correct expression of viral structural and enzymatic proteins, it is an important means of regulating viral translation levels upon ISG activation (Figure 2B).

Another ISG modulator of −1PRF in viral mRNA translation is Shiftless Antiviral Inhibitor of Ribosomal Frameshifting (Shiftless or SHFL), while its role in antiviral regulation is broad, including interaction with other factors such as PABPC1, LARP1, UPF1, and MOV10 [34,37,83,114,115], SHFL inhibits −1PRF through binding to RNA and the translating ribosome (Figure 2B) [35,116]. Interestingly, though SHFL is a strong −1PRF modulator, its ability to restrict viral replication is not limited to this activity since some of the viruses restricted do not possess −1PRF signals [117–119]. There is evidence that SHFL interacts with stalled ribosomes, signaling the recruitment of the eRF1-eRF3 complex to result in premature translation termination and mRNA decay [35,120]. SHFL is able to regulate the translation of numerous positive single-stranded RNA viruses, such as *Flaviviridae* (DENV, HCV, WNVKUN, ZIKV, JEV), *Togaviridae* (CHIKV and SINV), and *Picornaviridae* (EMCV), some DNA viruses such as members of *Herpesviridae* [104], many of the *Retroviridae* family (HIV-1, RSV, HTLV, MMuLV, HIV-2), and *Coronaviridae* (SARS-CoV-2) [35–38,121–123]. The diverse strategies employed by ISGs underscore the adaptability and complexity of host antiviral defenses. From global translational arrests to selective viral translation inhibition, these factors exemplify the precision of the host's response mechanisms.

5. Targeting the Template

Interacting directly with the translation machinery represents an effective way to inhibit viral replication by shutting down the entire protein synthesis pathway. Nevertheless, viral protein synthesis can also be impaired by viral mRNA targeting factors that sequester the mRNA template from translation by (1) physically separating the mRNA from active ribosomes, (2) chemically or structurally modifying the mRNA, or (3) degrading the mRNA (Figure 2C).

In mammalian cells, cytoplasmic membraneless compartments are known to form or expand in response to a viral infection. Due to the high concentration of proteins and RNA, these subcellular compartments are often referred to as ribonucleoprotein (RNP) granules, which comprise processing bodies (PB), stress granules (SG), and antiviral granules [124,125]. These granules share some common characteristics such as the absence of translational activity and the presence of non-translating mRNAs, as well as factors involved in translation inhibition and mRNA decay machinery such as the RNA-induced silencing complex (RISC) and its associated factors [126]. Furthermore, a significant portion of the proteins present in these granules have been identified as playing a pro- and/or antiviral function that overlaps with known ISGs (ADAR, APOBEC3F/G, DCP2, IGF2BP2, MOV10, PARP12, PATL1, XRN1, SMG7, ZAP, etc.) [127]. Thus, relocating vRNA into these granules is one of the ways to repress their expression. Importantly, ADAR, MOV10, PARP12, ZAP, and SHFL co-localize with vRNAs in these RNP granules to inhibit viral replication [85,128–132].

MOV10 is a helicase identified for the first time as a protein preventing infection by Moloney Murine leukemia virus (MMuLV). Since then, MOV10 has been shown to be implicated in the inhibition of many viruses, such as HIV-1, HCV, influenza A virus (IAV), or dengue virus (DENV), but also retroelements [80,81,83–85]. It functions as an RNA helicase, which regulates microRNA (miRNA) and mRNA generation, maturation, and degradation through RISC and thereby influences viral cycles at various steps [86]. MOV10 recognizes and sequesters HIV-1 gRNA in P-bodies (Figure 2C) [80], but also inhibits IAV RNP by mediating the vRNA degradation through miRNA pathways [84]. In the case of DENV, MOV10 associates with the antiviral protein SHFL to inhibit translation [83]. One of MOV10 modes of action is to bind the G-rich structures near miRNA recognition elements in order to unfold them with its helicase activity and induce their subsequent translational repression or RNA degradation by recruiting RISC [81]. However, some viruses, such as Hepatitis B virus (HBV) or Enterovirus 71 (EV71), managed to co-opt MOV10 to positively regulate their life cycle [79,82]. In addition, downregulation of MOV10 has been associated with defects in the miRNA machinery, which is required by HCV with miR122 [133]. Conversely, MOV10 overexpression also represses HCV replication through an unexplored mechanism [87].

Other factors associated with miRNA-mediated translation inhibition include the Poly ADP-Ribose Polymerase (PARP) family [134]. Among them, PARP12 and PARP13 (ZAP), can also be localized in cell compartments specialized in translational repression [69,131]. Unlike ZAP, PARP12 still possesses its PARP activity, which appears to be involved in the repression of vRNA translation by ADP-ribosylation of Argonaute 2 (Ago2) [68]. While PARP12 has been described for its antiviral activity by inducing viral protein degradation through MonoADP-Ribosylation (MARylation), it seems that its MARylation activity is also required for translational inhibition [67,69]. The importance of the PARP domain is further emphasized by its presence in ZAP-L. Mutations performed in the ZAP-L PARP domain led to a decrease in its antiviral activity, though as previously described, the antiviral activity of ZAP is not exclusive to the PARP domain [29,135]. One of the ways ZAP (-S and -L) and PARP12 repress translation is by binding to G-C-rich RNA sequences through conserved CCCH tandem zinc fingers. In the case of ZAP, it has been shown to bind as a homodimer before recruiting exonucleases, ultimately leading to the target mRNA degradation [136,137]. Additionally, PARP12 inhibits translation by binding to polysomes [67].

The Adenosine Deaminase Acting on RNA 1 (ADAR1) was implicated in the restriction of multiple viral genomes [78]. ADAR proteins are RNA editing enzymes that deaminate adenosines (A) to produce inosines (I) within double-stranded RNA (Figure 2C) [70]. There are two main isoforms of ADAR1 (p150 and p110) which are able to edit viral genomes. Their expression is under the control of an IFN-inducible promoter due to the IFN-stimulated response element (ISRE) located upstream of the p150 promoter region [138,139]. However, there is evidence indicating that p110 is constitutively expressed [77]. Importantly, it was shown that p150 is responsible for most A-to-I editing events targeting both cellular and viral RNAs [77]. Contrary to p110, which is predominantly nuclear, p150 is also present in the cytoplasm and localizes to RNP granules [76]. ADAR editing can be performed in both a highly selective (hepatitis delta virus—HDV; glutamate receptor subunit GRIA2) or nonselective (hypermutation of Measles virus, MeV genome) manner [78]. Cellular machineries, including ribosomes, recognize A-to-I modifications as A-to-G substitutions [140]. Besides its activities on viral genomes, ADAR is also able to edit cellular RNAs such as miRNAs. Both isoforms interact with Dicer and regulate miRNA maturation, as shown by the reduced expression of miRNAs when ADAR1 is absent [141]. Furthermore, ADAR1 editing of immature miRNAs also affects the translational landscape of cells, usually in a cell-type-dependent manner [71,72]. A-to-I substitutions in miRNA lead to a change in the targeted sequences, and thus a change in the regulation of a specific set of genes. In the end, these modifications have been shown to exhibit either positive or negative regulation depending on the specific context; thus, ADAR editing has a broad range of effects in the context of viral replication [73,78,142].

In eukaryotes, the degradation of vRNA serves as a crucial mechanism for controlling viral infections (Figure 2C). Many ISGs targeting vRNA possess an RNAse activity (ISG20) or activate an RNAse with a broad spectrum of targets (OAS—RNase L). In contrast with XRN1, which degrades RNA in RNP granules after decapping, RNase L and ISG20 do not require the mRNA to go through any such preliminary step [143]. A central player in the IFN antiviral pathway is the cellular endoribonuclease RNase L, activated through the 2′-5′-oligoadenylate synthetase (OAS)/RNase L system. Upon OAS activation, the synthesis of 2′-5′-linked oligoadenylates, a unique ligand for RNase L, occurs. RNase L features ankyrin repeats and a catalytic RNase domain, which mediates cleavage of RNA within single-stranded regions, leading to nonspecific degradation of both viral and host RNAs, including ribosomal RNA. This not only leads to the global downregulation of translation, but also eliminates invasive RNAs [20,21].

While ISG20 was briefly mentioned earlier for its role in activating IFIT1, it has also been shown to regulate over 100 genes, many of which are ISGs [51,52]. Mainly functioning as an RNA exonuclease with broad antiviral properties, recent studies challenge the prevailing mechanism of ISG20 by revealing its ability to target specific structures on the hepatitis B virus (HBV) and suggesting inhibition without viral RNA degradation. Using the vesicular stomatitis virus (VSV) as a model, it was demonstrated that ISG20 interferes with viral replication not by degrading viral RNA but by impairing its translation [53]. This translational control mechanism targets all RNAs originating from ectopically introduced genetic material, collectively defined here as "non-self", irrespective of their viral or non-viral origins. However, ISG20 does not affect the translation of endogenous mRNA transcripts, suggesting its ability to discriminate between the cell's own versus foreign genetic material [53,55].

Along with interferon-stimulated genes encoding for proteins, miRNA pathways are prevalent in vRNA silencing and can be found in RNP granules [144]. These miRNAs are generally associated with ISGs that regulate vRNA processing in these subcellular compartments [145,146]. One such pathway is mediated by the RNA interference factor Ago2 [147]. The binding of the miRNA to its target sequences requires a perfect base-pair complementarity to induce the cleavage by Ago2 and subsequent mRNA degradation [144]. Otherwise, the binding of both miRNA and Ago2 to target sites represses the translation of the transcript. Although Ago2 expression has not been shown to be upregulated upon viral infection and interferon stimulation, some Ago2-associated miRNAs are explored in [148–152]. The upregulation of these small non-coding RNAs could be responsible for vRNA silencing induced by Ago2 [148,153,154]. In fact, miRNAs were shown to interfere with the replication of multiple viruses such as HCV, PFV-1, VSV, HIV-1, and SARS-CoV-2 [148,154–156]. Furthermore, miRNA binding can change the structure of target mRNAs, which seems to be the case for miR-122 and HCV 5′ UTR region, but also for other mRNAs such as SARS-CoV-2 or the human pseudoknot of CCR5 mRNA [133,157–161]. Overall, miRNAs are responsible for the inhibition of translation in many cellular and viral RNAs [162].

6. Concluding Remarks

Although it has been essential to confirm the interactions between host factors and viruses, research focusing on specific virus–host factor interactions may carry an implicit bias, by the assumption that certain ISGs exclusively respond to specific viruses. In fact, ISGs are more likely to respond to multiple viral infections. For example, as ZAP research has progressed, the range of viral species modulated has expanded significantly to include Sindbis virus, MMuLV, HBV, Ebolavirus, IAV, and SARS-CoV-2 [163–168]. As with ZAP and other factors discussed in this review, it is likely that many other host-encoded factors have a broader phenotype than initially described, and thus likely inhibit a range of viruses upon infection. However, working with cell models that may not always express similar repertoires of antiviral factors, or probe for only specific interactions, we could be missing key information on host–pathogen interplay. Moving forward, it will become increasingly important to utilize techniques that explore virus–host interactions in a complete and unbiased way such as strategies involving large-scale and genome-wide screens, including interactome and multi-omics studies [36,169–171]. As we uncover the large complexity between viruses and hosts at a molecular level, advancements in live-cell and single-molecule RNA imaging also allow researchers to visualize vRNA replication and translation processes in infected cells, as well as track the heterogeneity of viral replication dynamics and antiviral responses [172–175].

In the interferon field, the rapid coevolution of viruses and hosts can be analogized to an arms race: a fight over the translation machinery. This is accurate in many ways, certainly involving the selective pressure of isoform-specific functions in host factors, while viruses continue to evolve new strategies for evasion [26]. However, recently it has been established that there are host factors that are vitally important for the virus life cycle [171].

Human evolution has benefitted from and depended on viral infection strategies. Human endogenous retroviruses are an integral part of the human genome, resulting from ancestral infections of human germline cells. Viruses have played important roles in human development such as the formation of the placenta, neuroprotective functions, and hormone-dependent organ function [176,177]. Thus, the interplay between viruses and host cells becomes even more complex—instead of an arms race to overcome an enemy, a model of coevolution that vitally depends on one another is perhaps more fitting. As we gain a better understanding of the continued dynamic coevolution between virus and host, we may uncover symbiosis where we originally perceived parasitism.

Deciphering the virus–host interplay will continue to be instrumental not only for identifying potential therapeutic targets but also in characterizing general insights into the dynamics of translation. For the development of new RNA-based therapeutics, future work is imperative to advance the understanding of how RNA elements modulate translation efficiency and accuracy in the context of immune responses.

Author Contributions: Conceptualization, N.C., A.S. and O.G.; writing—original draft preparation, A.S. and O.G.; writing—review and editing, N.C., A.S. and O.G.; visualization, A.S. and O.G.; supervision, N.C.; funding acquisition, N.C. All authors have read and agreed to the published version of the manuscript.

Funding: The work in Caliskan Laboratory is supported by Helmholtz Association and European Research Council (to N.C.; Grant number: 948636).

Acknowledgments: The authors would like to thank all of the researchers that continue to dedicate their time and energy to further the field of interferon-translation dynamics. We would also like to thank Redmond Smyth and the members of the Caliskan group, especially Ricarda Riegger and Rui Pedro (not the footballer) Marques da Costa, for their careful reading of this manuscript.

Conflicts of Interest: The authors declare no conflicts of interest.

References

1. Davey, N.E.; Travé, G.; Gibson, T.J. How Viruses Hijack Cell Regulation. *Trends Biochem. Sci.* **2011**, *36*, 159–169. [CrossRef] [PubMed]
2. Robinson, M.; Schor, S.; Barouch-Bentov, R.; Einav, S. Viral Journeys on the Intracellular Highways. *Cell. Mol. Life Sci.* **2018**, *75*, 3693–3714. [CrossRef] [PubMed]
3. Barnett, K.C.; Li, S.; Liang, K.; Ting, J.P.-Y. A 360° View of the Inflammasome: Mechanisms of Activation, Cell Death, and Diseases. *Cell* **2023**, *186*, 2288–2312. [CrossRef] [PubMed]
4. Wang, Y.; Liu, L. The Membrane Protein of Severe Acute Respiratory Syndrome Coronavirus Functions as a Novel Cytosolic Pathogen-Associated Molecular Pattern To Promote Beta Interferon Induction via a Toll-Like-Receptor-Related TRAF3-Independent Mechanism. *mBio* **2016**, *7*, e01872-15. [CrossRef] [PubMed]
5. Zhou, R.; Liu, L.; Wang, Y. Viral Proteins Recognized by Different TLRs. *J. Med. Virol.* **2021**, *93*, 6116–6123. [CrossRef] [PubMed]
6. Ivashkiv, L.B.; Donlin, L.T. Regulation of Type I Interferon Responses. *Nat. Rev. Immunol.* **2014**, *14*, 36–49. [CrossRef]
7. Borden, E.C.; Sen, G.C.; Uze, G.; Silverman, R.H.; Ransohoff, R.M.; Foster, G.R.; Stark, G.R. Interferons at Age 50: Past, Current and Future Impact on Biomedicine. *Nat. Rev. Drug Discov.* **2007**, *6*, 975–990. [CrossRef] [PubMed]
8. Patil, S.; Fribourg, M.; Ge, Y.; Batish, M.; Tyagi, S.; Hayot, F.; Sealfon, S.C. Single-Cell Analysis Shows That Paracrine Signaling by First Responder Cells Shapes the Interferon-β Response to Viral Infection. *Sci. Signal.* **2015**, *8*, ra16. [CrossRef]
9. Schoggins, J.W. Interferon-Stimulated Genes: What Do They All Do? *Annu. Rev. Virol.* **2019**, *6*, 567–584. [CrossRef]
10. Philips, R.L.; Wang, Y.; Cheon, H.; Kanno, Y.; Gadina, M.; Sartorelli, V.; Horvath, C.M.; Darnell, J.E.; Stark, G.R.; O'Shea, J.J. The JAK-STAT Pathway at 30: Much Learned, Much More to Do. *Cell* **2022**, *185*, 3857–3876. [CrossRef]
11. Feng, Q.; Langereis, M.A.; Van Kuppeveld, F.J.M. Induction and Suppression of Innate Antiviral Responses by Picornaviruses. *Cytokine Growth Factor Rev.* **2014**, *25*, 577–585. [CrossRef]
12. Rand, U.; Hillebrand, U.; Sievers, S.; Willenberg, S.; Köster, M.; Hauser, H.; Wirth, D. Uncoupling of the Dynamics of Host–Pathogen Interaction Uncovers New Mechanisms of Viral Interferon Antagonism at the Single-Cell Level. *Nucleic Acids Res.* **2014**, *42*, e109. [CrossRef] [PubMed]
13. Lei, X.; Xiao, X.; Wang, J. Innate Immunity Evasion by Enteroviruses: Insights into Virus-Host Interaction. *Viruses* **2016**, *8*, 22. [CrossRef]
14. Hotter, D.; Kirchhoff, F. Interferons and beyond: Induction of Antiretroviral Restriction Factors. *J. Leukoc. Biol.* **2018**, *103*, 465–477. [CrossRef] [PubMed]

15. Jouvenet, N.; Goujon, C.; Banerjee, A. Clash of the Titans: Interferons and SARS-CoV-2. *Trends Immunol.* **2021**, *42*, 1069–1072. [CrossRef] [PubMed]
16. Duggal, N.K.; Emerman, M. Evolutionary Conflicts between Viruses and Restriction Factors Shape Immunity. *Nat. Rev. Immunol.* **2012**, *12*, 687–695. [CrossRef]
17. Shuler, G.; Hagai, T. Rapidly Evolving Viral Motifs Mostly Target Biophysically Constrained Binding Pockets of Host Proteins. *Cell Rep.* **2022**, *40*, 111212. [CrossRef] [PubMed]
18. Balachandran, S.; Roberts, P.C.; Kipperman, T.; Bhalla, K.N.; Compans, R.W.; Archer, D.R.; Barber, G.N. Alpha/Beta Interferons Potentiate Virus-Induced Apoptosis through Activation of the FADD/Caspase-8 Death Signaling Pathway. *J. Virol.* **2000**, *74*, 1513–1523. [CrossRef] [PubMed]
19. Munir, M.; Berg, M. The Multiple Faces of Proteinkinase R in Antiviral Defense. *Virulence* **2013**, *4*, 85–89. [CrossRef]
20. Silverman, R.H. Viral Encounters with 2′,5′-Oligoadenylate Synthetase and RNase L during the Interferon Antiviral Response. *J. Virol.* **2007**, *81*, 12720–12729. [CrossRef]
21. Donovan, J.; Rath, S.; Kolet-Mandrikov, D.; Korennykh, A. Rapid RNase L–Driven Arrest of Protein Synthesis in the dsRNA Response without Degradation of Translation Machinery. *RNA* **2017**, *23*, 1660–1671. [CrossRef]
22. Chen, G.; Guo, X.; Lv, F.; Xu, Y.; Gao, G. P72 DEAD Box RNA Helicase Is Required for Optimal Function of the Zinc-Finger Antiviral Protein. *Proc. Natl. Acad. Sci. USA* **2008**, *105*, 4352–4357. [CrossRef]
23. Zhu, Y.; Gao, G. ZAP-Mediated mRNA Degradation. *RNA Biol.* **2008**, *5*, 65–67. [CrossRef]
24. Zhu, Y.; Chen, G.; Lv, F.; Wang, X.; Ji, X.; Xu, Y.; Sun, J.; Wu, L.; Zheng, Y.-T.; Gao, G. Zinc-Finger Antiviral Protein Inhibits HIV-1 Infection by Selectively Targeting Multiply Spliced Viral mRNAs for Degradation. *Proc. Natl. Acad. Sci. USA* **2011**, *108*, 15834–15839. [CrossRef]
25. Zhu, Y.; Wang, X.; Goff, S.P.; Gao, G. Translational Repression Precedes and Is Required for ZAP-Mediated mRNA Decay: ZAP-Mediated Translational Repression versus mRNA Decay. *EMBO J.* **2012**, *31*, 4236–4246. [CrossRef]
26. Schwerk, J.; Soveg, F.W.; Ryan, A.P.; Thomas, K.R.; Hatfield, L.D.; Ozarkar, S.; Forero, A.; Kell, A.M.; Roby, J.A.; So, L.; et al. RNA-Binding Protein Isoforms ZAP-S and ZAP-L Have Distinct Antiviral and Immune Resolution Functions. *Nat. Immunol.* **2019**, *20*, 1610–1620. [CrossRef]
27. Luo, X.; Wang, X.; Gao, Y.; Zhu, J.; Liu, S.; Gao, G.; Gao, P. Molecular Mechanism of RNA Recognition by Zinc-Finger Antiviral Protein. *Cell Rep.* **2020**, *30*, 46–52.e4. [CrossRef]
28. Lee, S.; Lee, Y.; Choi, Y.; Son, A.; Park, Y.; Lee, K.-M.; Kim, J.; Kim, J.-S.; Kim, V.N. The SARS-CoV-2 RNA Interactome. *Mol. Cell* **2021**, *81*, 2838–2850.e6. [CrossRef]
29. Zimmer, M.M.; Kibe, A.; Rand, U.; Pekarek, L.; Ye, L.; Buck, S.; Smyth, R.P.; Cicin-Sain, L.; Caliskan, N. The Short Isoform of the Host Antiviral Protein ZAP Acts as an Inhibitor of SARS-CoV-2 Programmed Ribosomal Frameshifting. *Nat. Commun.* **2021**, *12*, 7193. [CrossRef]
30. Yang, E.; Nguyen, L.P.; Wisherop, C.A.; Kan, R.L.; Li, M.M.H. The Role of ZAP and TRIM25 RNA Binding in Restricting Viral Translation. *Front. Cell. Infect. Microbiol.* **2022**, *12*, 886929. [CrossRef]
31. Li, M.M.H.; Lau, Z.; Cheung, P.; Aguilar, E.G.; Schneider, W.M.; Bozzacco, L.; Molina, H.; Buehler, E.; Takaoka, A.; Rice, C.M.; et al. TRIM25 Enhances the Antiviral Action of Zinc-Finger Antiviral Protein (ZAP). *PLoS Pathog.* **2017**, *13*, e1006145. [CrossRef] [PubMed]
32. Zheng, X.; Wang, X.; Tu, F.; Wang, Q.; Fan, Z.; Gao, G. TRIM25 Is Required for the Antiviral Activity of Zinc Finger Antiviral Protein. *J. Virol.* **2017**, *91*, e00088-17. [CrossRef] [PubMed]
33. Sanchez, J.G.; Sparrer, K.M.J.; Chiang, C.; Reis, R.A.; Chiang, J.J.; Zurenski, M.A.; Wan, Y.; Gack, M.U.; Pornillos, O. TRIM25 Binds RNA to Modulate Cellular Anti-Viral Defense. *J. Mol. Biol.* **2018**, *430*, 5280–5293. [CrossRef] [PubMed]
34. Suzuki, Y.; Chin, W.-X.; Han, Q.; Ichiyama, K.; Lee, C.H.; Eyo, Z.W.; Ebina, H.; Takahashi, H.; Takahashi, C.; Tan, B.H.; et al. Characterization of RyDEN (C19orf66) as an Interferon-Stimulated Cellular Inhibitor against Dengue Virus Replication. *PLoS Pathog.* **2016**, *12*, e1005357. [CrossRef] [PubMed]
35. Wang, X.; Xuan, Y.; Han, Y.; Ding, X.; Ye, K.; Yang, F.; Gao, P.; Goff, S.P.; Gao, G. Regulation of HIV-1 Gag-Pol Expression by Shiftless, an Inhibitor of Programmed -1 Ribosomal Frameshifting. *Cell* **2019**, *176*, 625–635.e14. [CrossRef] [PubMed]
36. Schmidt, N.; Lareau, C.A.; Keshishian, H.; Ganskih, S.; Schneider, C.; Hennig, T.; Melanson, R.; Werner, S.; Wei, Y.; Zimmer, M.; et al. The SARS-CoV-2 RNA–Protein Interactome in Infected Human Cells. *Nat. Microbiol.* **2020**, *6*, 339–353. [CrossRef] [PubMed]
37. Wang, X.; Zhang, A.-M. Functional Features of a Novel Interferon-Stimulated Gene SHFL: A Comprehensive Review. *Front. Microbiol.* **2023**, *14*, 1323231. [CrossRef]
38. Yu, D.; Zhao, Y.; Pan, J.; Yang, X.; Liang, Z.; Xie, S.; Cao, R. C19orf66 Inhibits Japanese Encephalitis Virus Replication by Targeting -1 PRF and the NS3 Protein. *Virol. Sin.* **2021**, *36*, 1443–1455. [CrossRef]
39. Guo, J.; Hui, D.J.; Merrick, W.C.; Sen, G.C. A New Pathway of Translational Regulation Mediated by Eukaryotic Initiation Factor 3. *EMBO J.* **2000**, *19*, 6891–6899. [CrossRef]
40. Hui, D.J.; Bhasker, C.R.; Merrick, W.C.; Sen, G.C. Viral Stress-Inducible Protein P56 Inhibits Translation by Blocking the Interaction of eIF3 with the Ternary Complex eIF2·GTP·Met-tRNAi. *J. Biol. Chem.* **2003**, *278*, 39477–39482. [CrossRef]
41. Wang, C.; Pflugheber, J.; Sumpter, R.; Sodora, D.L.; Hui, D.; Sen, G.C.; Gale, M. Alpha Interferon Induces Distinct Translational Control Programs To Suppress Hepatitis C Virus RNA Replication. *J. Virol.* **2003**, *77*, 3898–3912. [CrossRef] [PubMed]

42. Li, H.-T.; Su, Y.-P.; Cheng, T.-M.; Xu, J.-M.; Liao, J.; Chen, J.-C.; Ji, C.-Y.; Ai, G.-P.; Wang, J.-P. The Interaction between Interferon-Induced Protein with Tetratricopeptide Repeats-1 and Eukaryotic Elongation Factor-1A. *Mol. Cell Biochem.* **2010**, *337*, 101–110. [CrossRef] [PubMed]
43. Pichlmair, A.; Lassnig, C.; Eberle, C.-A.; Górna, M.W.; Baumann, C.L.; Burkard, T.R.; Bürckstümmer, T.; Stefanovic, A.; Krieger, S.; Bennett, K.L.; et al. IFIT1 Is an Antiviral Protein That Recognizes 5′-Triphosphate RNA. *Nat. Immunol.* **2011**, *12*, 624–630. [CrossRef] [PubMed]
44. Abbas, Y.M.; Pichlmair, A.; Górna, M.W.; Superti-Furga, G.; Nagar, B. Structural Basis for Viral 5′-PPP-RNA Recognition by Human IFIT Proteins. *Nature* **2013**, *494*, 60–64. [CrossRef] [PubMed]
45. Diamond, M.S.; Farzan, M. The Broad-Spectrum Antiviral Functions of IFIT and IFITM Proteins. *Nat. Rev. Immunol.* **2013**, *13*, 46–57. [CrossRef] [PubMed]
46. Feng, F.; Yuan, L.; Wang, Y.E.; Crowley, C.; Lv, Z.; Li, J.; Liu, Y.; Cheng, G.; Zeng, S.; Liang, H. Crystal Structure and Nucleotide Selectivity of Human IFIT5/ISG58. *Cell Res.* **2013**, *23*, 1055–1058. [CrossRef] [PubMed]
47. Vladimer, G.I.; GÃ3rna, M.W.; Superti-Furga, G. IFITs: Emerging Roles as Key Anti-Viral Proteins. *Front. Immunol.* **2014**, *5*, 94. [CrossRef] [PubMed]
48. Terenzi, F.; Hui, D.J.; Merrick, W.C.; Sen, G.C. Distinct Induction Patterns and Functions of Two Closely Related Interferon-Inducible Human Genes, ISG54 and ISG56. *J. Biol. Chem.* **2006**, *281*, 34064–34071. [CrossRef]
49. Sen, G.C.; Fensterl, V. Crystal Structure of IFIT2 (ISG54) Predicts Functional Properties of IFITs. *Cell Res.* **2012**, *22*, 1407–1409. [CrossRef]
50. Zhou, Z.; Wang, N.; Woodson, S.E.; Dong, Q.; Wang, J.; Liang, Y.; Rijnbrand, R.; Wei, L.; Nichols, J.E.; Guo, J.-T.; et al. Antiviral Activities of ISG20 in Positive-Strand RNA Virus Infections. *Virology* **2011**, *409*, 175–188. [CrossRef]
51. Leong, C.R.; Funami, K.; Oshiumi, H.; Mengao, D.; Takaki, H.; Matsumoto, M.; Aly, H.H.; Watashi, K.; Chayama, K.; Seya, T. Interferon-Stimulated Gene of 20 kDa Protein (ISG20) Degrades RNA of Hepatitis B Virus to Impede the Replication of HBV in Vitro and in Vivo. *Oncotarget* **2016**, *7*, 68179–68193. [CrossRef] [PubMed]
52. Weiss, C.M.; Trobaugh, D.W.; Sun, C.; Lucas, T.M.; Diamond, M.S.; Ryman, K.D.; Klimstra, W.B. The Interferon-Induced Exonuclease ISG20 Exerts Antiviral Activity through Upregulation of Type I Interferon Response Proteins. *mSphere* **2018**, *3*, e00209-18. [CrossRef] [PubMed]
53. Wu, N.; Nguyen, X.-N.; Wang, L.; Appourchaux, R.; Zhang, C.; Panthu, B.; Gruffat, H.; Journo, C.; Alais, S.; Qin, J.; et al. The Interferon Stimulated Gene 20 Protein (ISG20) Is an Innate Defense Antiviral Factor That Discriminates Self versus Non-Self Translation. *PLoS Pathog.* **2019**, *15*, e1008093. [CrossRef] [PubMed]
54. Imam, H.; Kim, G.-W.; Mir, S.A.; Khan, M.; Siddiqui, A. Interferon-Stimulated Gene 20 (ISG20) Selectively Degrades N6-Methyladenosine Modified Hepatitis B Virus Transcripts. *PLoS Pathog.* **2020**, *16*, e1008338. [CrossRef] [PubMed]
55. Deymier, S.; Louvat, C.; Fiorini, F.; Cimarelli, A. ISG20: An Enigmatic Antiviral RNase Targeting Multiple Viruses. *FEBS Open Bio* **2022**, *12*, 1096–1111. [CrossRef] [PubMed]
56. Lekmine, F.; Uddin, S.; Sassano, A.; Parmar, S.; Brachmann, S.M.; Majchrzak, B.; Sonenberg, N.; Hay, N.; Fish, E.N.; Platanias, L.C. Activation of the P70 S6 Kinase and Phosphorylation of the 4E-BP1 Repressor of mRNA Translation by Type I Interferons *. *J. Biol. Chem.* **2003**, *278*, 27772–27780. [CrossRef] [PubMed]
57. Lekmine, F.; Sassano, A.; Uddin, S.; Smith, J.; Majchrzak, B.; Brachmann, S.M.; Hay, N.; Fish, E.N.; Platanias, L.C. Interferon-γ Engages the P70 S6 Kinase to Regulate Phosphorylation of the 40S S6 Ribosomal Protein. *Exp. Cell Res.* **2004**, *295*, 173–182. [CrossRef] [PubMed]
58. Thyrell, L.; Hjortsberg, L.; Arulampalam, V.; Panaretakis, T.; Uhles, S.; Dagnell, M.; Zhivotovsky, B.; Leibiger, I.; Grandér, D.; Pokrovskaja, K. Interferon α-Induced Apoptosis in Tumor Cells Is Mediated through the Phosphoinositide 3-Kinase/Mammalian Target of Rapamycin Signaling Pathway *. *J. Biol. Chem.* **2004**, *279*, 24152–24162. [CrossRef] [PubMed]
59. Kaur, S.; Lal, L.; Sassano, A.; Majchrzak-Kita, B.; Srikanth, M.; Baker, D.P.; Petroulakis, E.; Hay, N.; Sonenberg, N.; Fish, E.N.; et al. Regulatory Effects of Mammalian Target of Rapamycin-Activated Pathways in Type I and II Interferon Signaling *. *J. Biol. Chem.* **2007**, *282*, 1757–1768. [CrossRef]
60. Burke, J.D.; Sonenberg, N.; Platanias, L.C.; Fish, E.N. Antiviral Effects of Interferon-β Are Enhanced in the Absence of the Translational Suppressor 4E-BP1 in Myocarditis Induced by Coxsackievirus B3. *Antivir. Ther.* **2011**, *16*, 577–584. [CrossRef]
61. Kroczynska, B.; Joshi, S.; Eklund, E.A.; Verma, A.; Kotenko, S.V.; Fish, E.N.; Platanias, L.C. Regulatory Effects of Ribosomal S6 Kinase 1 (RSK1) in IFNλ Signaling *. *J. Biol. Chem.* **2011**, *286*, 1147–1156. [CrossRef]
62. Wong, C.; Tavares-Ferreira, D.; Thörn Perez, C.; Sharif, B.; Uttam, S.; Amiri, M.; Lister, K.C.; Hooshmandi, M.; Nguyen, V.; Séguéla, P.; et al. 4E-BP1–Dependent Translation in Nociceptors Controls Mechanical Hypersensitivity via TRIM32/Type I Interferon Signaling. *Sci. Adv.* **2023**, *9*, eadh9603. [CrossRef]
63. Li, M.; Kao, E.; Gao, X.; Sandig, H.; Limmer, K.; Pavon-Eternod, M.; Jones, T.E.; Landry, S.; Pan, T.; Weitzman, M.D.; et al. Codon-Usage-Based Inhibition of HIV Protein Synthesis by Human Schlafen 11. *Nature* **2012**, *491*, 125–128. [CrossRef] [PubMed]
64. Stabell, A.C.; Hawkins, J.; Li, M.; Gao, X.; David, M.; Press, W.H.; Sawyer, S.L. Non-Human Primate Schlafen11 Inhibits Production of Both Host and Viral Proteins. *PLoS Pathog.* **2016**, *12*, e1006066. [CrossRef]
65. Malone, D.; Lardelli, R.M.; Li, M.; David, M. Dephosphorylation Activates the Interferon-Stimulated Schlafen Family Member 11 in the DNA Damage Response. *J. Biol. Chem.* **2019**, *294*, 14674–14685. [CrossRef]

66. Schmidt, S.V.; Schultze, J.L. New Insights into IDO Biology in Bacterial and Viral Infections. *Front. Immunol.* **2014**, *5*, 384. [CrossRef] [PubMed]
67. Atasheva, S.; Frolova, E.I.; Frolov, I. Interferon-Stimulated Poly(ADP-Ribose) Polymerases Are Potent Inhibitors of Cellular Translation and Virus Replication. *J. Virol.* **2014**, *88*, 2116–2130. [CrossRef]
68. Leung, A.K.L.; Vyas, S.; Rood, J.E.; Bhutkar, A.; Sharp, P.A.; Chang, P. Poly(ADP-Ribose) Regulates Stress Responses and microRNA Activity in the Cytoplasm. *Mol. Cell* **2011**, *42*, 489–499. [CrossRef] [PubMed]
69. Welsby, I.; Hutin, D.; Gueydan, C.; Kruys, V.; Rongvaux, A.; Leo, O. PARP12, an Interferon-Stimulated Gene Involved in the Control of Protein Translation and Inflammation. *J. Biol. Chem.* **2014**, *289*, 26642–26657. [CrossRef]
70. Bass, B.L. RNA Editing by Adenosine Deaminases That Act on RNA. *Annu. Rev. Biochem.* **2002**, *71*, 817–846. [CrossRef]
71. Lau, K.E.H.; Nguyen, N.T.; Kesavan, J.C.; Langa, E.; Fanning, K.; Brennan, G.P.; Sanz-Rodriguez, A.; Villegas-Salmerón, J.; Yan, Y.; Venø, M.T.; et al. Differential microRNA Editing May Drive Target Pathway Switching in Human Temporal Lobe Epilepsy. *Brain Commun.* **2023**, *6*, fcad355. [CrossRef] [PubMed]
72. Blow, M.; Grocock, R.; Van Dongen, S.; Enright, A.; Dicks, E.; Futreal, P.A.; Wooster, R.; Stratton, M. RNA Editing of Human microRNAs. *Genome Biol.* **2006**, *7*, R27. [CrossRef] [PubMed]
73. Toth, A.M.; Li, Z.; Cattaneo, R.; Samuel, C.E. RNA-Specific Adenosine Deaminase ADAR1 Suppresses Measles Virus-Induced Apoptosis and Activation of Protein Kinase PKR. *J. Biol. Chem.* **2009**, *284*, 29350–29356. [CrossRef] [PubMed]
74. Nishikura, K. Functions and Regulation of RNA Editing by ADAR Deaminases. *Annu. Rev. Biochem.* **2010**, *79*, 321–349. [CrossRef] [PubMed]
75. George, C.X.; Gan, Z.; Liu, Y.; Samuel, C.E. Adenosine Deaminases Acting on RNA, RNA Editing, and Interferon Action. *J. Interferon. Cytokine Res.* **2011**, *31*, 99–117. [CrossRef]
76. Weissbach, R.; Scadden, A.D.J. Tudor-SN and ADAR1 Are Components of Cytoplasmic Stress Granules. *RNA* **2012**, *18*, 462–471. [CrossRef] [PubMed]
77. Pfaller, C.K.; George, C.X.; Samuel, C.E. Adenosine Deaminases Acting on RNA (ADARs) and Viral Infections. *Annu. Rev. Virol.* **2021**, *8*, 239–264. [CrossRef] [PubMed]
78. Piontkivska, H.; Wales-McGrath, B.; Miyamoto, M.; Wayne, M.L. ADAR Editing in Viruses: An Evolutionary Force to Reckon With. *Genome Biol. Evol.* **2021**, *13*, evab240. [CrossRef] [PubMed]
79. Haussecker, D.; Cao, D.; Huang, Y.; Parameswaran, P.; Fire, A.Z.; Kay, M.A. Capped Small RNAs and MOV10 in Human Hepatitis Delta Virus Replication. *Nat Struct. Mol. Biol.* **2008**, *15*, 714–721. [CrossRef]
80. Burdick, R.; Smith, J.L.; Chaipan, C.; Friew, Y.; Chen, J.; Venkatachari, N.J.; Delviks-Frankenberry, K.A.; Hu, W.-S.; Pathak, V.K. P Body-Associated Protein Mov10 Inhibits HIV-1 Replication at Multiple Stages. *J. Virol.* **2010**, *84*, 10241–10253. [CrossRef]
81. Kenny, P.J.; Zhou, H.; Kim, M.; Skariah, G.; Khetani, R.S.; Drnevich, J.; Arcila, M.L.; Kosik, K.S.; Ceman, S. MOV10 and FMRP Regulate AGO2 Association with MicroRNA Recognition Elements. *Cell Rep.* **2014**, *9*, 1729–1741. [CrossRef]
82. Wang, H.; Chang, L.; Wang, X.; Su, A.; Feng, C.; Fu, Y.; Chen, D.; Zheng, N.; Wu, Z. MOV10 Interacts with Enterovirus 71 Genomic 5′UTR and Modulates Viral Replication. *Biochem. Biophys. Res. Commun.* **2016**, *479*, 571–577. [CrossRef]
83. Balinsky, C.A.; Schmeisser, H.; Wells, A.I.; Ganesan, S.; Jin, T.; Singh, K.; Zoon, K.C. IRAV (FLJ11286), an Interferon-Stimulated Gene with Antiviral Activity against Dengue Virus, Interacts with MOV10. *J. Virol.* **2017**, *91*. [CrossRef] [PubMed]
84. Li, J.; Hu, S.; Xu, F.; Mei, S.; Liu, X.; Yin, L.; Zhao, F.; Zhao, X.; Sun, H.; Xiong, Z.; et al. MOV10 Sequesters the RNP of Influenza A Virus in the Cytoplasm and Is Antagonized by Viral NS1 Protein. *Biochem. J.* **2019**, *476*, 467–481. [CrossRef] [PubMed]
85. Liu, D.; Ndongwe, T.P.; Puray-Chavez, M.; Casey, M.C.; Izumi, T.; Pathak, V.K.; Tedbury, P.R.; Sarafianos, S.G. Effect of P-body Component Mov10 on HCV Virus Production and Infectivity. *FASEB J.* **2020**, *34*, 9433–9449. [CrossRef] [PubMed]
86. Su, F.; Liu, X.; Jiang, Y. Roles of MOV10 in Animal RNA Virus Infection. *Front. Vet. Sci.* **2020**, *7*, 569737. [CrossRef] [PubMed]
87. Nawaz, A.; Shilikbay, T.; Skariah, G.; Ceman, S. Unwinding the Roles of RNA Helicase MOV10. *WIREs RNA* **2022**, *13*, e1682. [CrossRef] [PubMed]
88. Jaafar, Z.A.; Kieft, J.S. Viral RNA Structure-Based Strategies to Manipulate Translation. *Nat. Rev. Microbiol.* **2019**, *17*, 110–123. [CrossRef] [PubMed]
89. Sonenberg, N.; Hinnebusch, A.G. Regulation of Translation Initiation in Eukaryotes: Mechanisms and Biological Targets. *Cell* **2009**, *136*, 731–745. [CrossRef]
90. Leppek, K.; Das, R.; Barna, M. Functional 5′ UTR mRNA Structures in Eukaryotic Translation Regulation and How to Find Them. *Nat. Rev. Mol. Cell Biol.* **2018**, *19*, 158–174. [CrossRef]
91. Mailliot, J.; Martin, F. Viral Internal Ribosomal Entry Sites: Four Classes for One Goal. *WIREs RNA* **2018**, *9*, e1458. [CrossRef]
92. Ho, J.J.D.; Lee, S. A Cap for Every Occasion: Alternative eIF4F Complexes. *Trends Biochem. Sci.* **2016**, *41*, 821–823. [CrossRef]
93. Kahvejian, A.; Svitkin, Y.V.; Sukarieh, R.; M′Boutchou, M.-N.; Sonenberg, N. Mammalian Poly(A)-Binding Protein Is a Eukaryotic Translation Initiation Factor, Which Acts via Multiple Mechanisms. *Genes Dev.* **2005**, *19*, 104–113. [CrossRef]
94. Jackson, R.J.; Hellen, C.U.T.; Pestova, T.V. The Mechanism of Eukaryotic Translation Initiation and Principles of Its Regulation. *Nat. Rev. Mol. Cell Biol.* **2010**, *11*, 113–127. [CrossRef]
95. Brito Querido, J.; Díaz-López, I.; Ramakrishnan, V. The Molecular Basis of Translation Initiation and Its Regulation in Eukaryotes. *Nat. Rev. Mol. Cell Biol.* **2024**, *25*, 168–186. [CrossRef]
96. Yang, Y.; Wang, Z. IRES-Mediated Cap-Independent Translation, a Path Leading to Hidden Proteome. *J. Mol. Cell Biol.* **2019**, *11*, 911–919. [CrossRef]

97. Komar, A.A.; Hatzoglou, M. Cellular IRES-Mediated Translation. *Cell Cycle* **2011**, *10*, 229–240. [CrossRef]
98. Pelletier, J.; Sonenberg, N. Internal Initiation of Translation of Eukaryotic mRNA Directed by a Sequence Derived from Poliovirus RNA. *Nature* **1988**, *334*, 320–325. [CrossRef]
99. Jang, S.K.; Kräusslich, H.G.; Nicklin, M.J.; Duke, G.M.; Palmenberg, A.C.; Wimmer, E. A Segment of the 5′ Nontranslated Region of Encephalomyocarditis Virus RNA Directs Internal Entry of Ribosomes during in Vitro Translation. *J. Virol.* **1988**, *62*, 2636–2643. [CrossRef]
100. Tsukiyama-Kohara, K.; Iizuka, N.; Kohara, M.; Nomoto, A. Internal Ribosome Entry Site within Hepatitis C Virus RNA. *J. Virol.* **1992**, *66*, 1476–1483. [CrossRef]
101. Wilson, J.E.; Powell, M.J.; Hoover, S.E.; Sarnow, P. Naturally Occurring Dicistronic Cricket Paralysis Virus RNA Is Regulated by Two Internal Ribosome Entry Sites. *Mol. Cell Biol.* **2000**, *20*, 4990–4999. [CrossRef]
102. Rozman, B.; Fisher, T.; Stern-Ginossar, N. Translation—A Tug of War during Viral Infection. *Mol. Cell* **2023**, *83*, 481–495. [CrossRef]
103. Mears, H.V.; Sweeney, T.R. Better Together: The Role of IFIT Protein-Protein Interactions in the Antiviral Response. *J. Gen. Virol.* **2018**, *99*, 1463–1477. [CrossRef]
104. Diamond, M.S. IFIT1: A Dual Sensor and Effector Molecule That Detects Non-2′-O Methylated Viral RNA and Inhibits Its Translation. *Cytokine Growth Factor Rev.* **2014**, *25*, 543–550. [CrossRef]
105. Kumar, P.; Sweeney, T.R.; Skabkin, M.A.; Skabkina, O.V.; Hellen, C.U.T.; Pestova, T.V. Inhibition of Translation by IFIT Family Members Is Determined by Their Ability to Interact Selectively with the 5′-Terminal Regions of Cap0-, Cap1- and 5′ppp- mRNAs. *Nucleic Acids Res.* **2014**, *42*, 3228–3245. [CrossRef]
106. Fleith, R.C.; Mears, H.V.; Leong, X.Y.; Sanford, T.J.; Emmott, E.; Graham, S.C.; Mansur, D.S.; Sweeney, T.R. IFIT3 and IFIT2/3 Promote IFIT1-Mediated Translation Inhibition by Enhancing Binding to Non-Self RNA. *Nucleic Acids Res.* **2018**, *46*, 5269–5285. [CrossRef]
107. Richter, J.D.; Sonenberg, N. Regulation of Cap-Dependent Translation by eIF4E Inhibitory Proteins. *Nature* **2005**, *433*, 477–480. [CrossRef]
108. Ma, X.M.; Blenis, J. Molecular Mechanisms of mTOR-Mediated Translational Control. *Nat. Rev. Mol. Cell Biol.* **2009**, *10*, 307–318. [CrossRef]
109. Nunes, A.; Ribeiro, D.R.; Marques, M.; Santos, M.A.S.; Ribeiro, D.; Soares, A.R. Emerging Roles of tRNAs in RNA Virus Infections. *Trends Biochem. Sci.* **2020**, *45*, 794–805. [CrossRef]
110. Kim, E.T.; Weitzman, M.D. Schlafens Can Put Viruses to Sleep. *Viruses* **2022**, *14*, 442. [CrossRef]
111. Li, M.M.H.; Aguilar, E.G.; Michailidis, E.; Pabon, J.; Park, P.; Wu, X.; De Jong, Y.P.; Schneider, W.M.; Molina, H.; Rice, C.M.; et al. Characterization of Novel Splice Variants of Zinc Finger Antiviral Protein (ZAP). *J. Virol.* **2019**, *93*, e00715-19. [CrossRef]
112. Ficarelli, M.; Neil, S.J.D.; Swanson, C.M. Targeted Restriction of Viral Gene Expression and Replication by the ZAP Antiviral System. *Annu. Rev. Virol.* **2021**, *8*, 265–283. [CrossRef]
113. Yang, E.; Li, M.M.H. All About the RNA: Interferon-Stimulated Genes That Interfere With Viral RNA Processes. *Front. Immunol.* **2020**, *11*, 605024. [CrossRef]
114. Fonseca, B.D.; Zakaria, C.; Jia, J.-J.; Graber, T.E.; Svitkin, Y.; Tahmasebi, S.; Healy, D.; Hoang, H.-D.; Jensen, J.M.; Diao, I.T.; et al. La-Related Protein 1 (LARP1) Represses Terminal Oligopyrimidine (TOP) mRNA Translation Downstream of mTOR Complex 1 (mTORC1). *J. Biol. Chem.* **2015**, *290*, 15996–16020. [CrossRef]
115. Kinast, V.; Plociennikowska, A.; Anggakusuma; Bracht, T.; Todt, D.; Brown, R.J.P.; Boldanova, T.; Zhang, Y.; Brüggemann, Y.; Friesland, M.; et al. C19orf66 Is an Interferon-Induced Inhibitor of HCV Replication That Restricts Formation of the Viral Replication Organelle. *J. Hepatol.* **2020**, *73*, 549–558. [CrossRef]
116. Jäger, N.; Ayyub, S.A.; Peske, F.; Liedtke, D.; Bohne, J.; Hoffmann, M.; Rodnina, M.V.; Pöhlmann, S. The Inhibition of Gag-Pol Expression by the Restriction Factor Shiftless Is Dispensable for the Restriction of HIV-1 Infection. *Viruses* **2024**, *16*, 583. [CrossRef]
117. Kendra, J.A.; Advani, V.M.; Chen, B.; Briggs, J.W.; Zhu, J.; Bress, H.J.; Pathy, S.M.; Dinman, J.D. Functional and Structural Characterization of the Chikungunya Virus Translational Recoding Signals. *J. Biol. Chem.* **2018**, *293*, 17536–17545. [CrossRef]
118. Halma, M.T.J.; Ritchie, D.B.; Cappellano, T.R.; Neupane, K.; Woodside, M.T. Complex Dynamics under Tension in a High-Efficiency Frameshift Stimulatory Structure. *Proc. Natl. Acad. Sci. USA* **2019**, *116*, 19500–19505. [CrossRef]
119. Kelly, J.A.; Dinman, J.D. Shiftless Is a Novel Member of the Ribosome Stress Surveillance Machinery That Has Evolved to Play a Role in Innate Immunity and Cancer Surveillance. *Viruses* **2023**, *15*, 2296. [CrossRef]
120. Kobayashi, Y.; Zhuang, J.; Peltz, S.; Dougherty, J. Identification of a Cellular Factor That Modulates HIV-1 Programmed Ribosomal Frameshifting. *J. Biol. Chem.* **2010**, *285*, 19776–19784. [CrossRef]
121. Hanners, N.W.; Mar, K.B.; Boys, I.N.; Eitson, J.L.; De La Cruz-Rivera, P.C.; Richardson, R.B.; Fan, W.; Wight-Carter, M.; Schoggins, J.W. Shiftless Inhibits Flavivirus Replication in Vitro and Is Neuroprotective in a Mouse Model of Zika Virus Pathogenesis. *Proc. Natl. Acad. Sci. USA* **2021**, *118*, e2111266118. [CrossRef]
122. Napthine, S.; Hill, C.H.; Nugent, H.C.M.; Brierley, I. Modulation of Viral Programmed Ribosomal Frameshifting and Stop Codon Readthrough by the Host Restriction Factor Shiftless. *Viruses* **2021**, *13*, 1230. [CrossRef]
123. Rodriguez, W.; Mehrmann, T.; Hatfield, D.; Muller, M. Shiftless Restricts Viral Gene Expression and Influences RNA Granule Formation during Kaposi's Sarcoma-Associated Herpesvirus Lytic Replication. *J. Virol.* **2022**, *96*, e01469-22. [CrossRef] [PubMed]
124. Anderson, P.; Kedersha, N. RNA Granules: Post-Transcriptional and Epigenetic Modulators of Gene Expression. *Nat. Rev. Mol. Cell Biol.* **2009**, *10*, 430–436. [CrossRef] [PubMed]

125. Simpson-Holley, M.; Kedersha, N.; Dower, K.; Rubins, K.H.; Anderson, P.; Hensley, L.E.; Connor, J.H. Formation of Antiviral Cytoplasmic Granules during Orthopoxvirus Infection. *J. Virol.* **2011**, *85*, 1581–1593. [CrossRef]
126. Iwakawa, H.; Tomari, Y. Life of RISC: Formation, Action, and Degradation of RNA-Induced Silencing Complex. *Mol. Cell* **2022**, *82*, 30–43. [CrossRef]
127. Rusinova, I.; Forster, S.; Yu, S.; Kannan, A.; Masse, M.; Cumming, H.; Chapman, R.; Hertzog, P.J. INTERFEROME v2.0: An Updated Database of Annotated Interferon-Regulated Genes. *Nucleic Acids Res.* **2012**, *41*, D1040–D1046. [CrossRef]
128. Schoggins, J.W.; Wilson, S.J.; Panis, M.; Murphy, M.Y.; Jones, C.T.; Bieniasz, P.; Rice, C.M. A Diverse Range of Gene Products Are Effectors of the Type I Interferon Antiviral Response. *Nature* **2011**, *472*, 481–485. [CrossRef] [PubMed]
129. Kedersha, N.; Ivanov, P.; Anderson, P. Stress Granules and Cell Signaling: More than Just a Passing Phase? *Trends Biochem. Sci.* **2013**, *38*, 494–506. [CrossRef]
130. Luo, Y.; Na, Z.; Slavoff, S.A. P-Bodies: Composition, Properties, and Functions. *Biochemistry* **2018**, *57*, 2424–2431. [CrossRef]
131. Law, L.M.J.; Razooky, B.S.; Li, M.M.H.; You, S.; Jurado, A.; Rice, C.M.; MacDonald, M.R. ZAP's Stress Granule Localization Is Correlated with Its Antiviral Activity and Induced by Virus Replication. *PLoS Pathog.* **2019**, *15*, e1007798. [CrossRef] [PubMed]
132. Rodriguez, W.; Muller, M. Shiftless, a Critical Piece of the Innate Immune Response to Viral Infection. *Viruses* **2022**, *14*, 1338. [CrossRef] [PubMed]
133. Kunden, R.D.; Ghezelbash, S.; Khan, J.Q.; Wilson, J.A. Location Specific Annealing of miR-122 and Other Small RNAs Defines an Hepatitis C Virus 5′ UTR Regulatory Element with Distinct Impacts on Virus Translation and Genome Stability. *Nucleic Acids Res.* **2020**, *48*, 9235–9249. [CrossRef] [PubMed]
134. Malgras, M.; Garcia, M.; Jousselin, C.; Bodet, C.; Lévêque, N. The Antiviral Activities of Poly-ADP-Ribose Polymerases. *Viruses* **2021**, *13*, 582. [CrossRef] [PubMed]
135. Gläsker, S.; Töller, M.; Kümmerer, B.M. The Alternate Triad Motif of the Poly(ADP-Ribose) Polymerase-like Domain of the Human Zinc Finger Antiviral Protein Is Essential for Its Antiviral Activity. *J. Gen. Virol.* **2014**, *95*, 816–822. [CrossRef] [PubMed]
136. Guo, X.; Ma, J.; Sun, J.; Gao, G. The Zinc-Finger Antiviral Protein Recruits the RNA Processing Exosome to Degrade the Target mRNA. *Proc. Natl. Acad. Sci. USA* **2007**, *104*, 151–156. [CrossRef]
137. Chen, S.; Xu, Y.; Zhang, K.; Wang, X.; Sun, J.; Gao, G.; Liu, Y. Structure of N-Terminal Domain of ZAP Indicates How a Zinc-Finger Protein Recognizes Complex RNA. *Nat. Struct. Mol. Biol.* **2012**, *19*, 430–435. [CrossRef]
138. George, C.X.; Samuel, C.E. Human RNA-Specific Adenosine Deaminase *ADAR1* Transcripts Possess Alternative Exon 1 Structures That Initiate from Different Promoters, One Constitutively Active and the Other Interferon Inducible. *Proc. Natl. Acad. Sci. USA* **1999**, *96*, 4621–4626. [CrossRef]
139. Sun, T.; Yu, Y.; Wu, X.; Acevedo, A.; Luo, J.-D.; Wang, J.; Schneider, W.M.; Hurwitz, B.; Rosenberg, B.R.; Chung, H.; et al. Decoupling Expression and Editing Preferences of ADAR1 P150 and P110 Isoforms. *Proc. Natl. Acad. Sci. USA* **2021**, *118*, e2021757118. [CrossRef]
140. Deffit, S.N.; Hundley, H.A. To Edit or Not to Edit: Regulation of ADAR Editing Specificity and Efficiency. *WIREs RNA* **2016**, *7*, 113–127. [CrossRef]
141. Ota, H.; Sakurai, M.; Gupta, R.; Valente, L.; Wulff, B.-E.; Ariyoshi, K.; Iizasa, H.; Davuluri, R.V.; Nishikura, K. ADAR1 Forms a Complex with Dicer to Promote MicroRNA Processing and RNA-Induced Gene Silencing. *Cell* **2013**, *153*, 575–589. [CrossRef] [PubMed]
142. Ward, S.V.; George, C.X.; Welch, M.J.; Liou, L.-Y.; Hahm, B.; Lewicki, H.; De La Torre, J.C.; Samuel, C.E.; Oldstone, M.B. RNA Editing Enzyme Adenosine Deaminase Is a Restriction Factor for Controlling Measles Virus Replication That Also Is Required for Embryogenesis. *Proc. Natl. Acad. Sci. USA* **2011**, *108*, 331–336. [CrossRef] [PubMed]
143. Ng, C.S.; Kasumba, D.M.; Fujita, T.; Luo, H. Spatio-Temporal Characterization of the Antiviral Activity of the XRN1-DCP1/2 Aggregation against Cytoplasmic RNA Viruses to Prevent Cell Death. *Cell Death Differ.* **2020**, *27*, 2363–2382. [CrossRef] [PubMed]
144. Bushati, N.; Cohen, S.M. microRNA Functions. *Annu. Rev. Cell Dev. Biol.* **2007**, *23*, 175–205. [CrossRef] [PubMed]
145. Beckham, C.J.; Parker, R. P Bodies, Stress Granules, and Viral Life Cycles. *Cell Host Microbe* **2008**, *3*, 206–212. [CrossRef]
146. Buchan, J.R.; Parker, R. Eukaryotic Stress Granules: The Ins and Outs of Translation. *Mol. Cell* **2009**, *36*, 932–941. [CrossRef] [PubMed]
147. Bartel, D.P. Metazoan MicroRNAs. *Cell* **2018**, *173*, 20–51. [CrossRef]
148. Pedersen, I.M.; Cheng, G.; Wieland, S.; Volinia, S.; Croce, C.M.; Chisari, F.V.; David, M. Interferon Modulation of Cellular microRNAs as an Antiviral Mechanism. *Nature* **2007**, *449*, 919–922. [CrossRef] [PubMed]
149. Scagnolari, C.; Zingariello, P.; Vecchiet, J.; Selvaggi, C.; Racciatti, D.; Taliani, G.; Riva, E.; Pizzigallo, E.; Antonelli, G. Differential Expression of Interferon-Induced microRNAs in Patients with Chronic Hepatitis C Virus Infection Treated with Pegylated Interferon Alpha. *Virol. J.* **2010**, *7*, 311. [CrossRef]
150. Hecker, M.; Thamilarasan, M.; Koczan, D.; Schröder, I.; Flechtner, K.; Freiesleben, S.; Füllen, G.; Thiesen, H.-J.; Zettl, U. MicroRNA Expression Changes during Interferon-Beta Treatment in the Peripheral Blood of Multiple Sclerosis Patients. *IJMS* **2013**, *14*, 16087–16110. [CrossRef]
151. Forster, S.C.; Tate, M.D.; Hertzog, P.J. MicroRNA as Type I Interferon-Regulated Transcripts and Modulators of the Innate Immune Response. *Front. Immunol.* **2015**, *6*, 334. [CrossRef]
152. Mahesh, G.; Biswas, R. MicroRNA-155: A Master Regulator of Inflammation. *J. Interferon Cytokine Res.* **2019**, *39*, 321–330. [CrossRef]

153. Yang, C.; Li, K.; Pfeffer, S.; Pfeffer, L. The Type I IFN-Induced miRNA, miR-21. *Pharmaceuticals* **2015**, *8*, 836–847. [CrossRef]
154. Scheel, T.K.H.; Luna, J.M.; Liniger, M.; Nishiuchi, E.; Rozen-Gagnon, K.; Shlomai, A.; Auray, G.; Gerber, M.; Fak, J.; Keller, I.; et al. A Broad RNA Virus Survey Reveals Both miRNA Dependence and Functional Sequestration. *Cell Host Microbe* **2016**, *19*, 409–423. [CrossRef]
155. Arghiani, N.; Nissan, T.; Matin, M.M. Role of microRNAs in COVID-19 with Implications for Therapeutics. *Biomed. Pharmacother.* **2021**, *144*, 112247. [CrossRef]
156. Siniscalchi, C.; Di Palo, A.; Russo, A.; Potenza, N. Human MicroRNAs Interacting With SARS-CoV-2 RNA Sequences: Computational Analysis and Experimental Target Validation. *Front. Genet.* **2021**, *12*, 678994. [CrossRef]
157. Belew, A.T.; Meskauskas, A.; Musalgaonkar, S.; Advani, V.M.; Sulima, S.O.; Kasprzak, W.K.; Shapiro, B.A.; Dinman, J.D. Ribosomal Frameshifting in the CCR5 mRNA Is Regulated by miRNAs and the NMD Pathway. *Nature* **2014**, *512*, 265–269. [CrossRef]
158. Wang, J.; Zhang, T.; Yu, Z.; Tan, W.T.; Wen, M.; Shen, Y.; Lambert, F.R.P.; Huber, R.G.; Wan, Y. Genome-Wide RNA Structure Changes during Human Neurogenesis Modulate Gene Regulatory Networks. *Mol. Cell* **2021**, *81*, 4942–4953.e8. [CrossRef]
159. Cunningham, C.L.; Frye, C.J.; Makowski, J.A.; Kensinger, A.H.; Shine, M.; Milback, E.J.; Lackey, P.E.; Evanseck, J.D.; Mihailescu, M.-R. Effect of the SARS-CoV-2 Delta-Associated G15U Mutation on the S2m Element Dimerization and Its Interactions with miR-1307-3p. *RNA* **2023**, *29*, 1754–1771. [CrossRef]
160. Panigrahi, M.; Palmer, M.A.; Wilson, J.A. Enhanced Virus Translation Enables miR-122-Independent Hepatitis C Virus Propagation. *J. Virol.* **2023**, *97*, e00858-21. [CrossRef]
161. Scott, S.; Li, Y.; Bermek, O.; Griffith, J.D.; Lemon, S.M.; Choi, K.H. Binding of microRNA-122 to the Hepatitis C Virus 5′ Untranslated Region Modifies Interactions with Poly(C) Binding Protein 2 and the NS5B Viral Polymerase. *Nucleic Acids Res.* **2023**, *51*, 12397–12413. [CrossRef] [PubMed]
162. Girardi, E.; López, P.; Pfeffer, S. On the Importance of Host MicroRNAs During Viral Infection. *Front. Genet.* **2018**, *9*, 439. [CrossRef] [PubMed]
163. Gao, G.; Guo, X.; Goff, S.P. Inhibition of Retroviral RNA Production by ZAP, a CCCH-Type Zinc Finger Protein. *Science* **2002**, *297*, 1703–1706. [CrossRef] [PubMed]
164. Bick, M.J.; Carroll, J.-W.N.; Gao, G.; Goff, S.P.; Rice, C.M.; MacDonald, M.R. Expression of the Zinc-Finger Antiviral Protein Inhibits Alphavirus Replication. *J. Virol.* **2003**, *77*, 11555–11562. [CrossRef] [PubMed]
165. Müller, S.; Möller, P.; Bick, M.J.; Wurr, S.; Becker, S.; Günther, S.; Kümmerer, B.M. Inhibition of Filovirus Replication by the Zinc Finger Antiviral Protein. *J. Virol.* **2007**, *81*, 2391–2400. [CrossRef] [PubMed]
166. Mao, R.; Nie, H.; Cai, D.; Zhang, J.; Liu, H.; Yan, R.; Cuconati, A.; Block, T.M.; Guo, J.-T.; Guo, H. Inhibition of Hepatitis B Virus Replication by the Host Zinc Finger Antiviral Protein. *PLoS Pathog.* **2013**, *9*, e1003494. [CrossRef] [PubMed]
167. Liu, C.-H.; Zhou, L.; Chen, G.; Krug, R.M. Battle between Influenza A Virus and a Newly Identified Antiviral Activity of the PARP-Containing ZAPL Protein. *Proc. Natl. Acad. Sci. USA* **2015**, *112*, 14048–14053. [CrossRef] [PubMed]
168. Kmiec, D.; Lista, M.J.; Ficarelli, M.; Swanson, C.M.; Neil, S.J.D. S-Farnesylation Is Essential for Antiviral Activity of the Long ZAP Isoform against RNA Viruses with Diverse Replication Strategies. *PLoS Pathog.* **2021**, *17*, e1009726. [CrossRef]
169. Rebendenne, A.; Roy, P.; Bonaventure, B.; Chaves Valadão, A.L.; Desmarets, L.; Arnaud-Arnould, M.; Rouillé, Y.; Tauziet, M.; Giovannini, D.; Touhami, J.; et al. Bidirectional Genome-Wide CRISPR Screens Reveal Host Factors Regulating SARS-CoV-2, MERS-CoV and Seasonal HCoVs. *Nat. Genet.* **2022**, *54*, 1090–1102. [CrossRef]
170. Hou, J.; Wei, Y.; Zou, J.; Jaffery, R.; Sun, L.; Liang, S.; Zheng, N.; Guerrero, A.M.; Egan, N.A.; Bohat, R.; et al. Integrated Multi-Omics Analyses Identify Anti-Viral Host Factors and Pathways Controlling SARS-CoV-2 Infection. *Nat. Commun.* **2024**, *15*, 109. [CrossRef]
171. Steiner, S.; Kratzel, A.; Barut, G.T.; Lang, R.M.; Aguiar Moreira, E.; Thomann, L.; Kelly, J.N.; Thiel, V. SARS-CoV-2 Biology and Host Interactions. *Nat. Rev. Microbiol.* **2024**, *22*, 206–225. [CrossRef] [PubMed]
172. Boersma, S.; Khuperkar, D.; Verhagen, B.M.P.; Sonneveld, S.; Grimm, J.B.; Lavis, L.D.; Tanenbaum, M.E. Multi-Color Single-Molecule Imaging Uncovers Extensive Heterogeneity in mRNA Decoding. *Cell* **2019**, *178*, 458–472.e19. [CrossRef] [PubMed]
173. Boersma, S.; Rabouw, H.H.; Bruurs, L.J.M.; Pavlovič, T.; Van Vliet, A.L.W.; Beumer, J.; Clevers, H.; Van Kuppeveld, F.J.M.; Tanenbaum, M.E. Translation and Replication Dynamics of Single RNA Viruses. *Cell* **2020**, *183*, 1930–1945.e23. [CrossRef] [PubMed]
174. Lyon, K.; Aguilera, L.U.; Morisaki, T.; Munsky, B.; Stasevich, T.J. Live-Cell Single RNA Imaging Reveals Bursts of Translational Frameshifting. *Mol. Cell* **2019**, *75*, 172–183.e9. [CrossRef] [PubMed]
175. Bruurs, L.J.M.; Müller, M.; Schipper, J.G.; Rabouw, H.H.; Boersma, S.; Van Kuppeveld, F.J.M.; Tanenbaum, M.E. Antiviral Responses Are Shaped by Heterogeneity in Viral Replication Dynamics. *Nat. Microbiol.* **2023**, *8*, 2115–2129. [CrossRef] [PubMed]
176. Chuong, E.B. The Placenta Goes Viral: Retroviruses Control Gene Expression in Pregnancy. *PLoS Biol.* **2018**, *16*, e3000028. [CrossRef]
177. Luganini, A.; Gribaudo, G. Retroviruses of the Human Virobiota: The Recycling of Viral Genes and the Resulting Advantages for Human Hosts During Evolution. *Front. Microbiol.* **2020**, *11*, 1140. [CrossRef]

viruses MDPI

Review

Translational Control of Alphavirus–Host Interactions: Implications in Viral Evolution, Tropism and Antiviral Response

Iván Ventoso [1,*], Juan José Berlanga [1], René Toribio [2] and Irene Díaz-López [3]

[1] Centro de Biología Molecular "Severo Ochoa" (CSIC-UAM) and Departamento de Biología Molecular, Universidad Autónoma de Madrid (UAM), 28049 Madrid, Spain; jberlanga@cbm.csic.es
[2] Centro de Biotecnología y Genómica de Plantas, Universidad Politécnica de Madrid-Instituto Nacional de Investigación y Tecnología Agraria y Alimentaria (UPM-INIA), 28049 Madrid, Spain; toribio.francisco@upm.es
[3] MRC Laboratory of Molecular Biology, Cambridge CB2 0QH, UK; idiaz@mrc-lmb.cam.ac.uk
* Correspondence: iventoso@cbm.csic.es; Tel.: +34-911964809

Abstract: Alphaviruses can replicate in arthropods and in many vertebrate species including humankind, but only in vertebrate cells do infections with these viruses result in a strong inhibition of host translation and transcription. Translation shutoff by alphaviruses is a multifactorial process that involves both host- and virus-induced mechanisms, and some of them are not completely understood. Alphavirus genomes contain cis-acting elements (RNA structures and dinucleotide composition) and encode protein activities that promote the translational and transcriptional resistance to type I IFN-induced antiviral effectors. Among them, IFIT1, ZAP and PKR have played a relevant role in alphavirus evolution, since they have promoted the emergence of multiple viral evasion mechanisms at the translational level. In this review, we will discuss how the adaptations of alphaviruses to vertebrate hosts likely involved the acquisition of new features in viral mRNAs and proteins to overcome the effect of type I IFN.

Keywords: alphaviruses; translation initiation; eIF2α phosphorylation; RNA structure; antiviral response; interferon; evolution; host range

1. Alphavirus Replication, Tropism and Interference with Host Gene Expression

The *Alphavirus* genus includes over 30 viral species that are classified in seven complexes based on their antigenic characteristics (ICTV Taxonomy 2022 release) (Figure 1A). Most alphaviruses are transmitted between vertebrate hosts (e.g., nonhuman primates, birds, rodents and marsupials) by hematophagous arthropods, mainly mosquitoes [1,2]. In general, the host tropism in alphaviruses is broad, and many zoonotic transmissions to humans have been documented. Thus, the chikungunya virus (CHIKV) has evolved from a sylvatic cycle between primates and forest mosquitoes to an urbanized transmission that involves mosquitoes of the *Aedes* genus (e.g., tiger mosquito) and humans [3–5]. The emergence of Venezuelan equine encephalitis virus (VEEV) strains causing infection in humans is also frequent [6]. Historically, mosquito-borne alphaviruses have been divided into Old World and New World viruses according not only to their geographical distribution, but also to the clinical manifestations they cause in humans [1]. Thus, Old World alphaviruses include the CHIKV, Sindbis virus (SINV), Semliki Forest virus (SFV), Ross River virus (RRV) and o'nyong-nyong virus (ONNV), which were first isolated in Africa and Australia and mainly cause arthritis. New World alphaviruses include the Venezuelan (VEEV), eastern (EEEV) and western (WEEV) equine encephalitis viruses, which are endemic to North, Central and South America and cause diseases in horses and humans with neurological symptoms [7].

Alphaviruses are enveloped, and their genomes consist of a positive-sense single-stranded RNA (ssRNA) of approximately 12 kilobases that encodes four nonstructural proteins (nsP1-4) and five structural proteins (capsid, E3, E2, 6k and E1) in two large open reading frames that are translated as polyproteins (Figure 1B) [8,9]. The nonstructural polyprotein is translated directly from the genomic RNA (gmRNA), whereas the structural polyprotein is translated from a subgenomic mRNA (sgmRNA) that is transcribed by the viral replicase from an internal promotor. Both gmRNA and sgmRNA are 5′ capped (m7Gppp or cap0) and 3′ polyadenylated (Figure 1B) [10]. The nsp1 has both N-7 methyltransferase and guanyltransferase activities that are responsible for the capping of a fraction of both the gmRNA and sgmRNA during RNA transcription [11–13]. The nonstructural protein 2 (nsp2) is a multifunctional protein involved in viral RNA synthesis, polyprotein processing, the blockade of host gene expression and viral pathogenesis [9,14]. In addition to its role as the main viral protease that cleaves the nonstructural polyprotein precursor (p1234), nsp2 also contains a NTPase/RNA helicase domain involved in viral RNA synthesis and a SAM (S-adenosyl-methionine)-dependent methyltransferase-like domain that cooperates with the RNA helicase domain [9]. The role of nsp3 in alphavirus replication involves the recruitment of multiple host factors towards the viral replication complex (vRCs), including members of the Ras–GAP SH3 domain-binding proteins family (G3BP) in mammals and the corresponding ortholog in mosquito (Rasputin) [15,16]. Interestingly, since G3BP is involved in stress granule formation upon stress-induced eIF2α phosphorylation, G3BPs hijacking by nsp3 has been proposed to induce the rapid disassembly of stress granules that are formed early in response to an infection, although very recent reports have questioned this idea [17–19]. Nsp4 is the viral RNA-dependent RNA polymerase involved in negative- and positive-stranded RNA synthesis to generate gmRNA and sgmRNA [20]. The regulation of these activities involves the sequential processing of the nsP1-4 precursor by the nsp2 protease [21]. In the early stages of infection, the partially processed nsPs (P123 + nsp4) preferentially synthesize a negative RNA strand to form double-stranded RNA intermediates (dsRNA). Later, the fully processed nsPs produce gmRNA and sgmRNA [21,22]. Recent data suggest that viral RNA synthesis initiates in the discrete foci associated with the plasma membrane to further translocate into the cytoplasm to form bigger membrane-bound spherules (or cytopathic vacuoles) containing the mature vRCs [23–25]. Although dsRNA replicative forms tend to accumulate into the spherules to serve as templates for the synthesis of new gmRNA molecules, the fact that antiviral sensors such as PKR, RIG-I and MDA5 become activated after alphavirus infection suggests that some viral dsRNA molecules are exposed to these antiviral sensors (Figure 1B) [26].

Although alphavirus replication has been studied mainly in mammalian cells, comparative analyses also found similar vRCs associated with the membrane-bound spherules in insect cells infected with SINV and other alphaviruses, suggesting that the basic aspects of alphavirus replication are conserved among these hosts [28,29]. However, the outcome of virus replication in mammalian and insect cells is different, suggesting the existence of underlying differences in the way virus replication impacts the host cell physiology. Alphavirus replication in mammalian cells is highly productive, and it is generally associated with a strong cytopathic effect that precedes cell lysis. In mosquito cells, however, alphaviruses easily establish nonlytic, persistent infections, where an active viral replication occurs without signs of a cytopathic effect (Figure 2) [28,30,31]. This fact nicely reflects how alphaviruses and other arboviruses use insects both as reservoirs to persist in nature and as vectors to ensure transmission to vertebrate hosts. At a molecular level, the replication of alphaviruses in mammalian cells is associated with a profound inhibition of both host transcription and translation that ultimately contributes to the cytopathogenesis. Thus, mammalian cells infected with representative Old World and New World members almost exclusively synthesize viral RNAs and proteins at later times of infection. In insect cells, on the contrary, the synthesis of large amounts of viral RNAs and proteins is well tolerated and occurs without significant interference with host transcription and translation [32,33].

This suggests that instead of relying solely on the replication process, alphaviruses employ specific mechanisms to halt gene expression in mammalian cells.

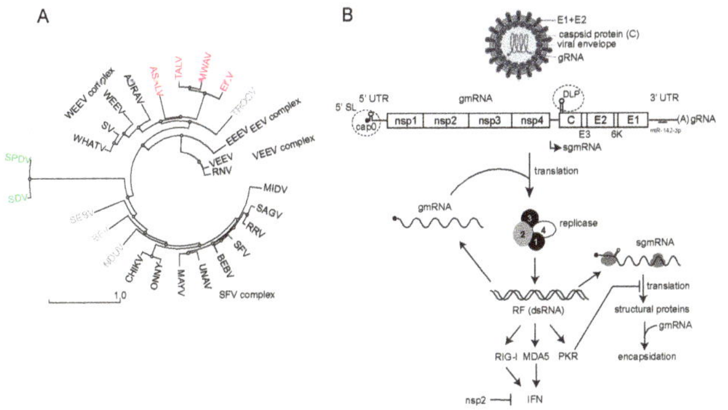

Figure 1. (**A**) Phylogenetic tree of some members of the *Alphavirus* genus. Structural polyprotein sequences (C-E3-E2-6K-E1) were aligned using Muscle, and the resulting distance-based phylogenetic tree was built using the tools available in the NGPhylogeny.fr suite. Some of the main complexes of the genus are shown. Insect-specific viruses close to the WEEV complex are in red. Aquatic alphaviruses infecting fish are in green. Members of other complexes are in light gray. (**B**) Schematic representation of alphavirus replication cycle including virion composition and genomic organization. The nonstructural (nsP1-3) and structural coding sequences (C-E3-E2-6K-E1) are indicated, as well as some of the cis-acting structures located in the 5′UTR of gmRNA and sgmRNA that are relevant to this review (dashed circles). Viral replicase is involved in the synthesis of (−) ssRNA, (+) gmRNA and sgmRNA from dsRNA intermediaries (RF). gmRNA and sgmRNA are translated independently to produce nonstructural and structural proteins, respectively. Capsid protein recruits and assembles gmRNA to produce new virus particles. The accumulation of dsRNA molecules can trigger the activation of antiviral sensors (PKR, RIG-I and MDA5) that results in an antiviral response and the synthesis of type I IFN. As a countermeasure, viral nsp2 blocks transcription of the IFN gene in mammalian cells. Binding of a specific microRNA (miR-142-3p) to the 3′ NC region of EEEV that restricts the replication of this virus in hematopoietic cells is also shown [27].

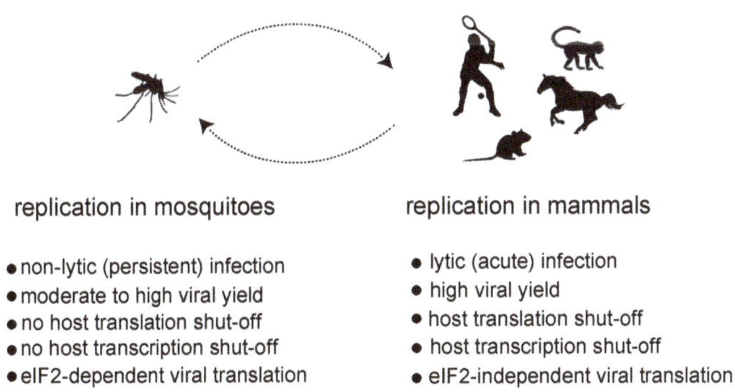

Figure 2. Comparison of the main characteristics associated with replication of alphaviruses in mosquitoes and mammalian hosts.

2. Mechanisms of Virus-Induced Host Translation Shutoff

Like many cytolytic viruses that infect mammals, an alphavirus infection often results in a strong blockade of host gene expression at both the transcriptional and translational levels [34]. The causative mechanisms of this interference or "shut-off" have been studied in both Old World (SINV and CHIKV) and New World (VEEV) members of the genus, using highly susceptible cell lines of human and murine origins [14,35–37]. Thus, the infection of BHK21 or murine embryonic fibroblasts (MEFs) with SINV and SFV at a high multiplicity of infection resulted in a complete shutoff of host translation at 3–4 h postinfection, concomitant with an almost exclusive translation of viral mRNAs, which can represent up to 30% of the total translation activity of uninfected cells. In other alphaviruses such as the VEEV and CHIKV, this effect is less dramatic in cell culture models [37–40]. Both the ongoing and de novo translation initiations of host mRNAs are abrogated in SINV-infected cells, suggesting that both the initiation and reinitiation (by ribosomal recycling) of mRNAs are blocked [35].

Different causative mechanisms have been proposed to explain the shutoff induced by alphaviruses [14,41,42]. Regarding mechanisms that impact the activities of translation factors, the strong phosphorylation of eIF2α by dsRNA-activated kinase (PKR) activation observed in cells infected with the SINV, SFV and VEEV could explain the translational block of the vast majority of mRNAs in infected cells [35,36,43]. eIF2α phosphorylation and host translation shutoff was also confirmed in mouse brain and organotypic cultures infected with the SINV [44]. However, SINV and CHIKV infections of PKR-knockout MEFs or PKR-knockdown human fibroblasts (HF), respectively, still induced a shutoff comparable to the control cells, suggesting the existence of an underlying mechanism(s) to halt host translation [35,36]. As discussed below, eIF2α phosphorylation in response to infection must be interpreted as an attempt of the host cell to prevent the translation of viral mRNAs in the context of an IFN response. More recently, a significant phosphorylation of elongation factor 2 (eEF2) has been detected in cells infected with many alphaviruses, including the CHIKV, SINV, SFV and VEEV [45]. eEF2 phosphorylation at the T56 residue reduces the activity of this factor, thus decreasing the rate of translation elongation [46]. The helicase activity associated with the NTPase domains of the CHIKV and VEEV nsp2s was sufficient to induce eEF2 phosphorylation and translation inhibition, although the contribution of this modification to the host shutoff induced by the CHIKV and other alphaviruses remains to be determined [45]. The activity of the nsp2s of Old World alphaviruses has also been linked to translation shutoff, since the SINV and CHIKV with point mutations in nsp2 that abrogated the host transcriptional inhibitory activity also failed to induce a complete inhibition of host translation in infected cells [14,36]. However, it is not clear if this defect in the translation shutoff could be an indirect consequence of a lesser impact of the mutated nsp2 on RNAPII activity that could make more mRNA available in the cytoplasm for translation.

Other indirect causes of shutoff have been proposed, although the precise mechanisms involved and their contributions to the shutoff phenomenon are still an open question. Thus, the large quantities of subgenomic mRNAs (sgmRNAs) accumulated in cells infected with the SINV and SFV could sequester translation components, such as ribosomes and translation factors, to redirect translation towards sgmRNAs [35,47]. Although sgmRNA is efficiently translated in infected cells by acting as a super competitor mRNA, the fact that replicons of the SINV and VEEV, which lack the entire sgmRNA, still repressed host translation, which seriously limits the contribution of sgmRNA accumulation to the translation shutoff by these viruses [48,49]. Related to this fact, the viral RNA synthesis directed by alphaviral replicases has been correlated to host translation inhibition in SINV-infected cells. Thus, the addition of RNA synthesis inhibitors such as 6-aza-uridine or ribavirin reduced the accumulation of both genomic RNA (gmRNA) and sgmRNA in SINV-infected cells and partially prevented or delayed the inhibition of host translation [47]. Some authors also reported the release of nuclear proteins to the cytoplasm in SINV-infected cells that could be interfering with translation by sequestering host mRNAs [47,50]. However, a direct causative correlation between these observations and the shutoff phenomenon is still lacking, so their contribution remains to be measured. Early reports also described an inhibitory effect of the capsid protein of the SFV on host translation [51]. Finally, alterations in the plasma membrane permeability observed in cells infected with alphaviruses and other animal viruses have also been proposed as a cause of translation shutoff some decades ago, but currently, these lines of investigation have been discontinued.

3. Interferon-Induced Genes That Block Translation of Alphaviral mRNAs: Roles of IFIT-1, ZAP and PKR

Alphaviruses induce strong innate and adaptive responses that greatly influence the outcome of infection in animal models [7,52,53]. Thus, type I interferon (IFN-α/β) signaling is critical for the protection against many alphaviruses, in that mice devoid of type I IFN receptor (IFNAR) easily succumbed to infection by the SINV, SFV, CHIKV and VEEV [54–56]. In cultured cells, all alphaviruses tested to date showed a sensitivity to IFN-α/β pretreatment [40,57–59]. The infection of both primary and immortalized cell lines with alphaviruses may result in the secretion of significant amounts of IFN-α/β, and both RIG-I and MDA-5 sensors are required for an optimal response against SINV and VEEV [26,37,60–62]. However, in general, alphaviruses are not considered strong inducers of IFN secretion. As discussed below, the inhibitory effects of viral nsp2 and eIF2α phosphorylation on host transcription and translation, respectively, limit the synthesis of IFN and other antiviral cytokines in alphavirus-infected cells.

IFN-α/β exert their antiviral effects through the activities of hundreds of genes whose expressions are stimulated by IFN signaling. Many of these interferon-stimulated genes (ISGs) encode proteins that are effectors of the antiviral response by blocking specific steps of the virus cycle such as decapsidation, RNA synthesis, viral translation or virus release [63]. Different combinations of these effectors can result in a specialized response against specific virus families, although some of them exert a broader antiviral activity [64]. Individual and combinatorial gene-targeting studies have identified a dozen ISGs that restrict alphavirus replication in cultured cells and in animal models [65,66]. Previous reports have shown that type I IFN pretreatment blocked the translation of alphaviruses, but not cellular mRNAs [67]. Interestingly, many of these IFN effectors exert their activity by blocking translation and/or by inducing the degradation of viral mRNAs, a finding that highlights the relevance of translation in controlling virus–host interactions and antiviral responses. These effectors include PKR, RNAseL/OAS, ISG20, ZAP, IFIT1 and IFIT3. In a recent loss-of-function screening analysis, the simple combination of three of these ISGs (ZAP, IFIT1 and IFIT3) accounted for the majority of the IFN-mediated restriction of VEEV and SINV replication in cultured cells [66].

IFIT1. Interferon-induced proteins with tetratricopeptide repeats (IFITs) are robustly induced by IFN-α/β treatment. IFIT1 is the prototype of this family, showing the strongest antiviral activity against several virus families including *Alphavirus* and *Coronavirus* [68,69]. The antiviral effects of IFITs depend on their ability to bind the 5′ extreme of mRNA to block the translation initiation [70]. However, the presence and type of cap structure greatly influences the affinity of the IFIT proteins for mRNAs. IFIT1 can interact with cap0-mRNAs, the simplest version of the cap structure, whereas IFI1B can also interact with cap1-mRNAs that contain an additional 2′-O-methylation at the first nucleotide of the mRNA [71]. Some groups also reported the interaction of IFIT1 with uncapped 5′-ppp-mRNA, although further studies suggested that only IFIT5 was able to interact with 5′-ppp-RNA in vitro [72]. In addition, IFIT3 can modulate the cap-binding affinity/specificity of IFIT1 [73,74]. Interestingly, further 2′-O-methylation at the second nucleotide of the mRNA (cap2) completely prevented the interaction of IFI1/1B with the cap structure, allowing the cellular mRNAs that generally contain cap2 structures to escape from IFIT-mediated inhibition [72]. Alphaviruses, like other viral families, encode their own capping enzyme (nsp1), but they lack 2′-O-methyl activity, so the resulting cap0-mRNAs are targets of IFIT1-mediated translation inhibition (Figure 3). Elegant experiments using chimeric nonreplicative alphaviruses showed that IFIT1 blocked the translation of incoming gmRNA upon IFN treatment [75]. However, as discussed below, not all the alphaviruses are equally sensitive to IFIT1 because of adaptive changes to evade the IFN response.

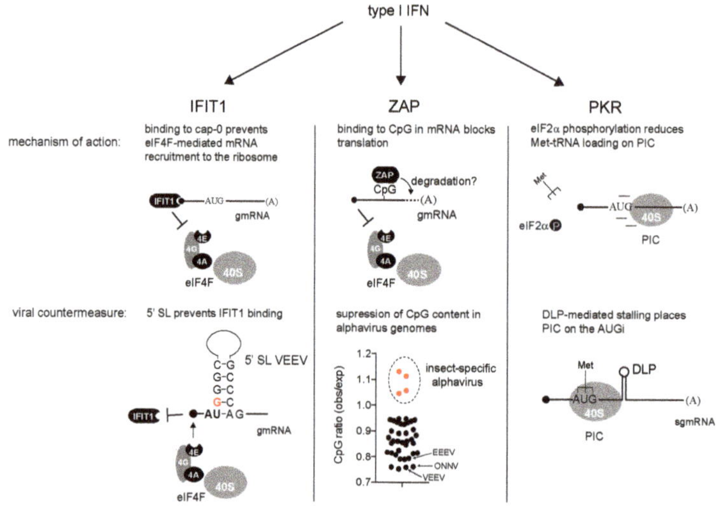

Figure 3. Mechanistic action of IFIT1, ZAP and PKR on the translation of alphaviral mRNA and some of the countermeasures used by alphaviruses to evade them. The 5′ stem loop in the gmRNA of VEEV (wt 3908 strain) is schematically shown. The G + 3, which is critical for IFIT1 resistance, is colored in red. To calculate the CpG ratio, the frequencies of CpG dinucleotides in full length genomes of 41 alphaviruses were divided by the product of corresponding C and G frequencies. Insect-specific alphaviruses are EILV, TALV, ASALV and MWAV. eIF2α phosphorylation not only prevents the location of AUGi of mRNA, but also reduces the formation of PIC (translation pre-initiation complex). For simplicity, the rest of components of the translation initiation complex were omitted. Black dots: alphaviruses infecting vertebrates. Red dots: insect-specific alphaviruses.

ZAP. The zinc finger antiviral protein (ZAP or ZC3HAV1) is able to inhibit the replication of many alphaviruses including the SINV, RRV, SFV and to a lesser extent, the VEEV, CHIKV and ONNV, as well members of other RNA and DNA virus families [65,76–79]. The ZAP is expressed as four splice isoforms that share an N-terminal domain with four CCH zinc fingers that are directly involved in RNA binding [80–83]. The large isoform (ZAPL) is constitutively expressed, thus providing the cells with basal ZAP activity, whereas the small isoform (ZAPS) is significantly induced by type I IFN treatment (2–3-fold increase) [81,82]. ZAP binding to viral mRNAs restricts virus replication by blocking translation and/or inducing the degradation of the target mRNAs (Figure 3) [77,84]. Structural and biochemical analyses showed that the RNA binding domains of mammalian ZAPs bind ssRNA molecules containing CpG dinucleotides with a high affinity [85–87]. This preference for CpG-rich regions in RNAs has been interpreted as the basis for ZAP specificity that could discriminate between self (host) and nonself (viral) mRNAs, since CpG is generally underrepresented in mammalian genomes [85,88]. Although ZAP binding to specific regions of the HIV-1 and SINV that are rich in CpG dinucleotides has been reported, the basis for this specificity has not yet been fully defined [80,87,89]. ZAP binding requires not only the proper spacing of CpGs, but also, the nucleotide composition surrounding CpGs influences the ZAP's affinity for target RNAs [86,90]. In alphaviruses, the ZAP blocked the translation initiation by preventing the assembly of the cap binding complex (eIF4F) on the incoming viral gmRNAs at early times of infection (Figure 3) [84,91]. However, since the ZAP preferentially binds to internal regions of the SINV gmRNA, it is not clear how ZAP binding prevents mRNA recruitment by the eIF4F complex, which occurs distally [80]. The destabilization of alphaviral mRNAs upon ZAP binding could also occur as described in other viruses, so that both mechanisms could be operating to restrict the replication of alphaviruses [92]. For alphaviruses, the longer ZAP isoforms showed a better antiviral effect [82]. A full ZAP activity requires cofactors such as the ubiquitin E3 ligase tripartite motif-containing protein (TRIM25) that also interacts with viral mRNAs [91,93].

PKR. The dsRNA-activated protein kinase (PKR) was the first ISG described, and it represents one of the four eIF2α kinases that respond to stress in mammals [94–97]. Although constitutive levels of PKR are present in most cell types, especially in fibroblasts, IFN treatment increases up to 10-fold the amount of PKR in the cell [95,98]. PKR is one of the main effectors of IFN-α/β, and animals devoid of PKR are much more susceptible to RNA and DNA viruses [97–99]. PKR plays a dual role as a sensor and IFN effector, owing to the presence of two dsRNA binding motifs (dsRBMs) at the N-terminal of the kinase [96]. The accumulation of dsRNA molecules as byproducts of viral RNA replication activates the PKR that phosphorylates the eIF2α at the S51 residue, thus preventing the recycling of this factor necessary to charge 40S subunits with new molecules of Met-tRNAi for translation initiation (Figure 3) [95,96]. Even low levels of eIF2α phosphorylation in response to PKR activation can have an amplified impact on translation, because phospho-eIF2 acts as a competitive inhibitor of eIF2 recycling, which involves the GTP-exchange factor eIF2B [100,101]. Since eIF2α phosphorylation leads to a general blockade of both viral and host translations, the effect of PKR activation can be considered less discriminative when compared to ZAP and IFIT1. Most pathogenic viruses express viral antagonists that prevent or counteract PKR activation in infected cells, thus making these viruses more resistant to IFN [96,102]. However, alphaviruses generally do not prevent the activation of dsRNA-activated kinase (PKR) in infected cells [35]. In return, they have acquired the ability to initiate translation independently of eIF2 as discussed below.

4. Adaptive Changes in Alphavirus mRNAs to Counteract the Antiviral Effect of IFN

As a result of thousands of years of virus–host coevolution, most pathogenic viruses adapted to counteract the antiviral activity of IFN [103]. In alphaviruses, these adaptive changes involved the acquisition of cis-acting elements in genomic and subgenomic mRNAs, changes in the dinucleotide composition and the occurrence of new activities in viral proteins that greatly influence viral translation and host gene expression. Collectively, these features and activities are aimed to promote the resistance to IFN (Figure 3).

IFIT1 resistance. Pioneer work by Hyde et al. reported a single mutation (G to A) at nucleotide 3 in the 5′ UTR of the gmRNA of the VEEV that drastically increased the sensitivity to IFIT1 and IFN in cell culture and animal models [69]. A structural analysis confirmed the existence of a short stem loop located at the 5′ extreme of gmRNA, whose stability (and/or topology) reduced the binding affinity of IFIT1 for VEEV gmRNA (Figure 3). Further studies found equivalent RNA structures in other alphaviruses, including the SINV, SFV, CHIKV and EEEV, that also promoted the resistance to IFIT1 at different extents depending on the virus [75]. A mutational analysis showed that, to generate IFIT1 resistance, the two first nucleotides of the gmRNA need to be AU (paired or not) and need to be followed by a G–C-rich stem likely involved in structure stabilization and/or topology (Figure 3) [75]. Interestingly, the existence of this structure was originally described in alphaviruses as a conserved element (CSE 1) acting as a promotor for the synthesis of both negative and positive RNA strands [8]. This raises the possibility that a previous structural element in the 5′ extreme involved in virus replication could also be adapted to shield the gmRNA against IFIT1, thus mimicking the protective effect of 2′ methylation at the first and second positions in cellular mRNA [69]. However, not all pathogenic alphaviruses are equally resistant to the antiviral effect of IFIT1. Thus, the EEEV shows only a partial resistance to IFIT1 that is probably compensated by the low levels of type I IFN induced by this virus in cultured cells and animals [53,75,104].

ZAP resistance. As described above, ZAP binds CpG-rich regions in mRNA to promote the translational repression and further degradation of alphaviral gmRNAs. However, since alphaviruses are not equally sensitive to ZAP-mediated restriction, some reports have analyzed the correlation between the CpG content of viral genome and the ZAP sensitivity. At the genomic scale, no obvious correlation was found between the sensitivity to the ZAP and the CpG composition among the main pathogenic alphaviruses analyzed. Therefore, some reports suggested that alphavirus evasion of the ZAP correlated with CpG suppression in specific regions of gmRNA, although this possibility remains to be confirmed [89]. Expanding the focus to include the insect-specific alphaviruses in the analysis could shed light on the phenomenon of CpG suppression in pathogenic alphaviruses. Insect-specific alphaviruses includes EILAT (EILV), TaiForest (TALV), AguaSalud (ASALV) and Mwinilunga (MWAV), which are close to the WEEV group, but they are unable to replicate in vertebrate cells [105–108]. The CpG frequency found in these viruses fits what is expected according to the C and G compositions of their genomes (Figure 3). In contrast, a clear overall CpG suppression was found in the rest of the vertebrate-adapted members of the genus, especially in the VEEV and ONNV, which showed the lowest sensitivity to the ZAP (Figure 3) [75]. Since no orthologs of the ZAP and IFN genes are found in insects, insect-specific alphaviruses have not been subjected to the selective pressures imposed by these genes, so that CpG suppression found in the rest of the alphaviruses could be interpreted as a footprint of ZAP evasion, as described previously for other viruses [85,109]. Then, a further suppression of the CpG content in specific regions of the viral genome targeted by the ZAP could explain the reduction in ZAP sensitivity found in some pathogenic members of the genus such as the VEEV and ONNV. Alternatively, viral resistance to the ZAP might also involve changes in the nucleotide sequences around the CpG, which are still to be deciphered, that could be modulating the ZAP affinity for target RNAs.

PKR resistance. Many Old World alphaviruses have acquired cis-acting RNA structures in sgmRNA to counteract PKR-mediated $\alpha\alpha$ phosphorylation in infected cells [110]. Genetic, functional and structural analyses have shown the existence of a stable stem–loop structure (DLP) located 27–30 nt downstream the AUG of sgmRNA in the SINV and SFV that confer resistance to PKR-mediated eIF2α phosphorylation (Figure 3) [35,110,111]. A functional analysis also showed the existence of equivalent structures in the RRV, Sagiyama virus (SAGV), Bebaru virus (BEBV), Una virus (UNAV), Middelburg virus (MIDV) and to a lesser extent, in the EEEV [110]. The DLP activity greatly depends on the RNA structure stability, especially at the base of the stem, whereas the helix topology seems to be less relevant. Thus, structure destabilization by introducing adenines (A) in the C–G-rich stretch of the stem abolished DLP activity [35]. The distance of the DLP to AUGi was found to be critical, so the DLP structure must be located in a 24–37 nt window downstream the AUG of the sgmRNA to promote resistance to eIF2α phosphorylation [110]. Crosslinking experiments of 48S-PIC assembled with the SINV and SFV sgmRNAs revealed multiple contacts of DLP with nucleotides of 18S rRNA in the ES6S region, suggesting the existence of a trapping mechanism that promotes 48S-PIC stalling in a way that allows the proper placement of AUGi in the P site during the scanning process (Figure 3) [110]. Thus, the DLP-mediated stalling of a translation initiation complex would increase the resistance to PKR by reducing the dependence of sgmRNA for eIF2 in mammalian cells, or by allowing an eIF2-independent mechanism for the recruitment of Met-tRNA. Interestingly, the SINV mutant lacking a functional DLP, or aura virus (AURAV) that naturally carries a suboptimal DLP structure located too close to the AUGi, both showed very low levels of replication in primary cells and mice, and a dramatic increase in their sensitivity to IFN in vitro [59]. Moreover, an SINV mutant lacking DLP rapidly evolved after serial passages in MEFs to partially restore the DLP structure, a finding that stresses the importance of the DLP structure in alphavirus replication and PKR resistance [32]. However, in other alphaviruses such as the CHIKV, ONNV and Mayaro virus (MAYV), the presence of this structure in the sgmRNA was not detected, suggesting that the acquisition of DLP was not a universal trait in alphavirus evolution. Interestingly, the productive infection of cultured cells with the CHIKV resulted in very low levels of eIF2α phosphorylation as compared to the SINV or SFV [19]. This suggests the existence of an alternative strategy in this virus to prevent eIF2α phosphorylation, perhaps by the presence of a viral protein with an anti-PKR activity as described in many other virus families [112].

Suppression of IFN synthesis by nsp2s and capsid proteins. The activities of Old World nsp2s and New World capsid proteins have been linked to the transcriptional shutoff observed in vertebrate cells as an effective mechanism to prevent the synthesis of IFN and other antiviral cytokines by pathogenic alphaviruses. In addition to their function in replication and virus encapsidation, a significant fraction of Old World nsp2s and New World capsid proteins accumulate in the nuclei of infected cells at early times of infection, where they suppress host transcription and/or nuclear trafficking [113–115].

An extensive genetic analysis has firmly established that the Old World nsp2 blocks type I IFN transcription by inducing the proteasomal degradation of Rpb1, a catalytic subunit of the RNAPII complex [33]. This prevents the expression of IFN and other antiviral genes in response to infection, highlighting the role of nsp2 in innate immune evasion [60,116]. Rpb1 degradation involves polyubiquitination and depends on the integrity of the helicase and SAM-dependent methyltransferase-like domains of nsp2, so that point mutation in these domains can abolish Rpb1 degradation and lead to the subsequent halt of host mRNA synthesis [33,116]. Consequently, the SINV and CHIKV viruses carrying some of these mutations in nsp2 became potent inducers of type I IFN in primary cell cultures that resulted in a marked attenuation of virus replication, especially in vivo [116]. Importantly, these mutant viruses also lost their pathogenic potential when inoculated in mice, highlighting the relevance of nsp2 activity in virus–host interactions [37,116].

Surprisingly, the transcriptional shutoff induced by New World alphaviruses is carried out by a capsid protein rather than by nsp2. However, contrary to nsp2, the VEEV capsid protein did not induce the degradation of any RNAPII subunit [113,114]. The expression of the VEEV capsid protein both individually, and in the context of viral infection, potently blocked the host transcription and nuclear import in mammalian but not in mosquito cells [115]. A short region located at the N-terminals of the capsid proteins of VEEV and EEEV promotes the formation of a tetrameric complex of the capsid protein with CRM1 and alpha/beta-importin that obstructs the nuclear pore, thus blocking the nucleocytoplasmic transport [117]. Moreover, the N-terminal region of the capsid protein is essential for the transcription suppression activity of the VEEV capsid protein, suggesting that the blockade of nucleocytoplasmic transport could be indirectly responsible for transcriptional suppression. Although this was further supported by the observation that the blockade of nuclear import precedes the transcriptional shutoff, a causal relationship between these two phenomena remains to be established.

The occurrence of transcriptional suppression activities in different proteins of Old World and New World alphaviruses constitutes a remarkable case of evolutionary convergence that stresses the central role of IFN evasion mechanisms in the natural story of alphaviruses in their two main geographic areas of distribution.

5. Conclusions and Future Directions

The interference of alphaviruses with host translation seems to be a multifactorial process that results from the intricate balance between host-induced measures and virus-induced countermeasures. The interplay between alphavirus-induced transcription and translation shutoff deserves to be explored in the future, as well as the impact that the cytoplasmic accumulation of many RNA binding proteins observed in infected cells could have on host translation. To what extent IFN evasion mechanisms have shaped the genome of alphavirus also deserves further investigation to obtain clues on some aspects of the origin and evolutionary history of alphavirus that still remain controversial. This will require a better understanding of ZAP specificity that will allow us to predict the existence of ZAP binding sites in viral genomes.

Author Contributions: I.V. conceptualized this work. I.V., J.J.B., R.T. and I.D.-L. wrote this review together. All authors have read and agreed to the published version of the manuscript.

Funding: This work was supported by a grant from the Spanish Ministry of Science and Innovation (PID2021-125844OB-I00).

Institutional Review Board Statement: Not applicable.

Informed Consent Statement: Not applicable.

Data Availability Statement: No specific data were generated to support reported results.

Conflicts of Interest: The authors declare no conflicts of interest.

References

1. Powers, A.M.; Brault, A.C.; Shirako, Y.; Strauss, E.G.; Kang, W.; Strauss, J.H.; Weaver, S.C. Evolutionary relationships and systematics of the alphaviruses. *J. Virol.* **2001**, *75*, 10118–10131. [CrossRef]
2. Weaver, S.C.; Barrett, A.D.T. Transmission cycles, host range, evolution and emergence of arboviral disease. *Nat. Rev. Microbiol.* **2004**, *2*, 789–801. [CrossRef]
3. Weaver, S.C.; Winegar, R.; Manger, I.D.; Forrester, N.L. Alphaviruses: Population genetics and determinants of emergence. *Antivir. Res.* **2012**, *94*, 242–257. [CrossRef]
4. Tsetsarkin, K.A.; Chen, R.; Sherman, M.B.; Weaver, S.C. Chikungunya virus: Evolution and genetic determinants of emergence. *Curr. Opin. Virol.* **2011**, *1*, 310–317. [CrossRef] [PubMed]
5. Tsetsarkin, K.A.; Chen, R.; Weaver, S.C. Interspecies transmission and chikungunya virus emergence. *Curr. Opin. Virol.* **2016**, *16*, 143–150. [CrossRef] [PubMed]
6. Zacks, M.A.; Paessler, S. Encephalitic alphaviruses. *Vet. Microbiol.* **2010**, *140*, 281–286. [CrossRef] [PubMed]
7. Kafai, N.M.; Diamond, M.S.; Fox, J.M. Distinct Cellular Tropism and Immune Responses to Alphavirus Infection. *Annu. Rev. Immunol.* **2022**, *40*, 615–649. [CrossRef]
8. Strauss, J.H.; Strauss, E.G. The alphaviruses: Gene expression, replication, and evolution. *Microbiol. Rev.* **1994**, *58*, 491–562. [CrossRef]
9. Ahola, T.; McInerney, G.; Merits, A. Alphavirus RNA replication in vertebrate cells. *Adv. Virus Res.* **2021**, *111*, 111–156. [CrossRef]
10. Pettersson, R.F.; Soderlund, H.; Kaariainen, L. The nucleotide sequences of the 5′-terminal T1 oligonucleotides of Semliki-Forest-virus 42-S and 26-S RNAs are different. *Eur. J. Biochem.* **1980**, *105*, 435–443. [CrossRef]
11. Ahola, T.; Kaariainen, L. Reaction in alphavirus mRNA capping: Formation of a covalent complex of nonstructural protein nsP1 with 7-methyl-GMP. *Proc. Natl. Acad. Sci. USA* **1995**, *92*, 507–511. [CrossRef]
12. Ahola, T.; Laakkonen, P.; Vihinen, H.; Kaariainen, L. Critical residues of Semliki Forest virus RNA capping enzyme involved in methyltransferase and guanylyltransferase-like activities. *J. Virol.* **1997**, *71*, 392–397. [CrossRef]
13. Decroly, E.; Ferron, F.; Lescar, J.; Canard, B. Conventional and unconventional mechanisms for capping viral mRNA. *Nat. Rev. Microbiol.* **2011**, *10*, 51–65. [CrossRef]
14. Gorchakov, R.; Frolova, E.; Frolov, I. Inhibition of transcription and translation in Sindbis virus-infected cells. *J. Virol.* **2005**, *79*, 9397–9409. [CrossRef]
15. Gorchakov, R.; Garmashova, N.; Frolova, E.; Frolov, I. Different types of nsP3-containing protein complexes in Sindbis virus-infected cells. *J. Virol.* **2008**, *82*, 10088–10101. [CrossRef]
16. Fros, J.J.; Geertsema, C.; Zouache, K.; Baggen, J.; Domeradzka, N.; van Leeuwen, D.M.; Flipse, J.; Vlak, J.M.; Failloux, A.; Pijlman, G.P. Mosquito Rasputin interacts with chikungunya virus nsP3 and determines the infection rate in *Aedes albopictus*. *Parasit. Vectors* **2015**, *8*, 464. [CrossRef]
17. Panas, M.D.; Varjak, M.; Lulla, A.; Eng, K.E.; Merits, A.; Karlsson Hedestam, G.B.; McInerney, G.M. Sequestration of G3BP coupled with efficient translation inhibits stress granules in Semliki Forest virus infection. *Mol. Biol. Cell* **2012**, *23*, 4701–4712. [CrossRef] [PubMed]
18. Jayabalan, A.K.; Adivarahan, S.; Koppula, A.; Abraham, R.; Batish, M.; Zenklusen, D.; Griffin, D.E.; Leung, A.K.L. Stress granule formation, disassembly, and composition are regulated by alphavirus ADP-ribosylhydrolase activity. *Proc. Natl. Acad. Sci. USA* **2021**, *118*, e2021719118. [CrossRef] [PubMed]
19. Frolova, E.I.; Palchevska, O.; Dominguez, F.; Frolov, I. Alphavirus-induced transcriptional and translational shutoffs play major roles in blocking the formation of stress granules. *J. Virol.* **2023**, *97*, e0097923-23. [CrossRef] [PubMed]
20. Pietila, M.K.; Hellstrom, K.; Ahola, T. Alphavirus polymerase and RNA replication. *Virus Res.* **2017**, *234*, 44–57. [CrossRef] [PubMed]
21. Lemm, J.A.; Rumenapf, T.; Strauss, E.G.; Strauss, J.H.; Rice, C.M. Polypeptide requirements for assembly of functional Sindbis virus replication complexes: A model for the temporal regulation of minus- and plus-strand RNA synthesis. *EMBO J.* **1994**, *13*, 2925–2934. [CrossRef]
22. Lulla, V.; Karo-Astover, L.; Rausalu, K.; Saul, S.; Merits, A.; Lulla, A. Timeliness of Proteolytic Events Is Prerequisite for Efficient Functioning of the Alphaviral Replicase. *J. Virol.* **2018**, *92*, e00151-18. [CrossRef]
23. Frolova, E.I.; Gorchakov, R.; Pereboeva, L.; Atasheva, S.; Frolov, I. Functional Sindbis virus replicative complexes are formed at the plasma membrane. *J. Virol.* **2010**, *84*, 11679–11695. [CrossRef]
24. Froshauer, S.; Kartenbeck, J.; Helenius, A. Alphavirus RNA replicase is located on the cytoplasmic surface of endosomes and lysosomes. *J. Cell Biol.* **1988**, *107*, 2075–2086. [CrossRef] [PubMed]

25. Laurent, T.; Kumar, P.; Liese, S.; Zare, F.; Jonasson, M.; Carlson, A.; Carlson, L. Architecture of the chikungunya virus replication organelle. *eLife* **2022**, *11*, e83042. [CrossRef] [PubMed]
26. Akhrymuk, I.; Frolov, I.; Frolova, E.I. Both RIG-I and MDA5 detect alphavirus replication in concentration-dependent mode. *Virology* **2016**, *487*, 230–241. [CrossRef] [PubMed]
27. Trobaugh, D.W.; Gardner, C.L.; Sun, C.; Haddow, A.D.; Wang, E.; Chapnik, E.; Mildner, A.; Weaver, S.C.; Ryman, K.D.; Klimstra, W.B. RNA viruses can hijack vertebrate microRNAs to suppress innate immunity. *Nature* **2014**, *506*, 245–248. [CrossRef] [PubMed]
28. Jose, J.; Taylor, A.B.; Kuhn, R.J. Spatial and Temporal Analysis of Alphavirus Replication and Assembly in Mammalian and Mosquito Cells. *mBio* **2017**, *8*, e02294-16. [CrossRef] [PubMed]
29. Gliedman, J.B.; Smith, J.F.; Brown, D.T. Morphogenesis of Sindbis virus in cultured Aedes albopictus cells. *J. Virol.* **1975**, *16*, 913–926. [CrossRef] [PubMed]
30. Davey, M.W.; Dalgarno, L. Semliki Forest virus replication in cultured Aedes albopictus cells: Studies on the establishment of persistence. *J. Gen. Virol.* **1974**, *24*, 453–463. [CrossRef] [PubMed]
31. Kim, K.H.; Rumenapf, T.; Strauss, E.G.; Strauss, J.H. Regulation of Semliki Forest virus RNA replication: A model for the control of alphavirus pathogenesis in invertebrate hosts. *Virology* **2004**, *323*, 153–163. [CrossRef] [PubMed]
32. Ventoso, I. Adaptive changes in alphavirus mRNA translation allowed colonization of vertebrate hosts. *J. Virol.* **2012**, *86*, 9484–9494. [CrossRef] [PubMed]
33. Akhrymuk, I.; Kulemzin, S.V.; Frolova, E.I. Evasion of the innate immune response: The Old World alphavirus nsP2 protein induces rapid degradation of Rpb1, a catalytic subunit of RNA polymerase II. *J. Virol.* **2012**, *86*, 7180–7191. [CrossRef] [PubMed]
34. Stern-Ginossar, N.; Thompson, S.R.; Mathews, M.B.; Mohr, I. Translational Control in Virus-Infected Cells. *Cold Spring Harb. Perspect. Biol.* **2019**, *11*, a033001. [CrossRef] [PubMed]
35. Ventoso, I.; Sanz, M.A.; Molina, S.; Berlanga, J.J.; Carrasco, L.; Esteban, M. Translational resistance of late alphavirus mRNA to eIF2alpha phosphorylation: A strategy to overcome the antiviral effect of protein kinase PKR. *Genes Dev.* **2006**, *20*, 87–100. [CrossRef] [PubMed]
36. Gorchakov, R.; Frolova, E.; Williams, B.R.G.; Rice, C.M.; Frolov, I. PKR-dependent and -independent mechanisms are involved in translational shutoff during Sindbis virus infection. *J. Virol.* **2004**, *78*, 8455–8467. [CrossRef] [PubMed]
37. Meshram, C.D.; Lukash, T.; Phillips, A.T.; Akhrymuk, I.; Frolova, E.I.; Frolov, I. Lack of nsP2-specific nuclear functions attenuates chikungunya virus replication both in vitro and in vivo. *Virology* **2019**, *534*, 14–24. [CrossRef] [PubMed]
38. Abraham, R.; Hauer, D.; McPherson, R.L.; Utt, A.; Kirby, I.T.; Cohen, M.S.; Merits, A.; Leung, A.K.L.; Griffin, D.E. ADP-ribosyl-binding and hydrolase activities of the alphavirus nsP3 macrodomain are critical for initiation of virus replication. *Proc. Natl. Acad. Sci. USA* **2018**, *115*, E10457–E10466. [CrossRef]
39. Dominguez, F.; Palchevska, O.; Frolova, E.I.; Frolov, I. Alphavirus-based replicons demonstrate different interactions with host cells and can be optimized to increase protein expression. *J. Virol.* **2023**, *97*, e0122523. [CrossRef]
40. Bhalla, N.; Sun, C.; Metthew Lam, L.K.; Gardner, C.L.; Ryman, K.D.; Klimstra, W.B. Host translation shutoff mediated by non-structural protein 2 is a critical factor in the antiviral state resistance of Venezuelan equine encephalitis virus. *Virology* **2016**, *496*, 147–165. [CrossRef]
41. Carrasco, L.; Sanz, M.A.; Gonzalez-Almela, E. The Regulation of Translation in Alphavirus-Infected Cells. *Viruses* **2018**, *10*, 70. [CrossRef]
42. Fros, J.J.; Pijlman, G.P. Alphavirus Infection: Host Cell Shut-Off and Inhibition of Antiviral Responses. *Viruses* **2016**, *8*, 166. [CrossRef]
43. McInerney, G.M.; Kedersha, N.L.; Kaufman, R.J.; Anderson, P.; Liljestrom, P. Importance of eIF2alpha phosphorylation and stress granule assembly in alphavirus translation regulation. *Mol. Biol. Cell* **2005**, *16*, 3753–3763. [CrossRef] [PubMed]
44. Toribio, R.; Ventoso, I. Inhibition of host translation by virus infection in vivo. *Proc. Natl. Acad. Sci. USA* **2010**, *107*, 9837–9842. [CrossRef] [PubMed]
45. Treffers, E.E.; Tas, A.; Scholte, F.E.M.; de Ru, A.H.; Snijder, E.J.; van Veelen, P.A.; van Hemert, M.J. The alphavirus nonstructural protein 2 NTPase induces a host translational shut-off through phosphorylation of eEF2 via cAMP-PKA-eEF2K signaling. *PLoS Pathog.* **2023**, *19*, e1011179. [CrossRef] [PubMed]
46. Carlberg, U.; Nilsson, A.; Nygard, O. Functional properties of phosphorylated elongation factor 2. *Eur. J. Biochem.* **1990**, *191*, 639–645. [CrossRef] [PubMed]
47. Sanz, M.A.; Garcia-Moreno, M.; Carrasco, L. Inhibition of host protein synthesis by Sindbis virus: Correlation with viral RNA replication and release of nuclear proteins to the cytoplasm. *Cell. Microbiol.* **2015**, *17*, 520–541. [CrossRef] [PubMed]
48. Frolov, I.; Schlesinger, S. Comparison of the effects of Sindbis virus and Sindbis virus replicons on host cell protein synthesis and cytopathogenicity in BHK cells. *J. Virol.* **1994**, *68*, 1721–1727. [CrossRef] [PubMed]
49. Sanz, M.A.; Castello, A.; Carrasco, L. Viral translation is coupled to transcription in Sindbis virus-infected cells. *J. Virol.* **2007**, *81*, 7061–7068. [CrossRef] [PubMed]
50. Barnhart, M.D.; Moon, S.L.; Emch, A.W.; Wilusz, C.J.; Wilusz, J. Changes in cellular mRNA stability, splicing, and polyadenylation through HuR protein sequestration by a cytoplasmic RNA virus. *Cell Rep.* **2013**, *5*, 909–917. [CrossRef]
51. van Steeg, H.; Kasperaitis, M.; Voorma, H.O.; Benne, R. Infection of neuroblastoma cells by Semliki Forest virus. The interference of viral capsid protein with the binding of host messenger RNAs into initiation complexes is the cause of the shut-off of host protein synthesis. *Eur. J. Biochem.* **1984**, *138*, 473–478. [CrossRef] [PubMed]

52. Frolov, I.; Akhrymuk, M.; Akhrymuk, I.; Atasheva, S.; Frolova, E.I. Early events in alphavirus replication determine the outcome of infection. *J. Virol.* **2012**, *86*, 5055–5066. [CrossRef] [PubMed]
53. Ryman, K.D.; Klimstra, W.B. Host responses to alphavirus infection. *Immunol. Rev.* **2008**, *225*, 27–45. [CrossRef] [PubMed]
54. Schilte, C.; Couderc, T.; Chretien, F.; Sourisseau, M.; Gangneux, N.; Guivel-Benhassine, F.; Kraxner, A.; Tschopp, J.; Higgs, S.; Michault, A.; et al. Type I IFN controls chikungunya virus via its action on nonhematopoietic cells. *J. Exp. Med.* **2010**, *207*, 429–442. [CrossRef]
55. Grieder, F.B.; Vogel, S.N. Role of interferon and interferon regulatory factors in early protection against Venezuelan equine encephalitis virus infection. *Virology* **1999**, *257*, 106–118. [CrossRef]
56. Ryman, K.D.; Klimstra, W.B.; Nguyen, K.B.; Biron, C.A.; Johnston, R.E. Alpha/beta interferon protects adult mice from fatal Sindbis virus infection and is an important determinant of cell and tissue tropism. *J. Virol.* **2000**, *74*, 3366–3378. [CrossRef]
57. Friedman, R.M.; Fantes, K.H.; Levy, H.B.; Carter, W.B. Interferon action on parental Semliki forest virus ribonucleic acid. *J. Virol.* **1967**, *1*, 1168–1173. [CrossRef]
58. Friedman, R.M. Inhibition of arbovirus protein synthesis by interferon. *J. Virol.* **1968**, *2*, 1081–1085. [CrossRef]
59. Toribio, R.; Diaz-Lopez, I.; Berlanga, J.J.; Molina-Jimenez, F.; Majano, P.; Ventoso, I. Naturally Occurring and Engineered Alphaviruses Sensitive to Double-Stranded-RNA-Activated Protein Kinase Show Restricted Translation in Mammalian Cells, Increased Sensitivity to Interferon, and Marked Oncotropism. *J. Virol.* **2020**, *94*, e01630-19. [CrossRef]
60. Frolova, E.I.; Fayzulin, R.Z.; Cook, S.H.; Griffin, D.E.; Rice, C.M.; Frolov, I. Roles of nonstructural protein nsP2 and Alpha/Beta interferons in determining the outcome of Sindbis virus infection. *J. Virol.* **2002**, *76*, 11254–11264. [CrossRef]
61. Atasheva, S.; Kim, D.Y.; Frolova, E.I.; Frolov, I. Venezuelan equine encephalitis virus variants lacking transcription inhibitory functions demonstrate highly attenuated phenotype. *J. Virol.* **2015**, *89*, 71–82. [CrossRef]
62. Her, Z.; Malleret, B.; Chan, M.; Ong, E.K.S.; Wong, S.; Kwek, D.J.C.; Tolou, H.; Lin, R.T.P.; Tambyah, P.A.; Renia, L.; et al. Active infection of human blood monocytes by Chikungunya virus triggers an innate immune response. *J. Immunol.* **2010**, *184*, 5903–5913. [CrossRef]
63. Sadler, A.J.; Williams, B.R.G. Interferon-inducible antiviral effectors. *Nat. Rev. Immunol.* **2008**, *8*, 559–568. [CrossRef]
64. Schoggins, J.W.; Wilson, S.J.; Panis, M.; Murphy, M.Y.; Jones, C.T.; Bieniasz, P.; Rice, C.M. A diverse range of gene products are effectors of the type I interferon antiviral response. *Nature* **2011**, *472*, 481–485. [CrossRef]
65. Zhang, Y.; Burke, C.W.; Ryman, K.D.; Klimstra, W.B. Identification and characterization of interferon-induced proteins that inhibit alphavirus replication. *J. Virol.* **2007**, *81*, 11246–11255. [CrossRef]
66. McDougal, M.B.; De Maria, A.M.; Ohlson, M.B.; Kumar, A.; Xing, C.; Schoggins, J.W. Interferon inhibits a model RNA virus via a limited set of inducible effector genes. *EMBO Rep.* **2023**, *24*, e56901. [CrossRef] [PubMed]
67. Tesfay, M.Z.; Yin, J.; Gardner, C.L.; Khoretonenko, M.V.; Korneeva, N.L.; Rhoads, R.E.; Ryman, K.D.; Klimstra, W.B. Alpha/beta interferon inhibits cap-dependent translation of viral but not cellular mRNA by a PKR-independent mechanism. *J. Virol.* **2008**, *82*, 2620–2630. [CrossRef] [PubMed]
68. Daffis, S.; Szretter, K.J.; Schriewer, J.; Li, J.; Youn, S.; Errett, J.; Lin, T.; Schneller, S.; Zust, R.; Dong, H.; et al. 2′-O methylation of the viral mRNA cap evades host restriction by IFIT family members. *Nature* **2010**, *468*, 452–456. [CrossRef] [PubMed]
69. Hyde, J.L.; Gardner, C.L.; Kimura, T.; White, J.P.; Liu, G.; Trobaugh, D.W.; Huang, C.; Tonelli, M.; Paessler, S.; Takeda, K.; et al. A viral RNA structural element alters host recognition of nonself RNA. *Science* **2014**, *343*, 783–787. [CrossRef] [PubMed]
70. Pichlmair, A.; Lassnig, C.; Eberle, C.; Gorna, M.W.; Baumann, C.L.; Burkard, T.R.; Burckstummer, T.; Stefanovic, A.; Krieger, S.; Bennett, K.L.; et al. IFIT1 is an antiviral protein that recognizes 5′-triphosphate RNA. *Nat. Immunol.* **2011**, *12*, 624–630. [CrossRef] [PubMed]
71. Abbas, Y.M.; Pichlmair, A.; Gorna, M.W.; Superti-Furga, G.; Nagar, B. Structural basis for viral 5′-PPP-RNA recognition by human IFIT proteins. *Nature* **2013**, *494*, 60–64. [CrossRef] [PubMed]
72. Kumar, P.; Sweeney, T.R.; Skabkin, M.A.; Skabkina, O.V.; Hellen, C.U.T.; Pestova, T.V. Inhibition of translation by IFIT family members is determined by their ability to interact selectively with the 5′-terminal regions of cap0-, cap1- and 5′ppp-mRNAs. *Nucleic Acids Res.* **2014**, *42*, 3228–3245. [CrossRef] [PubMed]
73. Johnson, B.; VanBlargan, L.A.; Xu, W.; White, J.P.; Shan, C.; Shi, P.; Zhang, R.; Adhikari, J.; Gross, M.L.; Leung, D.W.; et al. Human IFIT3 Modulates IFIT1 RNA Binding Specificity and Protein Stability. *Immunity* **2018**, *48*, 487–499.e5. [CrossRef] [PubMed]
74. Fleith, R.C.; Mears, H.V.; Leong, X.Y.; Sanford, T.J.; Emmott, E.; Graham, S.C.; Mansur, D.S.; Sweeney, T.R. IFIT3 and IFIT2/3 promote IFIT1-mediated translation inhibition by enhancing binding to non-self RNA. *Nucleic Acids Res.* **2018**, *46*, 5269–5285. [CrossRef] [PubMed]
75. Reynaud, J.M.; Kim, D.Y.; Atasheva, S.; Rasalouskaya, A.; White, J.P.; Diamond, M.S.; Weaver, S.C.; Frolova, E.I.; Frolov, I. IFIT1 Differentially Interferes with Translation and Replication of Alphavirus Genomes and Promotes Induction of Type I Interferon. *PLoS Pathog.* **2015**, *11*, e1004863. [CrossRef] [PubMed]
76. Bick, M.J.; Carroll, J.N.; Gao, G.; Goff, S.P.; Rice, C.M.; MacDonald, M.R. Expression of the zinc-finger antiviral protein inhibits alphavirus replication. *J. Virol.* **2003**, *77*, 11555–11562. [CrossRef]
77. Ficarelli, M.; Neil, S.J.D.; Swanson, C.M. Targeted Restriction of Viral Gene Expression and Replication by the ZAP Antiviral System. *Annu. Rev. Virol.* **2021**, *8*, 265–283. [CrossRef]
78. Goncalves-Carneiro, D.; Takata, M.A.; Ong, H.; Shilton, A.; Bieniasz, P.D. Origin and evolution of the zinc finger antiviral protein. *PLoS Pathog.* **2021**, *17*, e1009545. [CrossRef]

79. Gao, G.; Guo, X.; Goff, S.P. Inhibition of retroviral RNA production by ZAP, a CCCH-type zinc finger protein. *Science* **2002**, *297*, 1703–1706. [CrossRef]
80. Guo, X.; Carroll, J.N.; Macdonald, M.R.; Goff, S.P.; Gao, G. The zinc finger antiviral protein directly binds to specific viral mRNAs through the CCCH zinc finger motifs. *J. Virol.* **2004**, *78*, 12781–12787. [CrossRef]
81. Schwerk, J.; Soveg, F.W.; Ryan, A.P.; Thomas, K.R.; Hatfield, L.D.; Ozarkar, S.; Forero, A.; Kell, A.M.; Roby, J.A.; So, L.; et al. RNA-binding protein isoforms ZAP-S and ZAP-L have distinct antiviral and immune resolution functions. *Nat. Immunol.* **2019**, *20*, 1610–1620. [CrossRef]
82. Li, M.M.H.; Aguilar, E.G.; Michailidis, E.; Pabon, J.; Park, P.; Wu, X.; de Jong, Y.P.; Schneider, W.M.; Molina, H.; Rice, C.M.; et al. Characterization of Novel Splice Variants of Zinc Finger Antiviral Protein (ZAP). *J. Virol.* **2019**, *93*, e00715-19. [CrossRef]
83. Kerns, J.A.; Emerman, M.; Malik, H.S. Positive selection and increased antiviral activity associated with the PARP-containing isoform of human zinc-finger antiviral protein. *PLoS Genet.* **2008**, *4*, e21. [CrossRef]
84. Zhu, Y.; Wang, X.; Goff, S.P.; Gao, G. Translational repression precedes and is required for ZAP-mediated mRNA decay. *EMBO J.* **2012**, *31*, 4236–4246. [CrossRef]
85. Takata, M.A.; Goncalves-Carneiro, D.; Zang, T.M.; Soll, S.J.; York, A.; Blanco-Melo, D.; Bieniasz, P.D. CG dinucleotide suppression enables antiviral defence targeting non-self RNA. *Nature* **2017**, *550*, 124–127. [CrossRef]
86. Luo, X.; Wang, X.; Gao, Y.; Zhu, J.; Liu, S.; Gao, G.; Gao, P. Molecular Mechanism of RNA Recognition by Zinc-Finger Antiviral Protein. *Cell Rep.* **2020**, *30*, 46–52.e4. [CrossRef]
87. Meagher, J.L.; Takata, M.; Goncalves-Carneiro, D.; Keane, S.C.; Rebendenne, A.; Ong, H.; Orr, V.K.; MacDonald, M.R.; Stuckey, J.A.; Bieniasz, P.D.; et al. Structure of the zinc-finger antiviral protein in complex with RNA reveals a mechanism for selective targeting of CG-rich viral sequences. *Proc. Natl. Acad. Sci. USA* **2019**, *116*, 24303–24309. [CrossRef] [PubMed]
88. Simmonds, P.; Xia, W.; Baillie, J.K.; McKinnon, K. Modelling mutational and selection pressures on dinucleotides in eukaryotic phyla—Selection against CpG and UpA in cytoplasmically expressed RNA and in RNA viruses. *BMC Genom.* **2013**, *14*, 610. [CrossRef] [PubMed]
89. Nguyen, L.P.; Aldana, K.S.; Yang, E.; Yao, Z.; Li, M.M.H. Alphavirus Evasion of Zinc Finger Antiviral Protein (ZAP) Correlates with CpG Suppression in a Specific Viral nsP2 Gene Sequence. *Viruses* **2023**, *15*, 830. [CrossRef] [PubMed]
90. Goonawardane, N.; Nguyen, D.; Simmonds, P. Association of Zinc Finger Antiviral Protein Binding to Viral Genomic RNA with Attenuation of Replication of Echovirus 7. *mSphere* **2021**, *6*, e01138-20. [CrossRef] [PubMed]
91. Yang, E.; Nguyen, L.P.; Wisherop, C.A.; Kan, R.L.; Li, M.M.H. The Role of ZAP and TRIM25 RNA Binding in Restricting Viral Translation. *Front. Cell. Infect. Microbiol.* **2022**, *12*, 886929. [CrossRef]
92. Kozaki, T.; Takahama, M.; Misawa, T.; Matsuura, Y.; Akira, S.; Saitoh, T. Role of zinc-finger anti-viral protein in host defense against Sindbis virus. *Int. Immunol.* **2015**, *27*, 357–364. [CrossRef]
93. Li, M.M.H.; Lau, Z.; Cheung, P.; Aguilar, E.G.; Schneider, W.M.; Bozzacco, L.; Molina, H.; Buehler, E.; Takaoka, A.; Rice, C.M.; et al. TRIM25 Enhances the Antiviral Action of Zinc-Finger Antiviral Protein (ZAP). *PLoS Pathog.* **2017**, *13*, e1006145. [CrossRef]
94. Roberts, W.K.; Hovanessian, A.; Brown, R.E.; Clemens, M.J.; Kerr, I.M. Interferon-mediated protein kinase and low-molecular-weight inhibitor of protein synthesis. *Nature* **1976**, *264*, 477–480. [CrossRef]
95. Meurs, E.; Chong, K.; Galabru, J.; Thomas, N.S.; Kerr, I.M.; Williams, B.R.; Hovanessian, A.G. Molecular cloning and characterization of the human double-stranded RNA-activated protein kinase induced by interferon. *Cell* **1990**, *62*, 379–390. [CrossRef]
96. Garcia, M.A.; Gil, J.; Ventoso, I.; Guerra, S.; Domingo, E.; Rivas, C.; Esteban, M. Impact of protein kinase PKR in cell biology: From antiviral to antiproliferative action. *Microbiol. Mol. Biol. Rev.* **2006**, *70*, 1032–1060. [CrossRef] [PubMed]
97. Chaumont, L.; Collet, B.; Boudinot, P. Double-stranded RNA-dependent protein kinase (PKR) in antiviral defence in fish and mammals. *Dev. Comp. Immunol.* **2023**, *145*, 104732. [CrossRef] [PubMed]
98. Yang, Y.L.; Reis, L.F.; Pavlovic, J.; Aguzzi, A.; Schafer, R.; Kumar, A.; Williams, B.R.; Aguet, M.; Weissmann, C. Deficient signaling in mice devoid of double-stranded RNA-dependent protein kinase. *EMBO J.* **1995**, *14*, 6095–6106. [CrossRef] [PubMed]
99. Balachandran, S.; Roberts, P.C.; Brown, L.E.; Truong, H.; Pattnaik, A.K.; Archer, D.R.; Barber, G.N. Essential role for the dsRNA-dependent protein kinase PKR in innate immunity to viral infection. *Immunity* **2000**, *13*, 129–141. [CrossRef] [PubMed]
100. Gordiyenko, Y.; Llacer, J.L.; Ramakrishnan, V. Structural basis for the inhibition of translation through eIF2alpha phosphorylation. *Nat. Commun.* **2019**, *10*, 2640. [CrossRef] [PubMed]
101. Krishnamoorthy, T.; Pavitt, G.D.; Zhang, F.; Dever, T.E.; Hinnebusch, A.G. Tight binding of the phosphorylated alpha subunit of initiation factor 2 (eIF2alpha) to the regulatory subunits of guanine nucleotide exchange factor eIF2B is required for inhibition of translation initiation. *Mol. Cell. Biol.* **2001**, *21*, 5018–5030. [CrossRef] [PubMed]
102. Cesaro, T.; Michiels, T. Inhibition of PKR by Viruses. *Front. Microbiol.* **2021**, *12*, 757238. [CrossRef]
103. Garcia-Sastre, A. Ten Strategies of Interferon Evasion by Viruses. *Cell Host Microbe* **2017**, *22*, 176–184. [CrossRef] [PubMed]
104. Aguilar, P.V.; Adams, A.P.; Wang, E.; Kang, W.; Carrara, A.; Anishchenko, M.; Frolov, I.; Weaver, S.C. Structural and nonstructural protein genome regions of eastern equine encephalitis virus are determinants of interferon sensitivity and murine virulence. *J. Virol.* **2008**, *82*, 4920–4930. [CrossRef] [PubMed]
105. Torii, S.; Orba, Y.; Hang'ombe, B.M.; Mweene, A.S.; Wada, Y.; Anindita, P.D.; Phongphaew, W.; Qiu, Y.; Kajihara, M.; Mori-Kajihara, A.; et al. Discovery of Mwinilunga alphavirus: A novel alphavirus in Culex mosquitoes in Zambia. *Virus Res.* **2018**, *250*, 31–36. [CrossRef] [PubMed]

106. Hermanns, K.; Marklewitz, M.; Zirkel, F.; Overheul, G.J.; Page, R.A.; Loaiza, J.R.; Drosten, C.; van Rij, R.P.; Junglen, S. Agua Salud alphavirus defines a novel lineage of insect-specific alphaviruses discovered in the New World. *J. Gen. Virol.* **2020**, *101*, 96–104. [CrossRef] [PubMed]
107. Nasar, F.; Palacios, G.; Gorchakov, R.V.; Guzman, H.; Da Rosa, A.P.T.; Savji, N.; Popov, V.L.; Sherman, M.B.; Lipkin, W.I.; Tesh, R.B.; et al. Eilat virus, a unique alphavirus with host range restricted to insects by RNA replication. *Proc. Natl. Acad. Sci. USA* **2012**, *109*, 14622–14627. [CrossRef]
108. Hermanns, K.; Zirkel, F.; Kopp, A.; Marklewitz, M.; Rwego, I.B.; Estrada, A.; Gillespie, T.R.; Drosten, C.; Junglen, S. Discovery of a novel alphavirus related to Eilat virus. *J. Gen. Virol.* **2017**, *98*, 43–49. [CrossRef]
109. Odon, V.; Fiddaman, S.R.; Smith, A.L.; Simmonds, P. Comparison of CpG- and UpA-mediated restriction of RNA virus replication in mammalian and avian cells and investigation of potential ZAP-mediated shaping of host transcriptome compositions. *RNA* **2022**, *28*, 1089–1109. [CrossRef]
110. Toribio, R.; Diaz-Lopez, I.; Boskovic, J.; Ventoso, I. An RNA trapping mechanism in Alphavirus mRNA promotes ribosome stalling and translation initiation. *Nucleic Acids Res.* **2016**, *44*, 4368–4380. [CrossRef]
111. Frolov, I.; Schlesinger, S. Translation of Sindbis virus mRNA: Effects of sequences downstream of the initiating codon. *J. Virol.* **1994**, *68*, 8111–8117. [CrossRef]
112. Rathore, A.P.S.; Ng, M.; Vasudevan, S.G. Differential unfolded protein response during Chikungunya and Sindbis virus infection: CHIKV nsP4 suppresses eIF2alpha phosphorylation. *Virol. J.* **2013**, *10*, 36. [CrossRef]
113. Garmashova, N.; Gorchakov, R.; Volkova, E.; Paessler, S.; Frolova, E.; Frolov, I. The Old World and New World alphaviruses use different virus-specific proteins for induction of transcriptional shutoff. *J. Virol.* **2007**, *81*, 2472–2484. [CrossRef]
114. Garmashova, N.; Gorchakov, R.; Frolova, E.; Frolov, I. Sindbis virus nonstructural protein nsP2 is cytotoxic and inhibits cellular transcription. *J. Virol.* **2006**, *80*, 5686–5696. [CrossRef]
115. Atasheva, S.; Garmashova, N.; Frolov, I.; Frolova, E. Venezuelan equine encephalitis virus capsid protein inhibits nuclear import in Mammalian but not in mosquito cells. *J. Virol.* **2008**, *82*, 4028–4041. [CrossRef] [PubMed]
116. Akhrymuk, I.; Lukash, T.; Frolov, I.; Frolova, E.I. Novel Mutations in nsP2 Abolish Chikungunya Virus-Induced Transcriptional Shutoff and Make the Virus Less Cytopathic without Affecting Its Replication Rates. *J. Virol.* **2019**, *93*, e02062-18. [CrossRef] [PubMed]
117. Atasheva, S.; Fish, A.; Fornerod, M.; Frolova, E.I. Venezuelan equine Encephalitis virus capsid protein forms a tetrameric complex with CRM1 and importin alpha/beta that obstructs nuclear pore complex function. *J. Virol.* **2010**, *84*, 4158–4171. [CrossRef] [PubMed]

viruses

MDPI

Article

Tanapox Virus and Yaba Monkey Tumor Virus K3 Orthologs Inhibit Primate Protein Kinase R in a Species-Specific Fashion

Dewi Megawati [1,2], Jeannine N. Stroup [1], Chorong Park [1,†], Taylor Clarkson [1,‡], Loubna Tazi [1], Greg Brennan [1] and Stefan Rothenburg [1,*]

1 Department of Medical Microbiology and Immunology, School of Medicine, University of California, Davis, CA 95616, USA; amegawati@ucdavis.edu (D.M.); jnstroup15@gmail.com (J.N.S.); chorong.park@northwestern.edu (C.P.); ltazi@ucdavis.edu (L.T.)

2 Department of Microbiology and Parasitology, Faculty of Medicine and Health Sciences, Warmadewa University, Denpasar 80239, Bali, Indonesia

* Correspondence: rothenburg@ucdavis.edu

† Current Address: Department of Microbiology-Immunology, Feinberg School of Medicine, Northwestern University, Chicago, IL 60208, USA.

‡ Current Address: Colorado Department of Public Health and Environment, Denver, CO 80246, USA.

Abstract: Yaba monkey tumor virus (YMTV) and Tanapox virus (TPV) are members of the Yatapoxvirus genus and can infect humans and other primates. Despite the threat posed by yatapoxviruses, the factors determining their host range are poorly understood. In this study, we analyzed the ability of YMTV and TPV orthologs of vaccinia virus K3 (called 012 in YMTV and TPV), which share 75% amino acid identity with one another, to inhibit PKR from 15 different primate species. We first used a luciferase-based reporter, and found that YMTV and TPV K3 orthologs inhibited PKR in a species-specific manner and showed distinct PKR inhibition profiles. TPV 012 inhibited PKR from 11 primates, including humans, substantially better than YMTV 012. In contrast, both K3 orthologs inhibited the other four primate PKRs comparably well. Using YMTV 012 and TPV 012 hybrids, we mapped the region responsible for the differential PKR inhibition to the C- terminus of the K3 orthologs. Next, we generated chimeric vaccinia virus strains to investigate whether TPV K3 and YMTV K3 orthologs could rescue the replication of a vaccinia virus strain that lacks PKR inhibitors K3L and E3L. Virus replication in primate-derived cells generally correlated with the patterns observed in the luciferase-based assay. Together, these observations demonstrate that yatapoxvirus K3 orthologs have distinct PKR inhibition profiles and inhibit PKR in a species-specific manner, which may contribute to the differential susceptibility of primate species to yatapoxvirus infections.

Keywords: protein kinase R; poxviruses; K3L; yatapoxvirus; translational regulation

1. Introduction

Yatapoxviruses are a small group of *Chordopoxvirinae* that are pathogenic to primates, including humans. The Yatapoxvirus genus includes Yaba monkey tumor virus (YMTV) and Tanapox virus (TPV) [1]. Previously, Yaba-like disease virus (YLDV) was considered a third member; however, its genome shares 98.68% sequence identity with TPV, and can be considered a different variant of the same virus [2]. The genomes of YMTV and TPV share 75% sequence identity [3]. TPV was first isolated from human skin biopsies during a TPV outbreak in 1957 in Tana River Valley, Kenya [4,5]. YMTV was first isolated from an outbreak of subcutaneous tumors in monkeys in Yaba, Nigeria, in 1958 [6]. Yatapoxvirus genomes are A+T rich and range in size from 135 to 145 kb [2,3,7,8]. TPV and YMTV exhibit differences in disease presentations. TPV infection is characterized by vesicular skin lesions, whereas YMTV infection produces a localized histiocyte-filled tumor (histiocytomas) [9].

While both TPV and YMTV infect a wide range of primates, they appear to exhibit host specificity. Serological surveys have indicated that TPV is endemic in African and

Malaysian monkeys, but not in Indian rhesus macaques or New World monkeys [4,5,10]. Human Tanapox disease is considered endemic to several regions of Africa and causes febrile illness and vesicular skin lesions similar to those produced in non-human primates. Sporadic cases have been identified in 30 locations spanning 6000 km from Sierra Leone to Tanzania, with larger outbreaks occurring from time to time [11]. Similarly, antibodies against YMTV have been found in great apes and Old World monkeys but not in New World monkeys or Indian rhesus macaques [10]. A serological survey of 456 primate sera including 26 chimpanzees, 326 Old World monkeys (African green monkeys, patas monkeys, baboons, colobus monkeys, and rhesus macaques), and 104 New World monkeys (spider monkeys, squirrel monkeys, owl monkeys, marmosets, and capuchin monkeys) indicated that antibodies against YMTV were evident in chimpanzees and Old World monkeys but not in any of the New World monkeys [10].

Following experimental subcutaneous inoculation with YMTV, tumor-like masses were detected in Asiatic rhesus macaques (*Macaca mulatta*, *Macaca irus*, and *Macaca speciosa*) but not in African green monkeys, mangabey monkeys, patas monkeys, mice, rats, rabbits, or guinea pigs [6,9,12], indicating host specificity. While no natural infection of YMTV in humans has been reported, the infection of human volunteers as well as an accidental infection of a laboratory worker have resulted in the development of mild histiocytomas [13]. Despite the threat posed to human health, the factors determining the host range of yatapoxviruses are poorly understood. In addition, vaccination with vaccinia virus does not appear to protect against yatapoxvirus infections [13,14]. Thus, it is important to study factors that determine the host range of yatapoxviruses.

A major determinant for the tropism of many viruses is the attachment and entry via host-specific cell receptors [15]. However, as poxviruses use ubiquitous cell surface receptors for entry, it is believed that the evasion of the innate immune response post entry is a more important determinant of poxvirus tropism [16,17]. Protein Kinase R (PKR) is a prominent host restriction factor against poxvirus infection (reviewed in [18]). PKR exists in a monomeric inactive form and is activated upon binding to double-stranded RNA (dsRNA) produced during infection by most viruses, which leads to PKR dimerization and autophosphorylation. Activated PKR phosphorylates the alpha subunit of eukaryotic initiation factor 2 (eIF2) [19]. Phosphorylated eIF2α has a high binding affinity for the regulatory core of the guanine nucleotide exchange factor eIF2B and prevents eIF2B from catalyzing GDP-GTP exchange to eIF2α [20]. As GTP-bound eIF2 is required for translation initiation, the resulting low availability of GTP-eIF2 leads to the shut-off of eIF2-dependent protein synthesis, including that of viral proteins. To overcome the antiviral effects of PKR activity, most poxviruses encode two PKR inhibitors, named E3 and K3 in vaccinia virus. E3 binds dsRNA and prevents PKR dimerization, whereas K3 inhibits PKR by acting as a pseudosubstrate for the eIF2α binding site [21–24].

It was previously shown that the inhibition of host PKR by K3 orthologs of several poxvirus families was a key determinant of host specificity [22,25–28]. Vaccinia virus lacking the E3L gene (VACVΔE3L) but maintaining the K3L gene was unable to replicate in HeLa cells but remained replication-competent in BHK21 cells, suggesting a role for K3 in determining host range [22]. This finding was supported by recent studies on leporipoxvirus, capripoxvirus, and orthopoxvirus K3 orthologs. The myxoma virus (MYXV) K3 ortholog M156 specifically inhibited rabbit PKR but failed to inhibit other PKR species [25]. Similarly, capripoxvirus K3 orthologs strongly inhibited human, goat, and sheep PKR, but only weakly inhibited mouse and cow PKR [26]. Our studies on the inhibition of PKR from a panel of mammalian species by orthopoxvirus K3 orthologs exhibited distinct inhibition profiles. Here, we used a luciferase-based assay and infections of mammalian cell lines to examine the ability of TPV and YMTV K3 orthologs to inhibit PKR derived from 15 primate species. Our results demonstrate that TPV and YMTV K3 orthologs have distinct PKR inhibition profiles and that they inhibit primate PKR in a species-specific manner.

2. Materials and Methods

2.1. Cell Lines

Tert immortalized gibbon fibroblasts (gibbon Tert-GF), HeLa cells (human, ATCC #CCL-2), HeLa PKR-knock-out (PKR^{ko}) cells, and BSC-40 cells (ATCC CRL-2761) were kindly provided by Dr. Adam Geballe [29]. RK13 cells (rabbit) expressing E3 and K3 (designated RK13+E3L+K3L) were previously described [30]. The cells were grown in Dulbecco's Modified Eagle's Medium (DMEM, Life Technologies, Carlsbad, CA, USA) supplemented with 5% fetal bovine serum (FBS, Lonza Bioscience, Walkersville, MD, USA) or 10% FBS and 100 IU/mL penicillin/streptomycin (Gibco, Life Technologies, Carlsbad, CA, USA). The RK13+E3L+K3L cell culture medium contained 500 µg/mL geneticin and 300 µg/mL zeocin (Life Technologies).

2.2. Plasmids

A library of primate PKR expressing plasmids were kindly provided by Dr. Nels Elde [31]. Primate PKR and viral antagonist genes VACV K3L (YP_232916.1), TPV 012 (EF420156.1), YMTV 012 (AY386371.1) were subcloned into the pSG5 expression vector for luciferase-based reporter assays. pGL3 luciferase reporter vector was purchased from Promega, Madison, WI, USA. Cloning was carried out using Gibson assembly techniques (New England Biolabs, Ipswich, MA, USA) [32]. To monitor VACV K3L, TPV 012, and YMTV 012 gene expressions, these genes were cloned into C-terminus DYK-tagged pSG5 vector for transient transfection assay. To generate hybrid $_{N}$TPV-$_{C}$YMTV and $_{N}$YMTV-$_{C}$TPV K3 orthologs, we PCR-amplified $_{N}$YMTV 012 (nucleotide 1–99), $_{C}$YMTV 012 (nucleotide 100–267), $_{N}$TPV 012 (nucleotide 1–99), and $_{C}$TPV 012 (nucleotide 100–267) and cloned the TPV 012 and YMTV 012 hybrid genes into pSG5 vector using Gibson assembly. To generate recombinant VACV expressing TPV 012 and YMTV 012 genes, the yatapoxvirus K3 ortholog genes were cloned into p837-GOI-mCherry-E3L, as previously described [33].

2.3. Luciferase-Based Reporter Assays

Luciferase-based assays were performed as previously described [34]. Briefly, 5×10^4 HeLa PKR^{ko} cells were seeded per well in 24-well plates overnight. The HeLa PKR^{ko} cells were co-transfected with 200 ng of the indicated PKR expression vector, 200 ng of each viral antagonist expression vector, and 50 ng of pGL3 firefly luciferase expression vector (Promega), following the manufacturer's protocol. Cells were lysed with mammalian lysis buffer (GE Healthcare, Chicago, IL, USA) at 48 h post transfection. Luciferase activity was measured using a GloMax luminometer (Promega) by adding luciferin (Promega) reagent to the cell lysates, as per the manufacturer's recommendations. Data are presented as relative luciferase activity in which all data were normalized to pSG5 empty vector. Experiments were conducted in triplicate for each of the three independent experiments.

2.4. Virus and Infections

Vaccinia virus variant vP872 [35] was kindly provided by Dr. Bertram Jacobs. VC-R4+VACV K3L and VC-R4+sheeppox 011 (SPPV 011), as well as the generation of VC-R4 from vP872 variant, were previously described [26,33]. The generation of chimeric vaccinia virus, VC-R4+TPV 012, and VC-R4+YMTV 012, was carried out through the scarless integration of the open reading frames of the yatapoxvirus K3L orthologs into the E3L locus [33]. The chimeric viruses were plaque-purified twice, and the K3L ortholog gene integrations were confirmed by means of Sanger sequencing. The viruses were purified by means of zonal sucrose gradient centrifugation, and the virus titer was determined on confluent 12-well plates of RK13+E3L+K3L cells. Plaque assays were performed with confluent 6-well plates of the indicated cell lines, which were infected with 50 plaque-forming units (pfu) of each indicated virus. One hour post infection, the media were replaced with DMEM containing 5% FBS, 1% carboxymethylcellulose (CMC), and 100 U P/S. After 48 h, cells were stained with 0.1% crystal violet, and the excess staining was washed with water. The plates were imaged using an iBright Imaging System (Invitrogen,

Waltham, MA, USA). Virus infection assays were performed in confluent 6-well plates of the indicated cells. The cells were infected with each indicated virus at a MOI of 0.01 to assess replication in multiple-cycle replication assays. Cells and supernatants were collected at 30 h post infection and subjected to three rounds of freezing at −80 °C and thawing at 37 °C. Lysates were sonicated twice for 15 s, at 50% amplitude (Qsonica Q500, Newtown, CT, USA). Virus titers were measured in plaque-forming units per mL (pfu/mL) on RK13+E3L+K3L cells. Virus titer data were analyzed using one-way ANOVA followed by Tukey's multiple comparisons test (GraphPad Prism version 10.2.3).

2.5. PCR of Viral Genomic DNA

HeLa PKR^{ko} cells were seeded in 10 cm dishes to a confluency of 90–100%. The cells were then infected with the indicated viruses at a MOI of 0.1 for 24 h. Viral genomic DNA (gDNA) extraction was performed as previously described [36]. About 100 ng of the isolated viral gDNA was used as a template in a PCR targeting K3L ortholog genes using Phusion High Fidelity DNA polymerase (NEB #M0530L). The forward primer sequence was 5′ GACGAACCACCAGAGGATGATG 3′ and the reverse primer sequence was 5′ AGTACTACAATTTGTATTTTTTAATCTATCTCA 3′. PCR products were gel-purified with Monarch DNA Gel Extraction Kit (NEB #T1020) and Sanger sequencing was performed to confirm the correct integration of the K3L orthologs.

2.6. Immunoblot Analyses

To determine the level of K3 ortholog expression in a transient transfection system, 4×10^5 HeLa PKR^{ko} cells were seeded in a 6-well plate. On the next day, the cells were transfected with 1 µg of pSG5 empty vector, DYK-tagged VACV K3L, DYK-tagged TPV 012, and DYK-tagged YMTV 012, using a Genjet to DNA ratio of 2:1. At 48 h post transfection, cells were washed with PBS, lysed with 1% sodium dodecyl sulfate (SDS) in DPBS, and sonicated at 50% amplitude for 5 s twice. About 10 µg of proteins were run on 12% SDS polyacrylamide gels (Biorad, Hercules, CA, USA) and transferred to polyvinyl difluoride (PVDF, GE Healthcare) membranes. The membranes were blocked with SuperBlock blocking buffer (Thermofisher, Waltham, MA, USA) for 1 h and probed with primary antibodies against Flag M2 (Sigma F1804, St. Louis, MO, USA) and beta-actin (Sigma A1978) at a dilution of 1:2000 in SuperBlock blocking buffer (Thermofisher) for 1 h at room temperature. After washing with Tris Buffered Saline 0.1% Tween® 20 (TBST, Fisher Bioreagents), the membranes were probed with secondary antibody horseradish peroxidase-conjugated donkey anti-mouse (Jackson ImmunoResearch Laboratories Inc., 715-035-150, West Grove, PA, USA) at a dilution of 1:10,000 in 1% (*w*/*v*) non-fat milk dissolved in TBST for 1 h. The membranes were washed three times for 10 min, and proteins were detected with Amersham™ ECL™ (GE Healthcare). Images were taken using the iBright Imaging System (Invitrogen).

To determine the level of K3L ortholog expression by the chimeric vaccinia virus, 8×10^5 HeLa PKR^{ko} cells were seeded in a 6-well plate. On the next day, the cells were infected with VC-R4, VC-R4+VACV K3L, VC-R4+TPV 012, and VC-R4+YMTV 012 at a MOI of 3 to ensure that the vast majority of cells were infected. Protein lysates were collected at 24 h post infection. A total of 12 µg of protein was electrophoresed using 12% SDS polyacrylamide gels and transferred to polyvinyl difluoride (PVDF, GE Healthcare) membranes. The membranes were probed with primary antibodies for beta-actin (Sigma A1978) at a dilution of 1:2000 in Thermofisher Superblock blocking buffer, primary antibodies against TPV 012 (Genscript, Piscataway, NJ, USA) at a dilution of 1:1000 in 5% milk dissolved in TBST, and VACV-K3 at a dilution of 1:500 in 5% milk dissolved in TBST overnight at 4 °C. Anti-TPV 012 (cVQVIRTDKLKGYVDVRHIT) and anti-VACV K3 (cKVIRVDYTKGYIDVNYKRM) were custom-produced using peptide–KLH conjugate in New Zealand rabbit (GenScript). After being washed with TBST three times, the membranes were probed with horseradish peroxidase-conjugated donkey anti-rabbit IgG (H+L) secondary antibody (Invitrogen, A16023) at a dilution of 1:10,000 in 1% milk for anti-TPV 012 and anti-VACV K3, and with horseradish peroxidase-conjugated donkey anti-mouse IgG (H+L) (Jackson

ImmunoResearch Laboratories Inc., 715-035-150) at a 1:10,000 dilution in TBST containing 1% (*w*/*v*) non-fat milk for beta-actin. The membranes were washed three times for 10 min, and proteins were detected with Amersham™ ECL™ (GE Healthcare). Images were taken using the iBright Imaging System (Invitrogen).

2.7. Phylogenetic Analysis

The amino-acid sequences of PKR from 21 different primate species (Table 1) were used to generate the phylogenetic tree. The multiple sequence alignment was performed using MUSCLE [37], and the resulting alignment was used to construct a phylogenetic tree using the maximum likelihood approach, as implemented in the program PhyML 3.0 [38]. The nodal support of the phylogenetic tree was assessed by means of a bootstrap analysis of 100 replicates. The resulting phylogenetic tree was visualized using FigTree v1.4.4 (available at: http://tree.bio.ed.ac.uk/software/figtree/).

Table 1. Accession numbers of genes used for the phylogenetic analysis.

Common Name	Scientific Name	GenBank Accession
Human	*Homo sapiens*	NM_002759
Gorilla	*Gorilla gorilla*	EU733258.1
Bonobo	*Pan paniscus*	EU733255.1
Chimpanzee	*Pan troglodytes*	EU733256.1
Orangutan	*Pongo pygmaeus pgmaeus*	EU733259.1
White-cheeked gibbon	*Nomascus leucogenys*	EU733257.1
White-bearded gibbon	*Hylobates albibarbis*	EU733270.1
Siamang	*Symphalangus syndactylus*	EU733271.1
Colobus monkey	*Colobus guereza*	EU733267.1
François' leaf monkey	*Trachypithecus francoisi*	EU733268.1 *
Baboon	*Papio anubis*	XM_009184002.4 *
Sooty mangabey	*Cercocebus torquatus atys*	EU733262.1
Rhesus macaque	*Macaca mulatta*	EU733261.1
Talapoin monkey	*Miopithecus talapoin talapoin*	EU733269.1
African green monkey	*Clorocebus aethiops*	EU733254.1
Patas monkey	*Erythrocebus patas*	EU733260.1
Black-handed spider monkey	*Ateles geoffroyi*	EU733263.1
Woolly monkey	*Lagothrix lagotricha*	EU733266.1
Tamarin	*Saguinus labiatus*	EU733264.1
Dusky titi	*Callicebus moloch*	EU733265.1
Owl monkey	*Aotus trivirgatus*	FJ374685.1
Rabbit	*Oryctolagus cuniculus*	NM_001082213.1

* Nucleotide differences between the PKR sequences we used and the GenBank sequences: François leaf PKR and EU733268.1: all synonymous: T208 > C (L70L), C861 > T (I287I), G1326 > A (G442G), T1446C (H482H). Baboon PKR and XM_009184002.4: one synonymous A465G (Q155Q) and four non-synonymous G1489A (E497K), A1498C (K500Q), G1511A (G504T), and A1558G (K520E).

3. Results

3.1. Species-Specific Inhibition of Primate PKRs by Yatapoxvirus K3 Orthologs

The inhibition of PKR by K3 orthologs has been shown to be a key determinant for the host range of poxviruses [24–28,34]. In this study, we examined the role of the K3 orthologs TPV 012 and YMTV 012 in inhibiting PKR derived from a diverse set of primate species. The comparison of the amino acid sequences between TPV and YMTV 012 revealed 22 amino acid differences (75% sequence identity). Yatapoxvirus K3 orthologs and VACV K3 showed about 36% sequence identity (Figure 1A,B). We cloned these three orthologs into the pSG5 expression vector with two C-terminal FLAG tags and transfected them into HeLa PKR^{ko} cells. The Western blot analysis of whole cell lysates showed that vaccinia K3 and TPV 012 were expressed at comparable levels, whereas YMTV 012 was expressed at higher level than the former (Figure 1C). We also raised polyclonal antibodies against a conserved epitope of TPV 012 (Figure 1A). Western blot HeLa PKR^{ko} cell lysates transiently transfected with untagged K3L orthologs showed a higher expression of YMTV 012 (Figure 1D). A previously generated antibody raised against VACV K3 detected both VACV K3 and YMTV 012 (Figure 1D), which could be explained by shared epitopes.

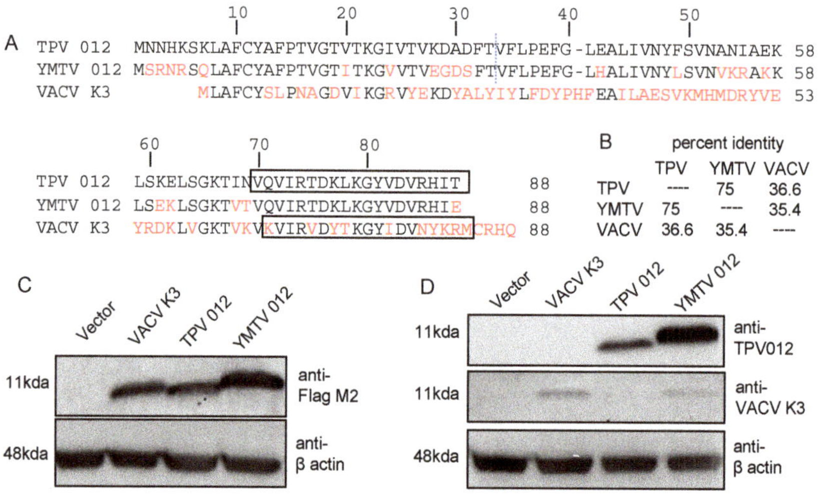

Figure 1. Amino acid differences among TPV 012, YMTV 012, and VACV K3. (**A**). Residues differing from TPV 012 are highlighted in red. The regions to the left and right of the dotted line were considered the N-terminal and C-terminal regions of the inhibitors, respectively. Sequences used to generate antibodies against TPV 012 and VACV K3 are boxed. (**B**). Percent identities between tested K3 orthologs were calculated from the multiple sequence alignment using Clustal Omega [39]. (**C,D**). HeLa PKR^{ko} cells were transfected with 1 μg of either pSG5 empty vector, pSG5 encoding FLAG-tag VACV K3, FLAG-tag TPV 012, or FLAG-tag YMTV 012. Proteins were harvested at 48 h post transfection, and protein expressions were detected with anti-Flag M2, anti-TPV 012, anti-VACV K3, and anti-β actin antibodies as indicated.

We next tested the sensitivities of 15 different primate species to VACV K3 and the yatapoxvirus K3 orthologs in an established luciferase-based expression assay in which PKR-deficient HeLa cells are co-transfected with a luciferase reporter plasmid, PKR, and PKR inhibitors [25,34]. In this assay, PKR is activated by dsRNA that is formed by overlapping transcripts generated from the transfected plasmids [40] and suppresses the translation of reporter mRNA, which serves as a proxy to quantify relative translation. Luciferase activity is thus inversely correlated with PKR activity, and the inhibition of PKR leads to elevated luciferase activity. We first co-transfected cells with PKRs from 15 primate species representing apes (human, gorilla, chimpanzee, orangutan, and white-cheeked gibbon), Old World monkeys (colobus monkey, François' leaf monkey, baboon, sooty mangabey, rhesus macaques, talapoin monkey, African green monkey, and patas monkey) and New World monkeys (tamarin and dusky titi), together with the luciferase reporter plasmid to assess PKR activity. All PKRs showed a strong and comparable inhibition of luciferase activity (Figure 2A). We next co-transfected each PKR with either K3 ortholog and the luciferase reporter plasmid (Figure 2B). Most primate PKRs tested were largely resistant to VACV K3 (less than 2-fold inhibition). Orangutan and white-cheeked gibbon PKR were weakly inhibited (about 2-fold), while rhesus macaque and tamarin PKR were inhibited at intermediate low levels (about 4-fold). In total, 11 of the 15 primate PKRs were inhibited substantially better by TPV 012 than by YMTV 012. PKRs from chimpanzee, white-cheeked gibbon, colobus monkey, and François' leaf monkey were inhibited comparably well by TPV 012 and YMTV 012. In addition to different inhibition profiles when comparing TPV 012 with YMTV 012, species-specific differences in PKR susceptibility were observed, with orangutan, François leaf monkey, and rhesus macaque PKR being most strongly inhibited, and nine other PKRs being intermediately inhibited by TPV 012. In contrast, nine tested PKRs were only weakly inhibited by YMTV 012.

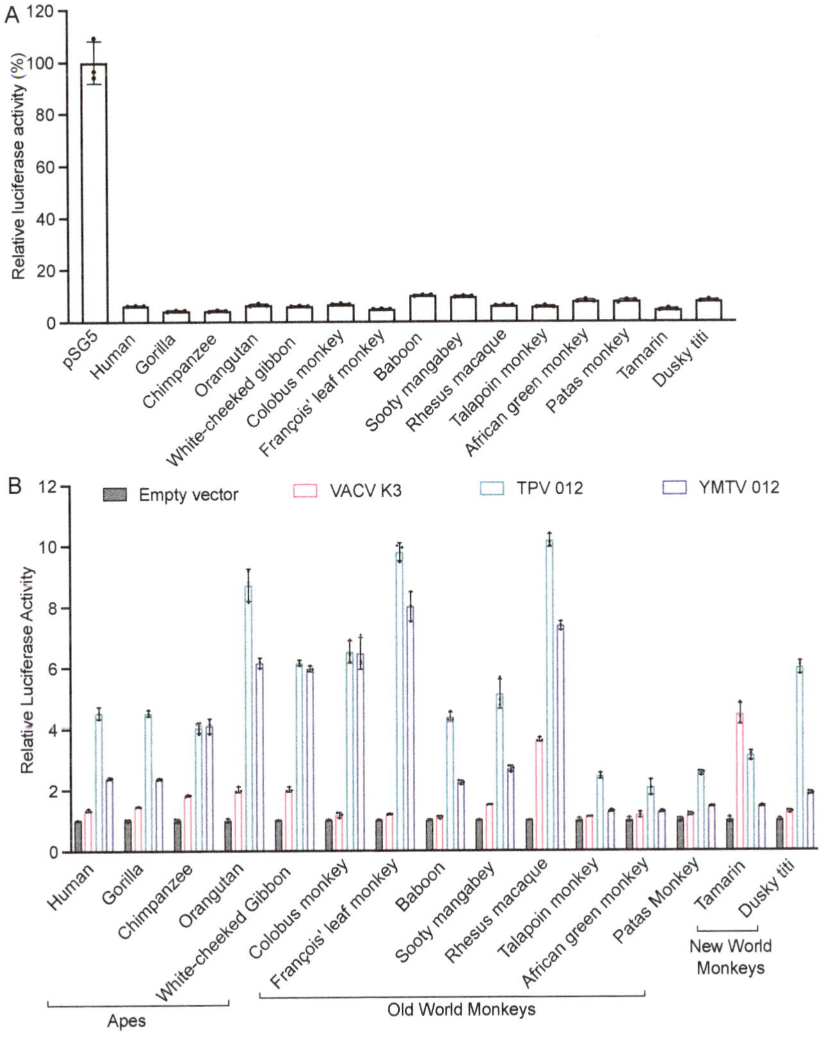

Figure 2. Differential sensitivities of tested PKRs to VACV K3, TPV 012, and YMTV 012 orthologs. (**A**). HeLa PKR^{ko} cells were co-transfected with expression vectors encoding firefly luciferase (50 ng) and with either empty vector or PKR (200 ng) from the indicated species. Luciferase activities were measured 48 h after transfection and normalized to empty vector transfected cells to obtain relative luciferase activities. (**B**). HeLa PKR^{ko} cells were transfected with expression vectors encoding firefly luciferase (50 ng), PKR (200 ng), and PKR inhibitor (200 ng) from the indicated species. Luciferase activities were measured 48 h after transfection and normalized to PKR-only transfected cells to obtain relative luciferase activities. Error bars represent the standard deviations from three independent transfections. The luciferase assay results shown are representative of at least three independent experiments.

We also projected the relative sensitivities of the tested PKRs to the K3 orthologs on a phylogenetic tree generated with PKR amino acid sequences from a panel of 21 primate species using rabbit PKR as an outgroup to visualize the relatedness among the primate species PKRs (Figure 3). This tree recapitulates the relatedness of primates PKRs from an earlier study. The closer relatedness of human PKR to gorilla PKR than to chimpanzee PKR, which is different from the generally accepted primate phylogeny, likely is a result of positive selection during PKR evolution [31]. We used a previously described scale to grade relative PKR inhibition [28], with 1 for no or very weak inhibition (<2-fold inhibition), 2 for weak inhibition (2- to 3-fold inhibition), 3 for intermediate low inhibition (3- to 5-fold inhibition), 4 for intermediate high inhibition (5- to 7-fold inhibition), 5 for high inhibition (7- to 10-fold inhibition), and 6 for very high inhibition (>10-fold inhibition).

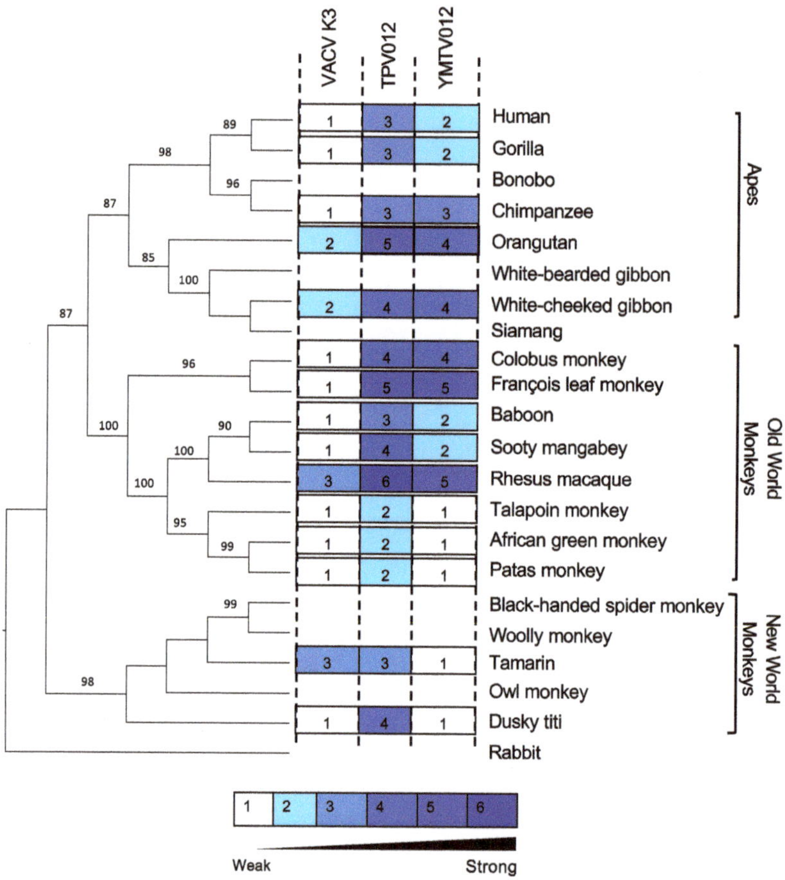

Figure 3. Sensitivities of tested PKRs projected on a PKR phylogenetic tree. We generated a phylogenetic tree with PKR protein sequences from a panel of 21 primate species and used rabbit PKR as an outgroup. The tree was built by PhyML with 100 bootstrap replicates and based on the maximum-likelihood method. Relative sensitivities of PKRs to either VACV K3, TPV 012, or YMTV 012 from multiple luciferase-based reporter assays are shown on a scale from 1 to 6. Scale 1 = no or very weak inhibition (≤2-fold increase in luciferase readout); 2 = weak inhibition (2- to 3-fold); 3 = intermediate low inhibition (3- to 5-fold); 4 = intermediate high inhibition (5- to 7-fold); 5 = high inhibition (7- to 10-fold); 6=very high inhibition (≥10-fold).

3.2. Differential PKR Inhibition Was Governed by the C-Terminus of Yatapoxvirus K3 Orthologs

The amino acid sequence alignment of YMTV and TPV K3 orthologs identified 22 amino acid differences between the two orthologs, which are distributed throughout the protein sequence. To investigate which regions are important for the differential PKR inhibition by yatapoxvirus K3 orthologs, we generated constructs encoding hybrid YMTV-TPV K3 orthologs in which the C-terminal regions of the two inhibitors were swapped. We took advantage of the shared region in the middle part of the gene (FTVFLPEFG) to separate the N-and C-termini, which leaves 11 amino acid differences in each terminus (Figure 4A). We analyzed the ability of these hybrid proteins to inhibit human, colobus monkey, and dusky titi PKR using the luciferase assay. The hybrid $_{N}$TPV-$_{C}$YMTV exhibited inhibition profiles that were more similar to YMTV 012 for each PKR tested (Figure 4B). Correspondingly, the $_{N}$YMTV-$_{C}$TPV hybrid showed a similar inhibition pattern to wild-type TPV 012. These results indicate that the C-terminus of TPV 012 and YMTV 012 is important for the observed differential PKR inhibition, although we cannot exclude a contribution of the N-terminus.

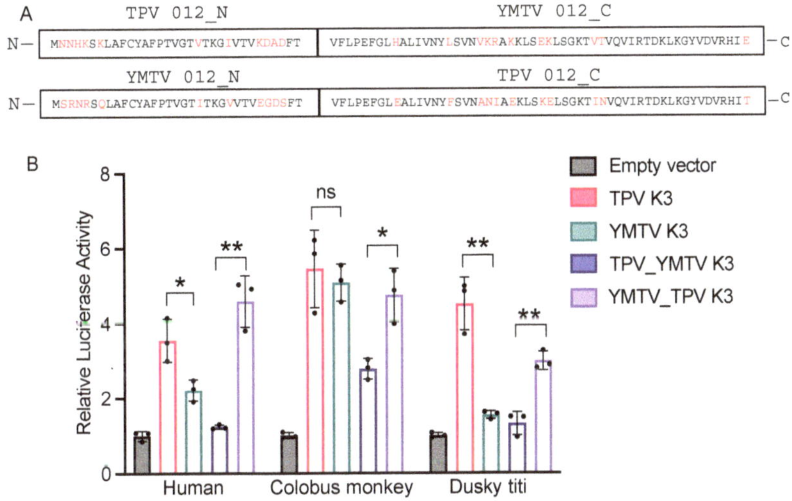

Figure 4. Identification of region of TPV 012 and YMTV 012 orthologs that confer differential PKR inhibition. (**A**). The predicted amino acid sequences of TPV 012 and YMTV 012 hybrid proteins. Constructs were designed such that the C-terminal regions of the two inhibitors were swapped to generate hybrid proteins. Different residues among K3 orthologs are highlighted in red. (**B**). Human HeLa-PKR^{ko} cells were transfected with expression vector encoding firefly luciferase, indicated PKRs, and either hybrid $_{N}$TPV-$_{C}$YMTV or $_{N}$YMTV-$_{C}$TPV inhibitors. The inhibition of indicated PKRs by TPV 012 and YMTV 012 serves as control. Luciferase activity was measured 48 h after transfection. Luciferase light units were normalized to PKR-only transfected cells to obtain relative luciferase activities. Error bars represent triplicate transfections and data are representative of three independent experiments. The data were analyzed using the unpaired *t*-test (GraphPad). *p*-values are indicated as follows: ns (non-significant) $p > 0.05$, * $p < 0.05$, ** $p < 0.01$.

3.3. Chimeric Viruses Expressing TPV 012 or YMTV 012 Displayed Cell-Type-Specific Differences in Plaque Formation and Virus Replication

Next, we investigated whether the ability of TPV 012 and YMTV 012 orthologs to inhibit primate PKRs correlated with the ability of the viral genes to rescue the replication of a VACV strain that lacks PKR inhibitors in primate-derived cell lines. The VACV strain VC-R4 (VACVΔE3LΔK3L) is a highly attenuated vaccinia virus variant that can only replicate in PKR-deficient cell lines or cell lines expressing PKR antagonists [33]. We inserted either TPV 012 or YMTV 012 into VC-R4 at the E3L locus using the scarless integration method as previously described, designating the chimeric viruses VC-R4-TPV 012 or VC-R4-YMTV 012, respectively [33]. To analyze the expression of the K3 orthologs by these chimeric viruses, we infected HeLa PKR^{ko} cells and performed Western blots using custom-made antibodies at 24 h post infection. In agreement with the transfection data (Figure 1), YMTV 012 was detected at a higher level compared to TPV 012 (Figure 5A). We next examined the plaque formation of these viruses in comparison to VACV vP872 (which lacks K3L but contains the second VACV PKR inhibitor E3), VC-R4+VACV K3L, and VC-R4+SPPV 011 which expresses the sheeppox virus K3 ortholog [26], in different primate-derived cell lines: HeLa (human), gibbon Tert-GF (gibbon), and BSC-40 and PRO1190 (African green monkey). As control, we used RK13+K3L+E3L cells (rabbit), which express VACV E3 and K3 and in which VC-R4 is permissive. We infected these cells with 50 PFU/well. As expected, all viruses formed plaques in permissive RK13+K3L+E3L cells. VC-R4 only formed plaques in RK13+K3L+E3L cells, and vP872 and VC-R4+SPPV 011 formed plaques in all cells tested (Figure 5B).

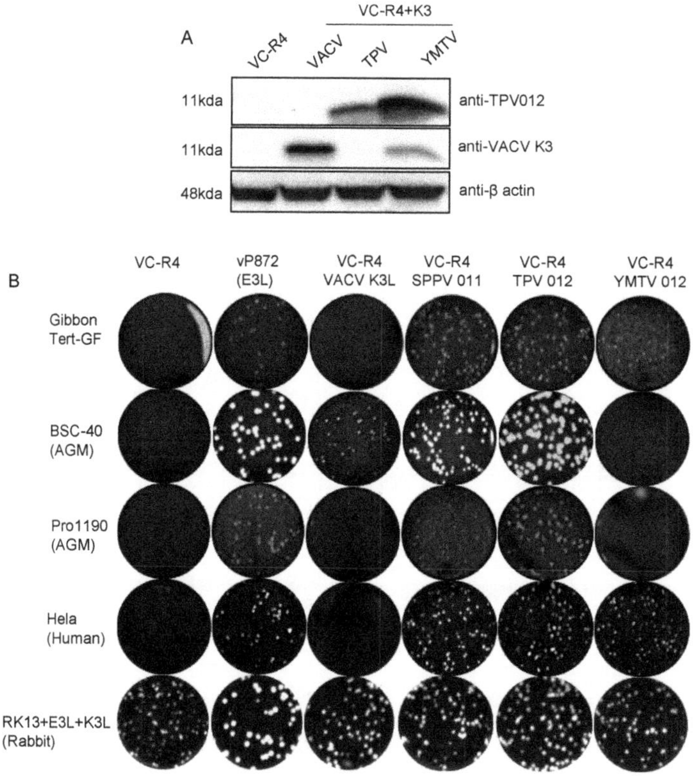

Figure 5. Generation of chimeric VC-R4 expressing yatapoxvirus K3 orthologs. (**A**). HeLa PKR^{ko} were infected with VC-R4, VC-R4+VACV K3, VC-R4+SPPV 011, VC-R4+TPV 012, or VC-R4+YMTV 012. Protein lysates were collected at 24 h post infection. The expression of K3 orthologs was determined by immunoblotting with the indicated antibodies. (**B**). Plaque formation of VC-R4 expressing K3 orthologs. A variety of primate-derived cells and RK13+E3L+K3L were infected with 50 pfu/mL of the indicated viruses, and plaque formation was visualized with crystal violet at 48 h after infection.

VC-R4+VACV K3L developed small plaques only in BSC-40 cells. VC-R4+TPV 012 infection led to the formation of plaques in all tested cells with comparable sizes to plaques formed by both VC-R4+SPPV 011 and vP872. Relative to VC-R4+TPV 012, VC-R4+YMTV 012 formed smaller plaques in HeLa and gibbon Tert-GF cells and failed to cause plaque formation in both African green monkey cells. In general, based on the plaque assay, TPV 012 rescued VC-R4 replication substantially better than YMTV 012 in primate-derived cell lines, despite the higher expression of YMTV 012.

We next measured virus titers to determine the capacity of yatapoxvirus K3 orthologs to restore the replication of VC-R4. We infected RK13+K3L+E3L and primate-derived cell lines HeLa (human), gibbon Tert-GF, BSC-40 (African green monkeys), and PRO1190 (African green monkeys) cells with either VC-R4, vP872, VC-R4+VACV K3L, VC-R4+SPPV 11, VC-R4+TPV 012, or VC-R4+YMTV 012 at a MOI of 0.01. Cell lysates were collected at 30 h post infection, and the virus titer was determined in RK13+K3L+E3L cells. Consistent with the plaque assays and luciferase assays, all viruses replicated to comparable titers in RK13+K3L+E3L cells (Figure 6). VC-R4 did not replicate in any cells tested (virus titer below 103 pfu/mL), except in RK13+K3L+E3L cells. VACV K3L rescued VC-R4 virus replication in all cells tested to variable extents, with virus titers about 10- to 100-fold lower than vP872. In HeLa and gibbon Tert-GF cells, SPPV 011 and TPV 012 both restored the replication of VC-R4 to a level comparable to vP872. In BSC-40 and PRO1190 cells, TPV 012 and SPPV 011 rescued VC-R4 replication comparably. Consistent with the plaque assays, VC-R4+YMTV 012 replicated to slightly lower levels than VC-R4+TPV 012 in HeLa cells, although this difference was not significant. In contrast, VC-R4+YMTV 012 significantly replicated with less efficiency than VC-R4+TPV 012 in gibbon Tert-GF cells (11-fold lower), BSC-40 cells (1394-fold lower), and PRO1190 cells (121-fold lower).

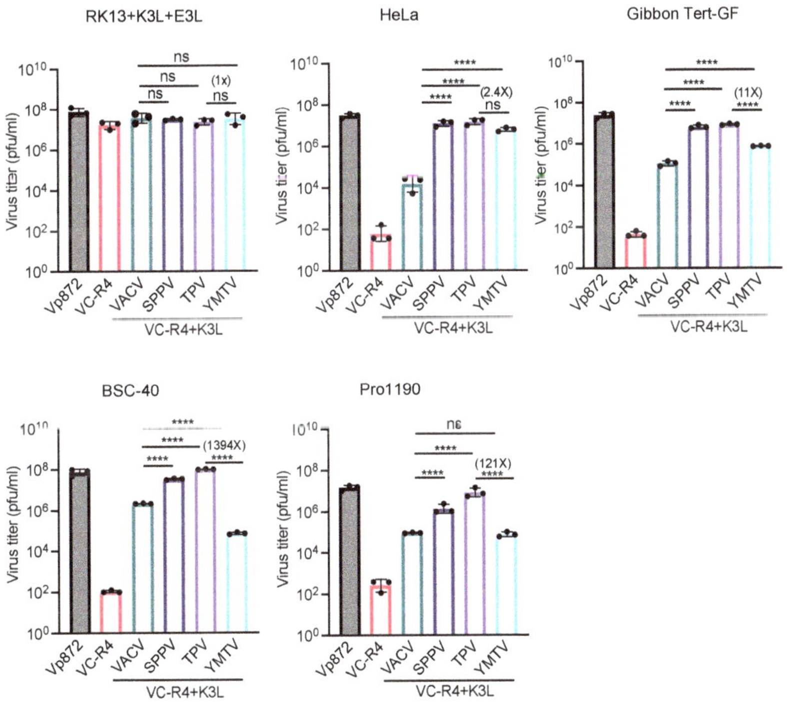

Figure 6. Virus titer of chimeric vaccinia viruses in primate-derived cell lines. A variety of primate-derived cells and RK13+E3L+K3L were infected with indicated viruses at a MOI of 0.01. Cell lysates were harvested at 30 h post infection and virus titers were determined in RK13+E3L+K3L cells. Error bars represent the standard deviations from three independent infections. Fold changes in virus titer between VC-R4+TPV 012 and VC-R4+YMTV 012 are shown. Data were analyzed using one-way ANOVA followed by Tukey's multiple comparisons test. p-values are marked as follows: ns = $p > 0.05$, **** $p < 0.0001$.

4. Discussion

Vaccinia virus K3 is a poxvirus host range factor that acts by inhibiting the host restriction factor PKR. Homology between the K3 and the S1 domain of eIF2α allows K3 to function as a pseudosubstrate of PKR, competitively inhibiting eIF2α phosphorylation [19]. This interaction between PKR and K3 has been studied using yeast and mammalian cell assay systems [41,42]. Previously, Elde et al. have studied the interaction of 10 primate PKR variants with VACV K3 and showed that PKR from Old World monkeys and New World monkeys were generally susceptible to VACV K3, whereas PKR from hominoids were resistant to VACV K3, indicating a species-specific inhibition of primate PKR by VACV K3 [31]. We recently described that K3 orthologs from myxoma virus, which infects rabbits and hares, and capripoxviruses, which infect sheep, goats, and cattle, also inhibit host PKRs in a species-specific manner [25,26,43]. Our recent studies revealed that K3 inhibition could not be predicted by the phylogenetic relatedness of PKR alone [28]. Therefore, in this study, we experimentally determined whether K3 orthologs of the Yatapoxvirus genus also exhibit host-specific PKR inhibition.

Yatapoxviruses infect primates and can cause zoonotic infections in humans. Here, we analyzed whether TPV and YMTV K3 orthologs inhibit PKR from 15 different primate species in comparison to VACV K3. Our results confirm and extend the results of species-specific VACV K3 inhibition previously reported in both yeast-based [31] and luciferase-based assays [28]. White-cheeked gibbon PKR was largely resistant to VACV K3, whereas VACV K3 strongly inhibited tamarin PKR and rhesus macaque PKR. Interestingly, dusky titi PKR, which was previously shown to be susceptible to VACV K3 in the yeast-based assay, appeared to be largely resistant to VACV K3 in our luciferase-based assays. Although the yeast assay can be used to predict the inhibition of PKR by viral antagonists, yeast expression systems have very different post-translational modifications relative to mammalian cells, which might affect protein function [44]. In general, our data showed that TPV 012 and YMTV 012 exhibited distinct PKR inhibition profiles and that these K3 orthologs inhibited primate PKR in a species-specific manner. All tested primate PKRs were sensitive to TPV 012, although to different degrees. We found a distinct profile of inhibition by YMTV 012 within the Old World monkeys in which talapoin monkey, African green monkey, and patas monkey PKRs were largely resistant to YMTV 012 inhibition. In contrast, rhesus macaque PKR was largely sensitive, supporting that the interaction between K3 orthologs and PKR could only be partially inferred by phylogenetic relatedness. This finding is in line with subcutaneous inoculation studies, showing that YMTV induced tumor formation in rhesus monkeys, whereas no tumors were observed in sooty mangabey, patas monkey, and African green monkeys [6,9,12].

TPV 012 and YMTV 012 differ by 22 amino acids out of 88 residues (Figure 1). K3 orthologs from the orthopoxvirus genus share a higher sequence identity (80.7–100%) among their members [28] than TPV 012 and YMTV 012. We found that residues within the C-terminus region are important for the differential PKR inhibition by these orthologs. There are two critical conserved motifs in the C-terminus of the K3 orthologs: the PKR recognition motif and helix insert region [19]. The PKR recognition motif, which includes residues K74, Y76, and D78, provides important determinants for high-affinity binding to PKR [19]. TPV 012 and YMTV 012 orthologs have identical residues in the PKR recognition motif "KGYVD", one amino acid difference compared to the PKR recognition motif "KGYID" of VACV K3. The helix insert of eIF2α appears to undergo conformational change upon binding to PKR, resulting in the Ser51 site of eIF2α becoming fully accessible to the phosphoacceptor binding site of PKR [45]. The K3 helix insert region is structurally homologous to the region adjacent to Ser51 in eIF2α. The unique conformation of the helix insert region of K3L functions as an inhibitor of PKR activation by preventing PKR trans-autophosphorylation and acting as a pseudosubstrate of PKR [45]. In addition, we and others have shown that one or more amino acid residues in orthopoxvirus K3 orthologs are responsible for the differential PKR inhibition and that the amino acid variants are predominantly located in the helix insert region [27,28].

In this study, the function of yatapoxvirus K3 proteins was further examined in the context of virus infection. The E3L and K3L double-knockout vaccinia virus strain VC-R4 (VACVΔE3LΔK3L) can only replicate in PKR-deficient cell lines or cell lines that express PKR antagonists. By inserting K3L orthologs from different poxviruses in the E3L locus of VC-R4, we can analyze the chimeric viruses' ability to replicate in PKR-competent cell lines. The advantage of using the chimeric virus system is that it allows us to assess the ability of K3L orthologs in rescuing the virus in PKR-competent cells. As expected, VC-R4 could not replicate in all tested cells, except RK13+E3L+K3L cells. VC-R4 formed smaller plaques than vP872, which can be explained by incomplete PKR inhibition by the ectopically expressed E3 and K3, and is consistent with the slightly less efficient replication of VC-R4 than vP872 in this cell line [46]. VC-R4 expressing SPPV 011 and TPV 012 replicated as efficiently as vP872 in all tested cell lines, as shown in plaque assays and virus titer. In contrast, VC-R4+VACV K3L and VC-R4+YMTV 012 formed smaller plaques in all tested cell lines, except in RK13+E3L+K3L where these viruses formed comparable plaque size with other viruses. Together these results indicate that VACV K3L and YMTV 012 are generally weaker inhibitors for the PKR orthologs that we tested. This result was recapitulated in the virus replication assay. There was an 11-fold higher virus titer in VC-R4+TPV 012-infected gibbon Tert-GF cells than in VC-R4+YMTV 012-infected cells. The difference in virus titer was more pronounced in PRO1190 and BSC-40 cells, with about 100- to 1000-fold lower virus titer in VC-R4+YMTV 012 infections. Although both cell lines are derived from African green monkeys, BSC-40 are immortalized epithelial cells and PRO1190 are primary fibroblasts. Differential susceptibility to various VACV recombinants by these two cell lines is well established, and may in part be due to three non-synonymous variants in PRO1190 PKR relative to BSC-40 PKR, although other host proteins, e.g., RNase L, have been implicated in some studies, suggesting that this phenotype is likely multifactorial [47–49]. While the general patterns observed in the luciferase-based assay were recapitulated in the infection assay, we observed some differences in the magnitude of inhibition.

Overall, the data presented here support previous observations regarding species-specific PKR inhibition by VACV K3 orthologs and extend them to a new genus of poxviruses [27,28,31]. Within primate species PKR, TPV 012 exhibited a broader range of PKR inhibition than did YMTV 012 and VACV K3. VACV K3 appeared to have a narrow range of PKR inhibition within primate species but efficiently inhibited PKR from a wide range mammals PKR [28]. It is worth mentioning that the ability of a virus to replicate and disseminate efficiently in its host is dependent on the entire interactome between the host and the virus [50]. In addition to K3L, yatapoxviruses encode several host range genes, including E3L, C7L, M11L, Serpins, P28-like, and B5R [51]. The role of the E3L gene of YMTV and C7L of YLDV in host range factors has been previously described [52,53]. However, the role of other genes as host range factors needs to be experimentally confirmed.

5. Conclusions

The possible reemergence of smallpox-like diseases through the zoonosis of animal poxviruses is a major public health concern. Members of the Yatapoxvirus genus infect humans and other primates and have caused outbreaks throughout Africa for the past 60 years. In addition, human infections with yatapoxviruses have been reported in animal handlers at primate centers in the United States and among travelers who visited Africa. Despite the threat posed to human health, the factors determining the host range of these viruses remain poorly understood. Here, we describe the inhibitory profiles of TPV and YMTV K3 orthologs against primate PKRs, and show that they display virus and host species-specific PKR inhibition. The TPV and YMTV K3 orthologs exhibited distinct PKR inhibition profiles, with TPV 012 generally inhibiting PKRs more strongly than YMTV 012. The species-specific PKR inhibition by the TPV and YMTV K3 orthologs may contribute to the differential susceptibility of primate species to TPV and YMTV infections.

Author Contributions: Conceptualization, D.M. and S.R.; methodology, D.M., J.N.S., C.P., T.C., L.T., G.B. and S.R.; validation, D.M., J.N.S., C.P., T.C., L.T., G.B. and S.R.; formal analysis, D.M., J.N.S., C.P., T.C., L.T., G.B. and S.R.; resources, S.R.; data curation, D.M. and L.T.; writing—original draft preparation, D.M.; writing—review and editing, D.M., J.N.S., C.P., L.T., G.B. and S.R.; visualization, D.M., L.T. and S.R.; supervision, S.R.; project administration, S.R.; funding acquisition, S.R. All authors have read and agreed to the published version of the manuscript.

Funding: This work was supported by Grant R01 AI114851 (to S.R.) from the National Institute of Allergy and Infectious Diseases.

Institutional Review Board Statement: Not applicable.

Informed Consent Statement: Not applicable.

Data Availability Statement: All data necessary for the interpretation of the findings presented in this work are contained within the manuscript figures.

Acknowledgments: We are grateful to Jeffrey Hastings for technical assistance. We thank Nels Elde, Bertram Jacobs, Charles Samuel, and Adam Geballe for generously providing materials and reagents. Parts of this work were included in the PhD dissertation of D.M.

Conflicts of Interest: The authors declare no conflicts of interest.

References

1. Haller, S.L.; Peng, C.; McFadden, G.; Rothenburg, S. Poxviruses and the evolution of host range and virulence. *Infect. Genet. Evol.* **2014**, *21*, 15–40. [CrossRef] [PubMed]
2. Lee, H.J.; Essani, K.; Smith, G.L. The genome sequence of Yaba-like disease virus, a yatapoxvirus. *Virology* **2001**, *281*, 170–192. [CrossRef]
3. Brunetti, C.R.; Amano, H.; Ueda, Y.; Qin, J.; Miyamura, T.; Suzuki, T.; Li, X.; Barrett, J.W.; McFadden, G. Complete genomic sequence and comparative analysis of the tumorigenic poxvirus Yaba monkey tumor virus. *J. Virol.* **2003**, *77*, 13335–13347. [CrossRef]
4. Downie, A.W.; Taylor-Robinson, C.H.; Caunt, A.E.; Nelson, G.S.; Manson-Bahr, P.E.; Matthews, T.C. Tanapox: A new disease caused by a pox virus. *Br. Med. J.* **1971**, *1*, 363–368. [CrossRef]
5. Jezek, Z.; Arita, I.; Szczeniowski, M.; Paluku, K.M.; Ruti, K.; Nakano, J.H. Human tanapox in Zaire: Clinical and epidemiological observations on cases confirmed by laboratory studies. *Bull. World Health Organ.* **1985**, *63*, 1027–1035.
6. Bearcroft, W.G.; Jamieson, M.F. An outbreak of subcutaneous tumours in rhesus monkeys. *Nature* **1958**, *182*, 195–196. [CrossRef]
7. Knight, J.C.; Novembre, F.J.; Brown, D.R.; Goldsmith, C.S.; Esposito, J.J. Studies on Tanapox virus. *Virology* **1989**, *172*, 116–124. [CrossRef]
8. Nazarian, S.H.; Barrett, J.W.; Frace, A.M.; Olsen-Rasmussen, M.; Khristova, M.; Shaban, M.; Neering, S.; Li, Y.; Damon, I.K.; Esposito, J.J.; et al. Comparative genetic analysis of genomic DNA sequences of two human isolates of Tanapox virus. *Virus Res.* **2007**, *129*, 11–25. [CrossRef] [PubMed]
9. Sproul, E.E.; Metzgar, R.S.; Grace, J.T., Jr. The pathogenesis of Yaba virus-induced histiocytomas in primates. *Cancer Res.* **1963**, *23*, 671–675.
10. Downie, A.W. Serological evidence of infection with Tana and Yaba pox viruses among several species of monkey. *J. Hyg.* **1974**, *72*, 245–250. [CrossRef]
11. Monroe, B.P.; Nakazawa, Y.J.; Reynolds, M.G.; Carroll, D.S. Estimating the geographic distribution of human Tanapox and potential reservoirs using ecological niche modeling. *Int. J. Health Geogr.* **2014**, *13*, 34. [CrossRef] [PubMed]
12. Ambrus, J.L.; Strandström, H.V. Susceptibility of Old World monkeys to Yaba virus. *Nature* **1966**, *211*, 876. [CrossRef]
13. Grace, J.T., Jr.; Mirand, E.A. Human susceptibility to a simian tumor virus. *Ann. N. Y. Acad. Sci.* **1963**, *108*, 1123–1128. [CrossRef] [PubMed]
14. Stich, A.; Meyer, H.; Köhler, B.; Fleischer, K. Tanapox: First report in a European traveller and identification by PCR. *Trans. R. Soc. Trop. Med. Hyg.* **2002**, *96*, 178–179. [CrossRef] [PubMed]
15. Dimitrov, D.S. Virus entry: Molecular mechanisms and biomedical applications. *Nat. Reviews. Microbiol.* **2004**, *2*, 109–122. [CrossRef]
16. McFadden, G. Poxvirus tropism. *Nat. Reviews. Microbiol.* **2005**, *3*, 201–213. [CrossRef]
17. Moss, B. Poxvirus cell entry: How many proteins does it take? *Viruses* **2012**, *4*, 688–707. [CrossRef]
18. Jacobs, B.L.; Langland, J.O. When two strands are better than one: The mediators and modulators of the cellular responses to double-stranded RNA. *Virology* **1996**, *219*, 339–349. [CrossRef]
19. Dar, A.C.; Sicheri, F. X-ray crystal structure and functional analysis of vaccinia virus K3L reveals molecular determinants for PKR subversion and substrate recognition. *Mol. Cell* **2002**, *10*, 295–305. [CrossRef]
20. Adomavicius, T.; Guaita, M.; Zhou, Y.; Jennings, M.D.; Latif, Z.; Roseman, A.M.; Pavitt, G.D. The structural basis of translational control by eIF2 phosphorylation. *Nat. Commun.* **2019**, *10*, 2136. [CrossRef]

21. Romano, P.R.; Zhang, F.; Tan, S.L.; Garcia-Barrio, M.T.; Katze, M.G.; Dever, T.E.; Hinnebusch, A.G. Inhibition of double-stranded RNA-dependent protein kinase PKR by vaccinia virus E3: Role of complex formation and the E3 N-terminal domain. *Mol. Cell. Biol.* **1998**, *18*, 7304–7316. [CrossRef] [PubMed]
22. Langland, J.O.; Jacobs, B.L. The role of the PKR-inhibitory genes, E3L and K3L, in determining vaccinia virus host range. *Virology* **2002**, *299*, 133–141. [CrossRef] [PubMed]
23. Langland, J.O.; Jacobs, B.L. Inhibition of PKR by vaccinia virus: Role of the N- and C-terminal domains of E3L. *Virology* **2004**, *324*, 419–429. [CrossRef] [PubMed]
24. Haller, S.L.; Park, C.; Bruneau, R.C.; Megawati, D.; Zhang, C.; Vipat, S.; Peng, C.; Senkevich, T.G.; Brennan, G.; Tazi, L.; et al. Molecular basis for the host range function of the poxvirus PKR inhibitor E3. *bioRxiv* **2024**. [CrossRef]
25. Peng, C.; Haller, S.L.; Rahman, M.M.; McFadden, G.; Rothenburg, S. Myxoma virus M156 is a specific inhibitor of rabbit PKR but contains a loss-of-function mutation in Australian virus isolates. *Proc. Natl. Acad. Sci. USA* **2016**, *113*, 3855–3860. [CrossRef] [PubMed]
26. Park, C.; Peng, C.; Brennan, G.; Rothenburg, S. Species-specific inhibition of antiviral protein kinase R by capripoxviruses and vaccinia virus. *Ann. N.Y. Acad. Sci.* **2019**, *1438*, 18–29. [CrossRef] [PubMed]
27. Cao, J.; Varga, J.; Deschambault, Y. Poxvirus encoded eIF2α homolog, K3 family proteins, is a key determinant of poxvirus host species specificity. *Virology* **2020**, *541*, 101–112. [CrossRef] [PubMed]
28. Park, C.; Peng, C.; Rahman, M.J.; Haller, S.L.; Tazi, L.; Brennan, G.; Rothenburg, S. Orthopoxvirus K3 orthologs show virus- and host-specific inhibition of the antiviral protein kinase PKR. *PLoS Pathog.* **2021**, *17*, e1009183. [CrossRef] [PubMed]
29. Carpentier, K.S.; Esparo, N.M.; Child, S.J.; Geballe, A.P. A Single Amino Acid Dictates Protein Kinase R Susceptibility to Unrelated Viral Antagonists. *PLoS Pathog.* **2016**, *12*, e1005966. [CrossRef]
30. Rahman, M.M.; Liu, J.; Chan, W.M.; Rothenburg, S.; McFadden, G. Myxoma virus protein M029 is a dual function immunomodulator that inhibits PKR and also conscripts RHA/DHX9 to promote expanded host tropism and viral replication. *PLoS Pathog.* **2013**, *9*, e1003465. [CrossRef]
31. Elde, N.C.; Child, S.J.; Geballe, A.P.; Malik, H.S. Protein kinase R reveals an evolutionary model for defeating viral mimicry. *Nature* **2009**, *457*, 485–489. [CrossRef] [PubMed]
32. Gibson, D.G.; Young, L.; Chuang, R.Y.; Venter, J.C.; Hutchison, C.A., 3rd; Smith, H.O. Enzymatic assembly of DNA molecules up to several hundred kilobases. *Nat. Methods* **2009**, *6*, 343–345. [CrossRef] [PubMed]
33. Vipat, S.; Brennan, G.; Park, C.; Haller, S.L.; Rothenburg, S. Rapid, Seamless Generation of Recombinant Poxviruses using Host Range and Visual Selection. *J. Vis. Exp. JoVE* **2020**, *159*, e61049. [CrossRef]
34. Rothenburg, S.; Seo, E.J.; Gibbs, J.S.; Dever, T.E.; Dittmar, K. Rapid evolution of protein kinase PKR alters sensitivity to viral inhibitors. *Nat. Struct. Mol. Biol.* **2009**, *16*, 63–70. [CrossRef] [PubMed]
35. Beattie, E.; Tartaglia, J.; Paoletti, E. Vaccinia virus-encoded eIF-2 alpha homolog abrogates the antiviral effect of interferon. *Virology* **1991**, *183*, 419–422. [CrossRef] [PubMed]
36. Esposito, J.; Condit, R.; Obijeski, J. The preparation of orthopoxvirus DNA. *J. Virol. Methods* **1981**, *2*, 175–179. [CrossRef]
37. Edgar, R.C. MUSCLE: Multiple sequence alignment with high accuracy and high throughput. *Nucleic Acids Res.* **2004**, *32*, 1792–1797. [CrossRef] [PubMed]
38. Guindon, S.; Dufayard, J.F.; Lefort, V.; Anisimova, M.; Hordijk, W.; Gascuel, O. New algorithms and methods to estimate maximum-likelihood phylogenies: Assessing the performance of PhyML 3.0. *Syst. Biol.* **2010**, *59*, 307–321. [CrossRef] [PubMed]
39. Sievers, F.; Higgins, D.G. Clustal Omega for making accurate alignments of many protein sequences. *Protein Sci.* **2018**, *27*, 135–145. [CrossRef]
40. Nejepinska, J.; Malik, R.; Wagner, S.; Svoboda, P. Reporters transiently transfected into mammalian cells are highly sensitive to translational repression induced by dsRNA expression. *PLoS ONE* **2014**, *9*, e87517. [CrossRef]
41. Kawagishi-Kobayashi, M.; Silverman, J.B.; Ung, T.L.; Dever, T.E. Regulation of the protein kinase PKR by the vaccinia virus pseudosubstrate inhibitor K3L is dependent on residues conserved between the K3L protein and the PKR substrate eIF2alpha. *Mol. Cell. Biol.* **1997**, *17*, 4146–4158. [CrossRef] [PubMed]
42. Davies, M.V.; Chang, H.W.; Jacobs, B.L.; Kaufman, R.J. The E3L and K3L vaccinia virus gene products stimulate translation through inhibition of the double-stranded RNA-dependent protein kinase by different mechanisms. *J. Virol.* **1993**, *67*, 1688–1692. [CrossRef] [PubMed]
43. Yu, H.; Peng, C.; Zhang, C.; Stoian, A.M.M.; Tazi, L.; Brennan, G.; Rothenburg, S. Maladaptation after a virus host switch leads to increased activation of the pro-inflammatory NF-κB pathway. *Proc. Natl. Acad. Sci. USA* **2022**, *119*, e2115354119. [CrossRef] [PubMed]
44. Vieira Gomes, A.M.; Souza Carmo, T.; Silva Carvalho, L.; Mendonça Bahia, F.; Parachin, N.S. Comparison of Yeasts as Hosts for Recombinant Protein Production. *Microorganisms* **2018**, *6*, 38. [CrossRef] [PubMed]
45. Dar, A.C.; Dever, T.E.; Sicheri, F. Higher-order substrate recognition of eIF2alpha by the RNA-dependent protein kinase PKR. *Cell* **2005**, *122*, 887–900. [CrossRef]
46. Hand, E.S.; Haller, S.L.; Peng, C.; Rothenburg, S.; Hersperger, A.R. Ectopic expression of vaccinia virus E3 and K3 cannot rescue ectromelia virus replication in rabbit RK13 cells. *PLoS ONE* **2015**, *10*, e0119189. [CrossRef] [PubMed]
47. Brennan, G.; Kitzman, J.O.; Rothenburg, S.; Shendure, J.; Geballe, A.P. Adaptive gene amplification as an intermediate step in the expansion of virus host range. *PLoS Pathog.* **2014**, *10*, e1004002. [CrossRef] [PubMed]

48. Brennan, G.; Kitzman, J.O.; Shendure, J.; Geballe, A.P. Experimental Evolution Identifies Vaccinia Virus Mutations in A24R and A35R That Antagonize the Protein Kinase R Pathway and Accompany Collapse of an Extragenic Gene Amplification. *J. Virol.* **2015**, *89*, 9986–9997. [CrossRef] [PubMed]
49. Banerjee, S.; Smith, C.; Geballe, A.P.; Rothenburg, S.; Kitzman, J.O.; Brennan, G. Gene amplification acts as a molecular foothold to facilitate cross-species adaptation and evasion of multiple antiviral pathways. *Virus Evol.* **2022**, *8*, veac105. [CrossRef]
50. Rothenburg, S.; Brennan, G. Species-Specific Host-Virus Interactions: Implications for Viral Host Range and Virulence. *Trends Microbiol.* **2020**, *28*, 46–56. [CrossRef]
51. Bratke, K.A.; McLysaght, A.; Rothenburg, S. A survey of host range genes in poxvirus genomes. *Infect. Genet. Evol.* **2013**, *14*, 406–425. [CrossRef] [PubMed]
52. Myskiw, C.; Arsenio, J.; Hammett, C.; van Bruggen, R.; Deschambault, Y.; Beausoleil, N.; Babiuk, S.; Cao, J. Comparative analysis of poxvirus orthologues of the vaccinia virus E3 protein: Modulation of protein kinase R activity, cytokine responses, and virus pathogenicity. *J. Virol.* **2011**, *85*, 12280–12291. [CrossRef] [PubMed]
53. Meng, X.; Chao, J.; Xiang, Y. Identification from diverse mammalian poxviruses of host-range regulatory genes functioning equivalently to vaccinia virus C7L. *Virology* **2008**, *372*, 372–383. [CrossRef] [PubMed]

viruses

MDPI

Article

Human Betacoronavirus OC43 Interferes with the Integrated Stress Response Pathway in Infected Cells

Stacia M. Dolliver, Caleb Galbraith and Denys A. Khaperskyy *

Department of Microbiology and Immunology, Faculty of Medicine, Dalhousie University, Halifax, NS B3H 4R2, Canada
* Correspondence: d.khaperskyy@dal.ca

Abstract: Viruses evolve many strategies to ensure the efficient synthesis of their proteins. One such strategy is the inhibition of the integrated stress response—the mechanism through which infected cells arrest translation through the phosphorylation of the alpha subunit of the eukaryotic translation initiation factor 2 (eIF2α). We have recently shown that the human common cold betacoronavirus OC43 actively inhibits eIF2α phosphorylation in response to sodium arsenite, a potent inducer of oxidative stress. In this work, we examined the modulation of integrated stress responses by OC43 and demonstrated that the negative feedback regulator of eIF2α phosphorylation GADD34 is strongly induced in infected cells. However, the upregulation of GADD34 expression induced by OC43 was independent from the activation of the integrated stress response and was not required for the inhibition of eIF2α phosphorylation in virus-infected cells. Our work reveals a complex interplay between the common cold coronavirus and the integrated stress response, in which efficient viral protein synthesis is ensured by the inhibition of eIF2α phosphorylation but the GADD34 negative feedback loop is disrupted.

Keywords: betacoronavirus OC43; integrated stress response; PERK; PKR; GADD34; ATF4

1. Introduction

Eukaryotic cells can rapidly respond to adverse conditions by activating specific stress response mechanisms. One of the conserved signaling pathways activated by various types of stress, including viral infections, is the integrated stress response (ISR) [1]. Its defining feature is the phosphorylation of the alpha subunit of the eukaryotic translation initiation factor 2 (eIF2α) on Serine-51, which causes the transient inhibition of bulk protein synthesis to conserve energy and minimize damage [2]. In mammals, there are four eIF2α kinases that are activated by different types of stress: heme-regulated inhibitor (HRI) kinase is activated by oxidative stress and heat shock [3,4], general control nonderepressible 2 (GCN2) kinase is activated by amino acid starvation or ultraviolet damage [5,6]; double stranded RNA (dsRNA)-activated protein kinase (PKR) is activated by dsRNA [7]; and PKR-like endoplasmic reticulum (ER) kinase (PERK) is activated by ER stress [8]. Although PKR is the kinase primarily dedicated to sensing viral dsRNA replication intermediates, other eIF2α kinases can also be activated by viral infections that perturb cellular homeostasis [9]. Since viruses rely on host machinery for their protein synthesis, they must evolve mechanisms to block or bypass the ISR to ensure efficient replication.

Upon phosphorylation, eIF2α stably binds the guanine exchange factor eIF2B, which prevents the regeneration of the translation initiation-competent GTP-bound form of eIF2 [2]. This stalls bulk translation initiation, while simultaneously causing the increased translation of a subset of messenger RNAs (mRNAs), some of which contain short upstream open reading frames (uORFs) in front of ORFs that encode stress response factors [10,11]. The depletion of initiation-competent eIF2 complexes increases leaky scanning, or uORF bypass by the 48S preinitiation complex, which promotes the translation of downstream

ORFs [2]. Activating transcription factor 4 (ATF4) is one of the important ISR factors preferentially translated via uORF bypass [12]. This transcriptional activator induces the expression of additional stress response genes, many of which are also encoded by uORF-containing mRNAs. They include another transcription factor C/EBP homologous protein (CHOP) and its downstream target growth arrest and DNA damage-inducible protein 34 (GADD34) [12]. GADD34 (also named protein phosphatase 1 regulatory subunit 15A, PPP1R15A) complexes with protein phosphatase 1 (PP1) and recruits it to dephosphorylate eIF2α [13]. Thus, the main function of GADD34 in the ISR is the negative feedback that terminates phospho-eIF2α mediated translation arrest.

Coronaviruses (CoVs) are a family of enveloped single-stranded positive-sense RNA viruses. There are seven CoVs known to infect humans, presenting from common colds to more severe and sometimes fatal respiratory disease [14]. Common cold coronaviruses include HCoV-OC43 (OC43) and HCoV-229E, which have been circulating in the human population for decades, as well as more recently identified HCoV-NL63 and HCoV-HKU1 [15]. The most recent pathogenic CoV, the severe acute respiratory syndrome coronavirus 2 (CoV2), emerged in 2019 and became the cause of the devastating pandemic of coronavirus disease 2019 (COVID-19). CoV2, as well as OC43 and HCoV-HKU1, belong to the Betacoronavirus genus [14,15]. In recent years, OC43 emerged as the model Betacoronavirus for studying viral replication and testing the candidate anti-CoV drugs because, unlike CoV2, it is considered safer and can be handled in a more accessible containment level 2 environment [16–20].

The OC43 virion has five structural proteins. Membrane (M) and envelope (E) proteins are responsible for viral particle formation. Nucleocapsid (N) is an RNA-binding protein that coats the viral genome. There are also two surface glycoproteins: hemagglutinin esterase (HE) and spike (S) [15]. S is a receptor binding protein that binds sialic acid on the cell surface [21] and mediates viral entry [14]. Once the virus enters a host cell, its large 30 kb positive sense capped and polyadenylated RNA genome associates with host translation machinery to initiate the synthesis of viral non-structural proteins (Nsps) involved in genome replication, subgenomic mRNA production, and the inhibition of host antiviral responses [14]. The replication of the viral genome in the cytoplasm can generate double-stranded RNA (dsRNA) intermediates which may activate cytoplasmic sensors, including retinoic acid-inducible gene I (RIG-I), melanoma differentiation-associated gene 5 (MDA5), and dsRNA-activated protein kinase (PKR) [7,22]. However, in CoV-infected cells, dsRNA is largely shielded inside double membrane replication transcription compartments that exclude and limit the activation of cytosolic sensors [14,23]. These double membrane compartments are derived from the endoplasmic reticulum (ER) of the host cell by the concerted action of viral Nsp3, Nsp4, and Nsp6 proteins [24,25]. Newly generated viral sub-genomic mRNAs that encode structural proteins are released from these compartments through specific crown-shaped pores and then associate with either free or ER-associated ribosomes in the cytoplasm to be translated [26].

Genomic RNA and sub-genomic mRNAs of coronaviruses are capped and polyadenylated and therefore resemble host mRNAs in their translation initiation mechanism, which is sensitive to the disruption of the cap-directed assembly of the translation preinitiation complex or eIF2α phosphorylation [27,28]. Therefore, coronaviruses must limit ISR to ensure robust protein synthesis and replication. Our previous work revealed that OC43 infection actively inhibits the phosphorylation of eIF2α [29]. In our cell culture infection model, very little eIF2α phosphorylation was detected in infected cells, and when they were treated with a potent inducer of eIF2α phosphorylation sodium arsenite (As), they had significantly reduced levels of p-eIF2α compared to mock-infected cells [29]. As treatment induces oxidative stress and activates HRI kinase [4]. This usually causes very strong eIF2α phosphorylation and the formation of stress granules (SGs), cytoplasmic condensates of polysome-free messenger ribonucleoprotein complexes that accumulate following the inhibition of translation initiation [9]. OC43 infection also inhibits the formation of As-induced stress granules (SGs) [29].

In the present work, we aimed to further characterize the modulation of ISR by OC43. We demonstrate that the OC43 infection of human embryonic kidney 293A cells strongly induces GADD34, the negative regulator of ISR. Although OC43 infection causes ER stress and the activation of PERK, which negatively affects viral protein synthesis and replication, PERK-mediated ATF4 production is not responsible for GADD34 upregulation in infected cells. Remarkably, we also show that OC43 induced GADD34 protein does not function in mediating increased eIF2α dephosphorylation.

2. Materials and Methods

2.1. Cells and Viruses

Human embryonic kidney (HEK) 293A cells and green monkey kidney (Vero) cells were purchased from American Type Culture Collection (ATCC, Manassas, VA, USA) and cultured in Dulbecco's modified Eagle's medium (DMEM) supplemented with heat-inactivated 10% fetal bovine serum (FBS) and 2 mM L-glutamine (all purchased from Thermo Fisher Scientific (Thermo), Waltham, MA, USA). Human betacoronavirus OC43 was purchased from ATCC. Initial virus stocks were generated in Vero cells as described in [29]. Subsequent OC43 passages for the expansion of stocks were undertaken in 293A cells. The 293A cells were infected at the multiplicity of infection (MOI) <0.1 for 1 h in serum-free DMEM, and then the inoculum was replaced with DMEM supplemented with 1% FBS + 0.5% BSA and incubation continued at 37 °C. At 48 h post infection (hpi), the viral supernatant was harvested, cleared by centrifugation at 2500× g for 5 min, and then aliquoted and stored at −80 °C. Virus titers were determined by 50% tissue culture infectious dose (TCID50) assay in 293A cells in the same infection media as above and calculated using the Spearman–Kärber method.

2.2. Cell Treatments

For poly(I:C) transfection, 1.5 µL of FuGENE HD reagent (Promega, Madison, WI, USA) was combined with 0.5 µg of low MW poly(I:C) (InvivoGen, San Diego, CA, USA) in 50 µL Opti-MEM-I (Gibco, Waltham, MA, USA) for 5 min and then added directly to media for each 20 mm well of cells. Cells were returned to a 37 °C incubator for 2 h. To induce eIF2α phosphorylation, sodium arsenite (As) was added to the media to a final concentration of 500 µM and cells were returned to the 37 °C incubator for 50 min; thapsigargin (Tg) was added to the media to a final concentration of 1 µM and cells were returned to the 37 °C incubator for 1 h or 4 h, as indicated. The integrated stress response inhibitor (ISRIB, final concentration 200 nM) or the Inositol-requiring enzyme 1 (IRE1) inhibitor 4µ8C (concentration range 5–25 µM) was added to the culture media and cells were returned to the 37 °C incubator for 23 h (for infection experiments) or the length of Tg treatment, as indicated. All compounds and inhibitors were purchased from Millipore Sigma (Oakville, ON, Canada).

2.3. Virus Infections

The 75–80% confluent monolayers of 293A cells grown on 20 mm wells of 12-well cluster dishes with or without glass coverslips were washed briefly with PBS and 300 µL of virus inoculum diluted to the specified MOI in 1% FBS DMEM. Cells were placed at 37 °C for 1 h, with manual horizontal shaking every 10–15 min. Then, virus inoculum was aspirated from cells, cells were washed briefly with PBS, and 1 mL of fresh 1% FBS DMEM was added to each well and cells were returned to 37 °C until the specified time post-infection.

2.4. Gene Silencing

For GADD34 silencing, 293A cells were transfected with Ambion Silencer Select siRNAs (siGADD34#1 (s24269) and siGADD34#2 (s24268)) using Lipofectamine RNAiMAX (Invitrogen, Waltham, MA, USA) according to manufacturer's protocol and used in infection/treatment experiments at 48 h post-transfection. For non-targeting siRNA control, cells were transfected with Silencer Select Negative Control #2 siRNA (siNT; Ambion, #4390846). For ATF4 silencing, shRNA inserts targeting human ATF4 gene (shATF4-1, target sequence catgatccctcagtgcataaa; shATF4-2, target sequence cctaggtctcttagatgatta) or control human interleukin 2 (IL2) that is not expressed in epithelial cells (shIL2, target sequence gctacctattgtaactattat) were cloned into pLKO.1-TRC vector (Addgene plasmid #10878, a gift from David Root, [30]). For PERK silencing, a guide RNA insert targeting human PERK (sgRNA sequence gaatataccgaagttcaaag) was cloned with the lentiCRISPR-v2 vector (Addgene plasmid #52961, a gift from Feng Zhang, [31]). The shRNAs and gRNA were designed using Broad Institute GPP Web Portal tools (https://portals.broadinstitute.org/gpp/public/). Then, 293A cells were transduced with lentiviruses generated with the above vectors at an MOI of 1.0, and stably transduced cells were selected with 1 μg/mL puromycin for 48 h. Resistant cells were seeded onto 12-well cluster dishes and used in experiments the following day.

2.5. Western Blotting

Whole-cell lysates were prepared by the direct lysis of PBS-washed cell monolayers with 1× Laemmli sample buffer (50 mM Tris-HCl pH 6.8, 10% glycerol, 2% sodium dodecyl sulphate (SDS), 100 mM DTT, 0.005% Bromophenol Blue). Lysates were immediately placed on ice, homogenized by passing through a 21-gauge needle, and stored at −20 °C. Aliquots of lysates thawed on ice were incubated at 95 °C for 3 min, cooled on ice, separated using denaturing polyacrylamide gel electrophoresis, transferred onto polyvinylidene fluoride (PVDF) membranes using Trans Blot Turbo Transfer System with RTA Transfer Packs (Bio-Rad Laboratories, Hercules, CA, USA) according to the manufacturer's protocol, and analyzed by immunoblotting using antibody-specific protocols. Antibodies to the following targets were used: ATF4 (1:1000; rabbit, Abcam, ab270980); β-actin (1:2000; HRP-conjugated, mouse, Santa Cruz, sc-47778); eIF2α (1:1000; rabbit, Cell Signaling, #5324); GADD34 (1:1000; mouse, Proteintech, #10449-1-AP); OC43 N (1:1000; mouse, Millipore, MAB9012); OC43 Spike (1:1000; rabbit, Cusabio, CSB-PA336163EA-1HIY); PERK (1:1000; rabbit, Cell Signaling, #5683); phospho-S51-eIF2α (1:1000; rabbit, Cell Signaling, #3398); phospho-T451-PKR (1:1000; rabbit, Epitomics, 2283-1); PKR (1:1000; mouse, Santa Cruz, sc-6282); and XBP1s (1:1000; mouse, Cell Signaling, #12782). For band visualization, HRP-conjugated anti-rabbit IgG (goat, Cell Signaling, #7074) or anti-mouse IgG (horse, Cell Signaling, #7076) were used with Clarity Western ECL Substrate on the ChemiDoc Touch Imaging System (Bio-Rad Laboratories). For analyses of phospho-eIF2α band intensities, Western blot signals were quantified using Bio-Rad Image Lab 5.2.1 software and values normalized to the total eIF2α bands for each lane.

2.6. Immunofluorescence Microscopy

Cell fixation, methanol permeabilization, and immunofluorescence staining were performed according to the procedure described in [32]. Primary antibody staining was performed overnight at +4 °C with antibodies to OC43 N (1:500; mouse, Millipore, MAB9012) and/or TIAR (1:1000; rabbit, Cell Signaling, #8509). Alexa Fluor-conjugated secondary antibodies used were donkey anti-mouse IgG AF488 (Invitrogen, Burlington, Canada, A21202) and donkey anti-rabbit IgG AF555 (Invitrogen, A31572). Where indicated, nuclei were stained with Hoechst 33342 dye (Invitrogen, H3570). Slides were mounted with ProLong Gold Antifade Mountant (Thermo Fisher, Burlington, Canada) and imaged using a Zeiss AxioImager Z2 fluorescence microscope and Zeiss ZEN 2011 software.

2.7. RNA Isolation and RT-QPCR

Total RNA was isolated from cells using an RNeasy Plus Mini kit (Qiagen, Toronto, Canada) according to the manufacturer's protocol and 250 ng of RNA was used to synthesize cDNA using qScript cDNA SuperMix (Quanta). Quantitative PCR amplification was performed using PerfeCTa SYBR Green PCR master mix (Quanta) with the specific primers listed below: 18S–Left: cgttcttagttggtggagcg, Right: ccggacatctaagggcatca; ACTB–Left:

catccgcaaagacctgtacg, Right: cctgcttgctgatccacatc; GADD34–Left: ctcagcgccca-gaaac, Right: ggaaatggacagtgaccttc; GAPDH–Left: gagtcaacggatttggtcgt, Right: ttgattttggagggatctcg; and OC43-nsp15–Left: atggcgtagtggtggacaag, Right: actcccaggctgtcgaattg. Relative target levels were determined using the ΔΔCt method with normalization to 18S.

3. Results

3.1. The Targeting Subunit of eIF2α Holophosphatase GADD34 Is Strongly Upregulated in OC43-Infected Cells

Our previously published analyses of eIF2α phosphorylation in OC43-infected cells showed that at 24 and 48 h post-infection (hpi), OC43 causes a small but consistent increase in p-eIF2α levels [29]. Therefore, to begin characterizing the effects of the virus on ISR, we wanted to test the activation status of host factors upstream and downstream of phospho-eIF2α in the pathway. Among the eIF2α kinases, we focused on PKR because of its primary function in recognition of viral dsRNA replication intermediates, and PERK, because coronaviruses are known to induce ER stress [33,34]. We infected 293A cells with OC43 at a multiplicity of infection (MOI) = 1.0 and analyzed the phosphorylation statuses of PKR, PERK, and eIF2α at 24 hpi using Western blotting. For positive controls, we used transfection with polyinosinic-polycytidylic acid (poly(I:C)) dsRNA mimic for PKR activation and treatment with the sarcoendoplasmic reticulum calcium ATPase (SERCA) inhibitor thapsigargin (Tg) for the induction of ER stress and the activation of PERK (Figure 1A,B). In both mock- and OC43-infected cells, poly(I:C) transfection induced the activation of PKR (as measured by phosphorylation at Threonine-451) and strong eIF2α phosphorylation, while no phospho-PKR signal could be detected in untransfected mock- or OC43-infected cells (Figure 1A). These results indicate that in our infection model, OC43 successfully blocks the detection of its dsRNA molecules by PKR. By contrast, virus infection caused the activation of PERK (as indicated by the appearance of an upstream shifted band corresponding to phosphorylated PERK), similar to the activation observed in mock-infected cells upon 1 h Tg treatment (Figure 1B, lanes 1–3). As expected, 4 h Tg treatment led to the induction of ATF4 and GADD34 and decreased phospho-eIF2α in both mock- and OC43-infected cells (Figure 1B, lanes 5,6). However, even without Tg treatment, OC43 infection strongly induced GADD34 to levels exceeding those observed after 4 h Tg treatment in mock cells (Figure 1B, compare lanes 2 and 5). To test if GADD34 protein accumulation was caused by increased mRNA levels, we isolated total RNA from mock- and OC43-infected cells and performed a reverse transcription–quantitative polymerase chain reaction (RT-QPCR) analysis of the GADD34 transcript. This analysis showed that, similar to 4 h Tg treatment, OC43 infection caused the strong upregulation of GADD34 mRNA levels (Figure 1C). Control GAPDH transcript was not significantly affected by Tg treatment or infection (Figure 1D). To confirm minimal eIF2α phosphorylation and strong GADD34 upregulation by OC43 infection in a different cell type, we used Vero cells. Similar to the infection of 293A cells, OC43 caused a small increase in eIF2α phosphorylation in Vero cells and decreased the maximal levels of phospho-eIFα upon Tg treatment (Figure 1E). The virus also caused the significant upregulation of GADD34 mRNA levels in Vero cells (Figure 1F,G), confirming that this phenotype is not specific to the 293A infection model.

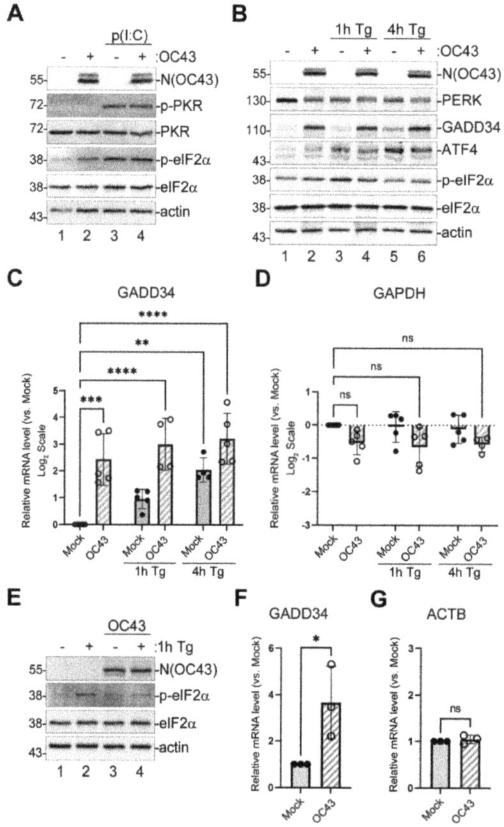

Figure 1. OC43 strongly induces GADD34 expression. 293A cells were infected with OC43 at MOI = 1.0. (**A**) Western blot analysis of lysates from mock- or OC43-infected cells collected at 24 h post-infection (hpi) with or without poly(I:C) transfection at 22 hpi (2 h poly(I:C) treatment). (**B**) Western blot analysis of lysates from mock- or OC43-infected cells collected at 24 hpi with or without Tg treatment for 1 or 4 h prior to harvesting. (**C**,**D**) Total RNA was isolated from mock- and infected cells treated as indicated and levels of GADD34 (**C**) and GAPDH (**D**) transcripts were determined by RT-QPCR. (**E**–**G**) Vero cells were infected with OC43 at MOI = 1.0. (**E**) Western blot analysis of lysates from mock- or OC43-infected Vero cells collected at 24 hpi with or without Tg treatment for 1 h prior to harvesting. (**F**,**G**) Total RNA was isolated from mock- and infected Vero cells and levels of GADD34 (**F**) and ACTB (**G**) transcripts were determined by RT-QPCR. On all plots, values were normalized to 18S and expressed as fold changes relative to mock untreated cells. Each data point represents an independent biological replicate ($N \geq 3$). Error bars = standard deviation. Two-way ANOVA and Dunnett's multiple comparisons tests were performed in C and D, and Student's *t*-Test in (**F**,**G**) to determine statistical significance (****, *p*-value < 0.0001; ***, *p*-value < 0.001; **, *p*-value < 0.01; *, *p*-value < 0.05; ns, non-significant).

3.2. The Activation of the Integrated Stress Response Pathway Is Not Responsible for GADD34 Induction in OC43-Infected Cells

Next, we wanted to determine whether ISR activation by OC43 was responsible for GADD34 upregulation and, consequently, the inhibition of As-induced eIF2α phosphorylation. To block phospho-eIF2α-mediated changes in translation initiation, we treated mock- and OC43-infected cells with ISR inhibitor (ISRIB, [35]). ISRIB does not affect eIF2α phosphorylation, but blocks its effects on the regeneration of the initiation-competent

GTP-bound form of eIF2 and prevents translation arrest [36]. In our infection model, ISRIB treatment did not alter the inhibition of As-induced SG formation by OC43, while it significantly decreased As-induced SG formation in mock-infected cells, as expected (Figure 2A,B). It also did not affect OC43-mediated GADD34 induction (Figure 2C), the inhibition of As-induced eIF2α phosphorylation by OC43 (Figure 2D,E), or virus replication (Figure 2F). To test whether the ISR-dependent ATF4 and GADD34 induction pathway is intact in 293A cells, we verified the effects of ISRIB on Tg-induced ISR in this model. As expected, the Tg-mediated induction of ATF4 and GADD34 was blocked by ISRIB (Figure 2G–I). Taken together, these results show that neither GADD34 induction nor the inhibition of eIF2α phosphorylation by OC43 are due to ISR activation by the virus.

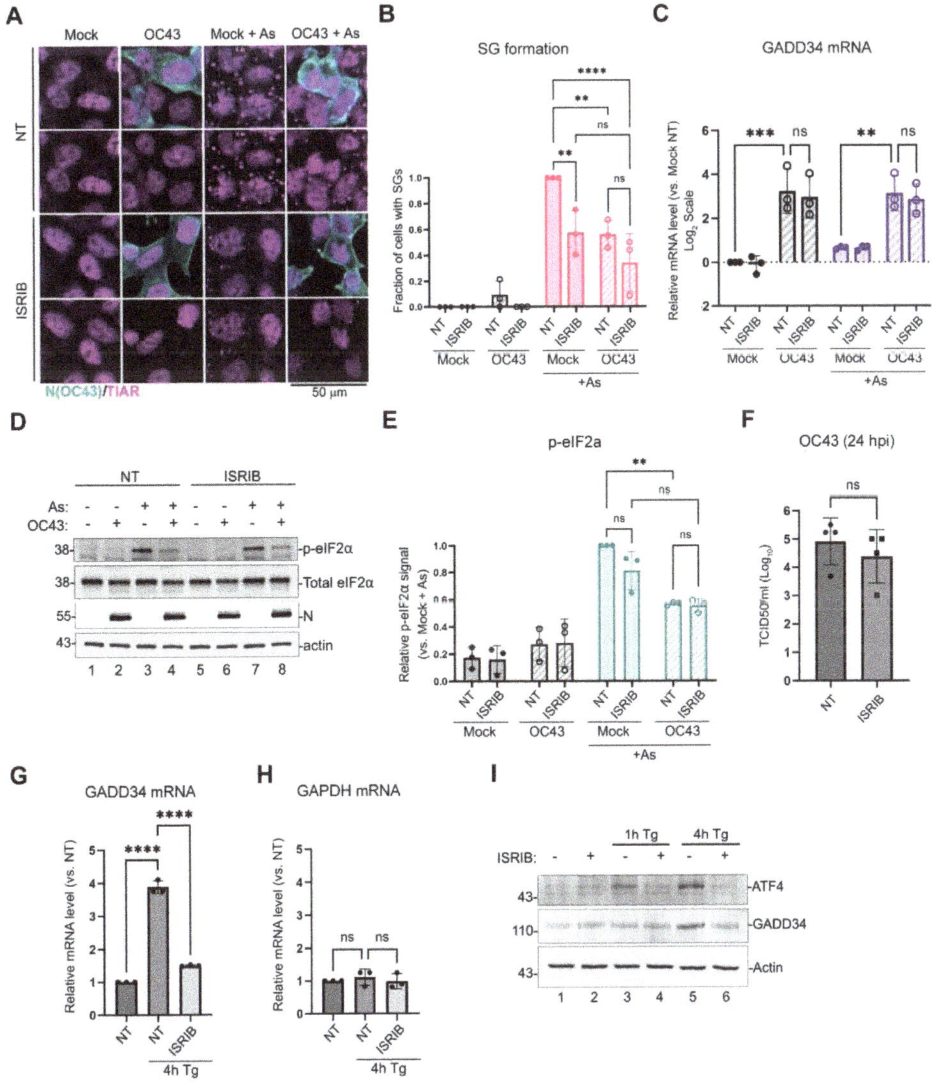

Figure 2. Integrated stress response inhibitor does not prevent GADD34 upregulation or the inhibition of eIF2α phosphorylation by OC43. The 293A cells were infected with OC43 at MOI = 1.0 and treated

with ISRIB at 1 hpi or left untreated. At 23 hpi, p-eIF2α phosphorylation and SG formation was induced by the sodium arsenite treatment (+As). (**A**) SG formation in infected and mock-infected cells was analyzed using immunofluorescence staining for SG marker TIAR (magenta). Infected cells were identified by staining for nucleoprotein (N(OC43), teal). (**B**) SG formation was quantified from immunofluorescence staining images represented in (**A**). (**C**) Total RNA was isolated from mock- and infected cells treated as indicated and levels of GADD34 transcripts were determined by RT-QPCR. Values were normalized to 18S and expressed as fold changes relative to mock untreated cells. (**D**) Western blot analysis of lysates from mock- or OC43-infected cells treated as indicated. (**E**) Relative level of eIF2α phosphorylation was quantified from Western blot images represented in (**D**). (**F**) Production of infectious virions in untreated and ISRIB-treated cells was measured in cell supernatants at 24 hpi. (**G–I**) ISRIB blocks Tg-induced ATF4 and GADD34 expression in 293A cells. Uninfected cells were treated with Tg with or without the addition of ISRIB as indicated. (**G,H**) Levels of GADD34 (**G**) and GAPDH (**H**) transcripts were determined by RT-QPCR. Values were normalized to 18S and expressed as fold changes relative to mock untreated cells. (**I**) Western blot analysis of lysates from cells treated as indicated. For all plots, each data point represents an independent biological replicate ($N \geq 3$). Error bars = standard deviation. Two-way ANOVA and Tukey multiple comparison tests or an unpaired *t*-Test (panel F only) were performed to determine statistical significance (****, p-value < 0.0001; ***, p-value < 0.001; **, p-value < 0.01; ns, non-significant).

The principal upstream transcription factor induced upon eIF2α phosphorylation by PERK is ATF4. To confirm that ATF4 is not responsible for the transcriptional upregulation of GADD34 in OC43-infected cells, we silenced its expression using two lentivirus-driven shRNA constructs. As a non-targeting control, we used shRNA against IL2, the cytokine that is not expressed in epithelial cells. Both shRNA constructs were effective in decreasing Tg-induced ATF4 expression to nearly undetectable levels in uninfected cells, as well as OC43-induced ATF4 expression (Figure 3A). At the same time, neither shRNA affected the induction of GADD34 protein by OC43 infection or Tg treatment (Figure 3A). ATF4 silencing also did not affect OC43 protein accumulation, virus replication, or GADD34 mRNA induction (Figure 3B–D). In a classical ISR pathway, ATF4 induces the CHOP transcription factor, which then induces GADD34. However, GADD34 can be induced by other transcription factors [37–39], and in addition to PERK, Tg and OC43 can activate other ER stress sensors. One ER stress sensor that is strongly activated by OC43 is inositol-requiring enzyme 1 (IRE1) [33]. IRE1 activation causes the non-canonical cytoplasmic splicing of mRNA-encoding transcription factor X-box binding protein 1 (XBP1), leading to the synthesis of XBP1s, an activator of transcription produced from spliced XBP1 mRNA [40]. Although the IRE1/XBP1s pathway has not been shown to upregulate GADD34, we tested whether IRE1 inhibition would perturb GADD34 expression in OC43-infected cells. As reported previously in another model [33], OC43 infection induced XBP1s in 293A cells (Figure 3E). The addition of specific IRE1 inhibitor 4μ8C inhibited XBP1s' induction triggered by OC43 infection or Tg treatment in a concentration-dependent manner (Figure 3E,F), but did not affect GADD34 induction.

3.3. *The PERK Arm of the Unfolded Protein Response Negatively Affects OC43 Protein Synthesis and Replication*

Having determined that OC43 infection activates PERK, we wanted to determine what role PERK plays in virus replication. First, we silenced PERK expression in 293A cells by transduction with the lentiviral-vector-encoding CRISPR/Cas9 system with guide RNA targeting PERK (gRNA-PERK). After 2 days, transduced cells expressed measurably lower PERK levels compared to parental untransduced cells (Figure 4A). When transduced cells were infected with OC43, they accumulated more viral proteins and GADD34 compared to parental cells (Figure 4A, lanes 3,4). Consistent with PERK being an upstream regulator of ATF4 expression, when transduced cells were treated with Tg, they induced lower levels of ATF4 protein compared to parental cells (Figure 4A, lanes 5,6). A similar effect was also observed in OC43-infected cells (Figure 4A, lanes 7,8). This suggests that PERK is not required for GADD34 upregulation by OC43 and that PERK-mediated ISR may have an antiviral role. Indeed, compared to parental cells, gRNA-PERK transduced cells generated significantly higher infectious virus titers (Figure 4B).

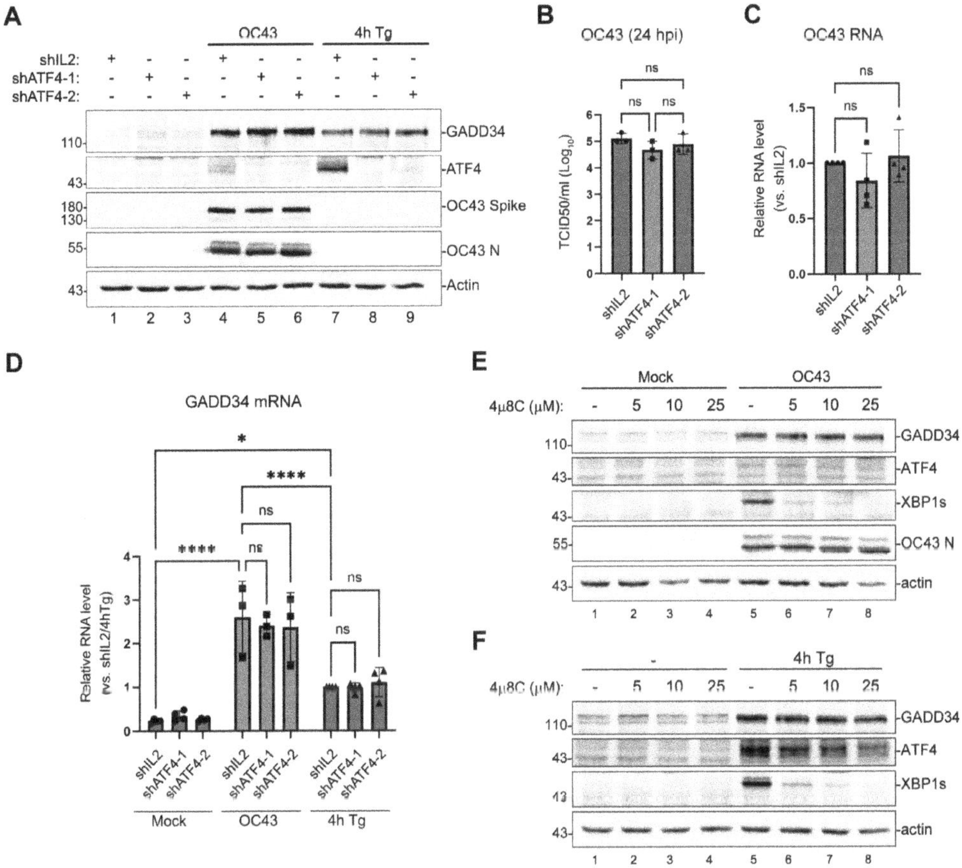

Figure 3. The upregulation of GADD34 expression by OC43 is ATF4-independent. (**A–D**) Cells transduced with lentiviruses encoding shRNAs against ATF4 (shATF4-1 and shATF4-2) or control IL-2 shRNA (shIL2) were infected with OC43 at multiplicity of infection (MOI) = 1.0 for 24 h or were treated with Tg for 4 h. (**A**) Western blot analysis of lysates from cells treated as indicated. (**B**) Production of infectious virions was measured in cell supernatants at 24 hpi. (**C**) Total RNA was isolated from infected cells and levels of OC43 genomic RNA were determined by RT-QPCR. Values were normalized to 18S and expressed as fold changes relative to shIL2 cells. (**D**) Levels of GADD34 transcripts in mock-infected, OC43-infected, and Tg-treated cells were determined by RT-QPCR. (**E**,**F**) IRE1 inhibition does not affect GADD34 upregulation in OC43-infected or Tg-treated cells. Western blot analysis of cells treated with the indicated concentrations of IRE1 inhibitor 4μ8C. (**E**) Mock- or OC43-infected cells analysed at 24 hpi. (**F**) Untreated (-) and Tg-treated cells. For all plots, each data point represents an independent biological replicate (N $\geq$ 3). Error bars = standard deviation. Two-way ANOVA and Tukey multiple comparisons tests were carried out to determine statistical significance (****, p-value < 0.0001; *, p-value < 0.05, ns, non-significant).

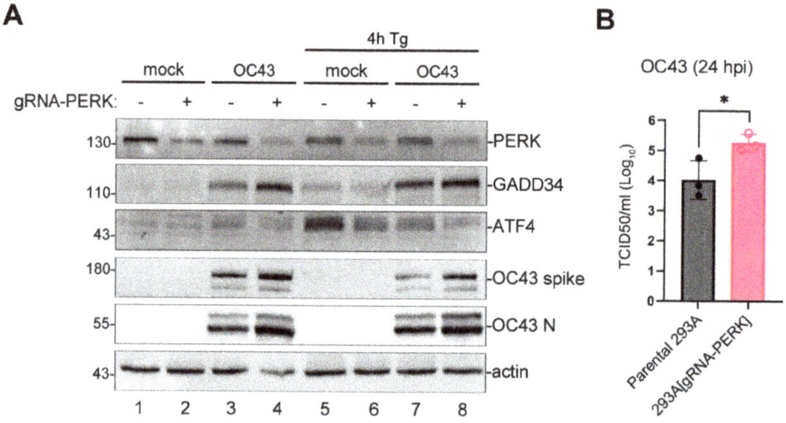

Figure 4. The silencing of PERK enhances OC43 protein synthesis and infectious virion production. The 293A cells transduced with lentivirus encoding Cas9 and guide RNA targeting PERK (gRNA-PERK) or parental untransduced cells were mock- or OC43-infected at MOI = 1.0. (**A**) Western blot analysis of cell lysates collected at 24 hpi. (**B**) The production of infectious virions was measured in cell supernatants at 24 hpi. Each data point represents an independent biological replicate (N = 3). Error bars = standard deviation. Student's t-Test was performed to determine statistical significance (*, p-value < 0.05).

3.4. OC43-Induced GADD34 Does Not Contribute to Decreased eIF2α Phosphorylation

To test if GADD34 upregulation is responsible for the inhibition of As-induced eIF2α phosphorylation in OC43 infected cells, and to determine if GADD34 induction plays a role in viral replication, we silenced its expression using transfection with two different GADD34-targeting siRNAs (siGADD34#1 and siGADD34#2). As a control, we used non-targeting siRNA (siNT). After 48 h, transfected cells were mock-infected or infected with OC43 virus. At 24 hpi, GADD34 expression and As-induced eIF2α phosphorylation were analyzed using Western blotting. Both siRNAs effectively silenced OC43-induced GADD34 expression without affecting the inhibition of eIF2α phosphorylation triggered by As (Figure 5A–D). Similarly, both siRNAs decreased GADD34 mRNA levels (Figure 5E). To determine viral replication kinetics, siRNA-transfected cells were infected with OC43 at MOI ~0.1 and viral replication was assessed at 12, 16, and 20 hpi using an immunofluorescence microscopy analysis of infected cells (Figure 5F), RT-QPCR for genomic RNA levels (Figure 5G), and the release of infectious virions (Figure 5H). These analyses revealed that in our infection model, GADD34 is not required for efficient OC43 replication, despite the strong induction of this host protein by the virus.

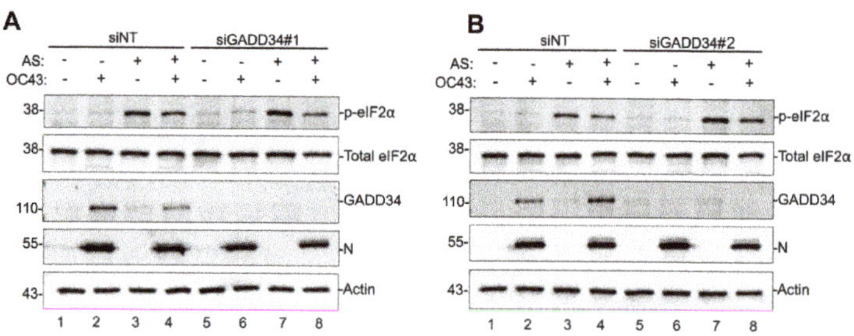

Figure 5. *Cont.*

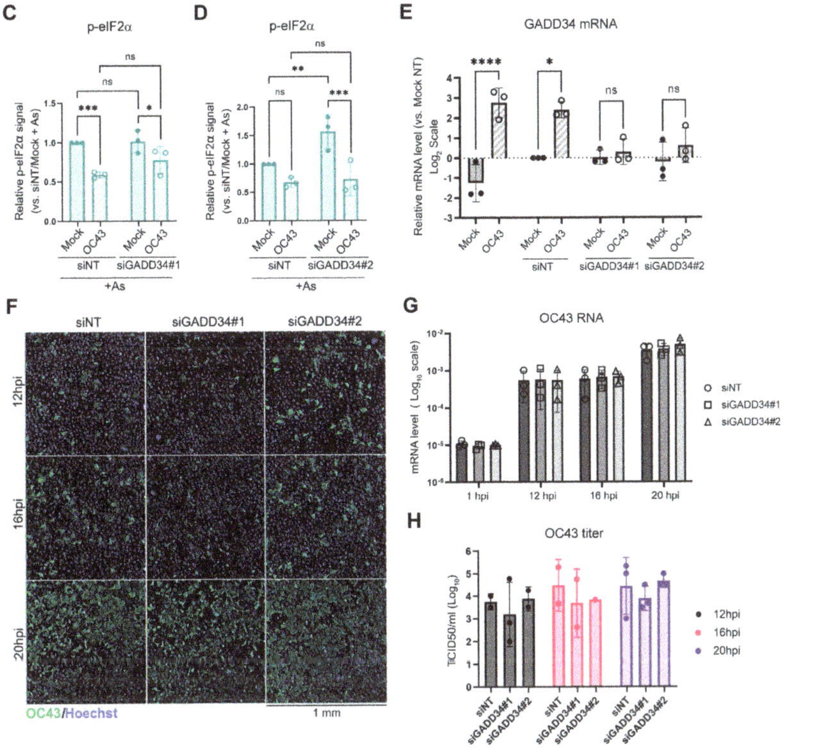

Figure 5. GADD34 is not responsible for the inhibition of eIF2α phosphorylation in OC43-infected cells. GADD34 expression was silenced in 293A cells using specific siRNAs (siGADD34#1 and siGADD34#2) or cells were transfected with non-targeting control siRNA (siNT). (**A**,**B**) Western blot analysis of untreated or As-treated mock- or OC43 infected cells performed at 24 hpi. (**C**,**D**) Relative level of eIF2α phosphorylation was quantified from Western blot images represented in (**A**,**B**). (**E**) Levels of GADD34 transcript were determined by RT-QPCR. (**F**) Immunofluorescence staining of OC43-infected cells pre-treated with the indicated siRNAs. Infected cells at 12, 16, and 20 hpi were visualised using the antibody to OC43 nucleoprotein (OC43, green). Nuclei were stained with Hoechst dye (blue). (**G**) Total RNA was isolated from infected cells at the indicated times post-infection and levels of OC43 genomic RNA were determined by RT-QPCR. Values were normalized to 18S. (**H**) The production of infectious virions was measured in cell supernatants. For all plots, each data point represents an independent biological replicate. Error bars = standard deviation. Two-way ANOVA and Tukey multiple comparisons tests were performed to determine statistical significance (****, p-value < 0.0001; ***, p-value < 0.001; **, p-value < 0.01; *, p-value < 0.05; ns, non-significant).

4. Discussion

In this work, we examined some of the key aspects of the ISR pathway in 293A cells infected with human common cold virus OC43 (Figure 6). Previously, we used the same infection model to demonstrate that OC43 infection proceeds efficiently in these cells and does not trigger strong eIF2α phosphorylation or SG formation [29]. Most importantly, OC43-infected cells become less responsive to As, a strong ISR inducer that triggers robust eIF2α phosphorylation and SG formation in mock- but not in OC43-infected cells. Due to its potency in triggering eIF2α phosphorylation and translation arrest, As is often used in research labs to induce ISR-dependent SG formation [4]. We probed the response to As in infected cells for the same reason; however, the level of oxidative stress induced by 50 μM As is not physiological and only serves as a probe to reveal potential eIF2α phosphorylation inhibition. Since coronavirus mRNAs are sensitive to ISR-induced translation arrest [28], they must evolve mechanisms to suppress ISR to ensure robust viral protein synthesis and replication. To characterize the upstream activation of eIF2α kinases by OC43, we focused on PKR and PERK, the two kinases that are often activated by virus infections [9]. For example, both kinases are activated by severe acute respiratory syndrome coronavirus and chicken gammacoronavirus infectious bronchitis virus [41,42], and PERK is shown to be activated by a number of different coronaviruses and can be involved in virus-induced cell death and inflammation [28,34,43,44]. In our infection model, we did not detect PKR activation by OC43. Moreover, we did not observe the inhibition of PKR activation by poly(I:C) transfection in OC43-infected cells compared to mock. This suggests that OC43 is efficient at avoiding the detection of its dsRNA molecules by PKR by shielding them inside the double membrane replication transcription compartments and/or by the action of the Nsp15 protein [45], and does not actively block PKR activation. By contrast, OC43 infection caused a detectable activation of PERK, consistent with its ability to trigger ER stress in infected cells [33,46,47]. Interestingly, PERK activation was previously shown to either inhibit (transmissible gastroenteritis virus, [28]) or promote (porcine epidemic diarrhea virus, [44]) coronavirus replication. In our study, we demonstrate that PERK is an antiviral factor for OC43, as the silencing of PERK expression in 293A cells enhanced viral protein accumulation and infectious virion release. Interestingly, despite PERK activation, eIF2α phosphorylation levels in OC43 remained low, confirming that it is inhibited in virus-infected cells.

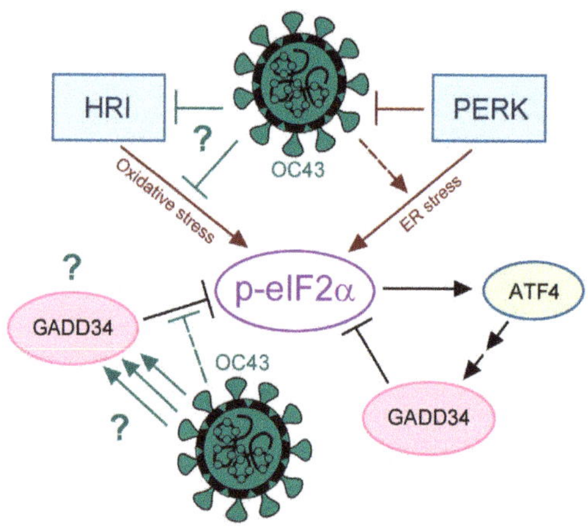

Figure 6. A schematic diagram representing the main findings of this work. Question marks indicate the main unanswered questions that should be addressed in future studies: How does OC43 inhibit HRI? What is the mechanism of GADD34 upregulation by OC43? What is the function of GADD34 in OC43-infected cells?

In our infection model, we demonstrate the robust induction of the downstream ISR effector GADD34 by OC43. GADD34 is a multifunctional protein involved in cellular stress responses. Its primary role in ISR is the recovery of protein synthesis after stress by forming a complex with PP1 to dephosphorylate eIF2α [13]. Beyond its involvement in the ISR, GADD34 plays a role in the modulation of cell death and cell cycle regulation, contributing to growth arrest under certain stress conditions [48–50]. The specific functions of GADD34 are context-dependent and vary based on the nature of the stress stimuli and cellular context [1,48,50]. Given robust GADD34 induction by OC43 infection, the inhibition of eIF2α phosphorylation by OC43 could be explained by GADD34-mediated recruitment of PP1 and increased eIF2α dephosphorylation rates. However, we were unable to alter the inhibition of As-induced eIF2α phosphorylation in OC43-infected cells by silencing

GADD34. In mock-infected cells, Tg induced lower levels of GADD34 compared to those induced by OC43 infection, and this induction resulted in the apparent dephosphorylation of eIF2α at 4 h post-treatment compared to its peak at 1 h post-treatment. These results indicate that most of the GADD34 induced by the virus is diverted to perform other functions, but that the ISR negative feedback loop is intact in 293A cells.

The induction of GADD34 by Tg followed eIF2α phosphorylation by PERK and the production of the ATF4 transcription factor that normally functions upstream of GADD34 induction in the ISR pathway. Our analyses showed that ATF4 is not required for GADD34 induction by ER stress in 293A cells, because GADD34 protein accumulation in response to Tg or OC43 infection was unaffected by ATF4 silencing. In addition to ATF4, GADD34 can be induced by transcription factors CHOP and ATF3 [1], as well as the interferon regulatory factor 3 (IRF3) [37]. Both CHOP and ATF3 are induced by ATF4; however, other stimuli can upregulate ATF3 [51,52]. IRF3 is the main transcription factor activated by cellular sensors of viral nucleic acids and an important driver of type I and type III interferon expression [53]. OC43 was shown to activate IRF3 and induce interferon response genes in lung fibroblast cell line MRC5 [54], but in another study OC43 was demonstrated to be very efficient at avoiding type I and type III interferon induction in primary human bronchial epithelial cells [55]. Future studies will determine if ATF3, IRF3, or other transcription factors implicated in GADD34 induction [38,39] are responsible for the upregulation of GADD34 in OC43-infected cells.

To summarize, our work revealed that OC43 infection activates the PERK arm of the ISR that suppresses viral protein synthesis and replication. At the same time, OC43 robustly induces GADD34, a multifunctional protein normally involved in eIF2α dephosphorylation. However, GADD34 silencing had no effect on the viral inhibition of As-induced eIF2α phosphorylation or virus replication. For our infection model, we used 293A cells, a kidney-derived cell line that can be productively infected by OC43 [29]. The 293A cells are a clone of the human embryonic kidney 293 cell line that contains a stably integrated copy of the adenovirus E1 gene that encodes the E1a and E1b proteins. These viral proteins can potentially skew the cellular responses to OC43 infection, as they are shown to attenuate interferon induction [56,57]. Our previous work showed that eIF2α phosphorylation and SG inhibition by OC43 occurs in multiple cell types, including the untransformed immortalized human upper airway BEAS2B cells [29]. In this study, we confirmed strong GADD34 upregulation by OC43 in Vero cells that do not express adenoviral E1 gene products. However, the use of transformed cell lines deficient in type I interferon responses represents a limitation of our study, and it is possible that GADD34 may perform an important proviral or antiviral function in other infection models. Another limitation of this study is that we only tested PKR and PERK activation by OC43. Importantly, while Tg-triggered eIF2α phosphorylation by PERK was partially inhibited by OC43, PKR-mediated phosphorylation of eIF2α in response to poly(I:C) transfection was not. It is possible that the mechanism of ISR inhibition by OC43 is stress- or kinase-specific. Given the threat of ISR and eIF2α phosphorylation to coronavirus mRNA translation, future research should examine the activation of HRI and GCN2 and address whether OC43 directly inhibits eIF2α phosphorylation by HRI or decreases oxidative stress in infected cells and how these mechanisms influence virus fitness and pathogenesis.

Author Contributions: Conceptualization, S.M.D. and D.A.K.; methodology, S.M.D.; validation, S.M.D. and D.A.K.; formal analysis, S.M.D. and C.G.; investigation, S.M.D. and C.G.; resources, D.A.K.; data curation, S.M.D.; writing—original draft preparation, S.M.D.; writing—review and editing, S.M.D., C.G. and D.A.K.; visualization, S.M.D.; supervision, D.A.K.; project administration, D.A.K.; funding acquisition, D.A.K. All authors have read and agreed to the published version of the manuscript.

Funding: This research was funded by the Canadian Institutes of Health Research (CIHR) Project Grant PJT-175130 to D.K.

Data Availability Statement: All data necessary for the interpretation of the findings presented in this work are contained within the manuscript figures.

Acknowledgments: We would like to thank members of the Khaperskyy lab for helpful discussions about the experimental design and their critical input on the draft manuscript.

Conflicts of Interest: The authors declare no conflicts of interest. The funders had no role in the design of the study, the collection, analyses, or interpretation of data, the writing of the manuscript, or in the decision to publish the results.

References

1. Pakos-Zebrucka, K.; Koryga, I.; Mnich, K.; Ljujic, M.; Samali, A.; Gorman, A.M. The Integrated Stress Response. *EMBO Rep.* **2016**, *17*, 1374–1395. [CrossRef]
2. Jackson, R.J.; Hellen, C.U.T.; Pestova, T.V. The Mechanism of Eukaryotic Translation Initiation and Principles of Its Regulation. *Nat. Rev. Mol. Cell Biol.* **2010**, *11*, 113–127. [CrossRef]
3. Lu, L.; Han, A.P.; Chen, J.J. Translation Initiation Control by Heme-Regulated Eukaryotic Initiation Factor 2alpha Kinase in Erythroid Cells under Cytoplasmic Stresses. *Mol. Cell Biol.* **2001**, *21*, 7971–7980. [CrossRef]
4. McEwen, E.; Kedersha, N.; Song, B.; Scheuner, D.; Gilks, N.; Han, A.; Chen, J.-J.; Anderson, P.; Kaufman, R.J. Heme-Regulated Inhibitor Kinase-Mediated Phosphorylation of Eukaryotic Translation Initiation Factor 2 Inhibits Translation, Induces Stress Granule Formation, and Mediates Survival upon Arsenite Exposure. *J. Biol. Chem.* **2005**, *280*, 16925–16933. [CrossRef]
5. Wek, S.A.; Zhu, S.; Wek, R.C. The Histidyl-tRNA Synthetase-Related Sequence in the eIF-2 Alpha Protein Kinase GCN2 Interacts with tRNA and Is Required for Activation in Response to Starvation for Different Amino Acids. *Mol. Cell Biol.* **1995**, *15*, 4497–4506. [CrossRef]
6. Deng, J.; Harding, H.P.; Raught, B.; Gingras, A.-C.; Berlanga, J.J.; Scheuner, D.; Kaufman, R.J.; Ron, D.; Sonenberg, N. Activation of GCN2 in UV-Irradiated Cells Inhibits Translation. *Curr. Biol.* **2002**, *12*, 1279–1286. [CrossRef]
7. García, M.A.; Meurs, E.F.; Esteban, M. The dsRNA Protein Kinase PKR: Virus and Cell Control. *Biochimie* **2007**, *89*, 799–811. [CrossRef]
8. Harding, H.P.; Zhang, Y.; Bertolotti, A.; Zeng, H.; Ron, D. Perk Is Essential for Translational Regulation and Cell Survival during the Unfolded Protein Response. *Mol. Cell* **2000**, *5*, 897–904. [CrossRef] [PubMed]
9. McCormick, C.; Khaperskyy, D.A. Translation Inhibition and Stress Granules in the Antiviral Immune Response. *Nat. Rev. Immunol.* **2017**, *17*, 647–660. [CrossRef] [PubMed]
10. Lee, Y.-Y.; Cevallos, R.C.; Jan, E. An Upstream Open Reading Frame Regulates Translation of GADD34 during Cellular Stresses That Induce eIF2alpha Phosphorylation. *J. Biol. Chem.* **2009**, *284*, 6661–6673. [CrossRef] [PubMed]
11. Palam, L.R.; Baird, T.D.; Wek, R.C. Phosphorylation of eIF2 Facilitates Ribosomal Bypass of an Inhibitory Upstream ORF to Enhance CHOP Translation. *J. Biol. Chem.* **2011**, *286*, 10939–10949. [CrossRef] [PubMed]
12. Harding, H.P.; Novoa, I.; Zhang, Y.; Zeng, H.; Wek, R.; Schapira, M.; Ron, D. Regulated Translation Initiation Controls Stress-Induced Gene Expression in Mammalian Cells. *Mol. Cell* **2000**, *6*, 1099–1108. [CrossRef] [PubMed]
13. Brush, M.H.; Weiser, D.C.; Shenolikar, S. Growth Arrest and DNA Damage-Inducible Protein GADD34 Targets Protein Phosphatase 1 Alpha to the Endoplasmic Reticulum and Promotes Dephosphorylation of the Alpha Subunit of Eukaryotic Translation Initiation Factor 2. *Mol. Cell Biol.* **2003**, *23*, 1292–1303. [CrossRef] [PubMed]
14. Hartenian, E.; Nandakumar, D.; Lari, A.; Ly, M.; Tucker, J.M.; Glaunsinger, B.A. The Molecular Virology of Coronaviruses. *J. Biol. Chem.* **2020**, *295*, 12910–12934. [CrossRef] [PubMed]
15. Woo, P.C.Y.; Huang, Y.; Lau, S.K.P.; Yuen, K.-Y. Coronavirus Genomics and Bioinformatics Analysis. *Viruses* **2010**, *2*, 1804–1820. [CrossRef] [PubMed]
16. Schirtzinger, E.E.; Kim, Y.; Davis, A.S. Improving Human Coronavirus OC43 (HCoV-OC43) Research Comparability in Studies Using HCoV-OC43 as a Surrogate for SARS-CoV-2. *J. Virol. Methods* **2022**, *299*, 114317. [CrossRef] [PubMed]
17. Duguay, B.A.; Herod, A.; Pringle, E.S.; Monro, S.M.A.; Hetu, M.; Cameron, C.G.; McFarland, S.A.; McCormick, C. Photodynamic Inactivation of Human Coronaviruses. *Viruses* **2022**, *14*, 110. [CrossRef] [PubMed]
18. Kim, M.I.; Lee, C. Human Coronavirus OC43 as a Low-Risk Model to Study COVID-19. *Viruses* **2023**, *15*, 578. [CrossRef]
19. Shen, L.; Yang, Y.; Ye, F.; Liu, G.; Desforges, M.; Talbot, P.J.; Tan, W. Safe and Sensitive Antiviral Screening Platform Based on Recombinant Human Coronavirus OC43 Expressing the Luciferase Reporter Gene. *Antimicrob. Agents Chemother.* **2016**, *60*, 5492–5503. [CrossRef]
20. Raymonda, M.H.; Ciesla, J.H.; Monaghan, M.; Leach, J.; Asantewaa, G.; Smorodintsev-Schiller, L.A.; Lutz, M.M.; Schafer, X.L.; Takimoto, T.; Dewhurst, S.; et al. Pharmacologic Profiling Reveals Lapatinib as a Novel Antiviral against SARS-CoV-2 in Vitro. *Virology* **2022**, *566*, 60–68. [CrossRef]
21. Künkel, F.; Herrler, G. Structural and Functional Analysis of the Surface Protein of Human Coronavirus OC43. *Virology* **1993**, *195*, 195–202. [CrossRef] [PubMed]
22. Rehwinkel, J.; Gack, M.U. RIG-I-like Receptors: Their Regulation and Roles in RNA Sensing. *Nat. Rev. Immunol.* **2020**, *20*, 537–551. [CrossRef] [PubMed]

23. Brian, D.A.; Baric, R.S. Coronavirus Genome Structure and Replication. *Curr. Top. Microbiol. Immunol.* **2005**, *287*, 1–30. [CrossRef] [PubMed]
24. Snijder, E.J.; Limpens, R.W.A.L.; de Wilde, A.H.; de Jong, A.W.M.; Zevenhoven-Dobbe, J.C.; Maier, H.J.; Faas, F.F.G.A.; Koster, A.J.; Bárcena, M. A Unifying Structural and Functional Model of the Coronavirus Replication Organelle: Tracking down RNA Synthesis. *PLoS Biology* **2020**, *18*, e3000715. [CrossRef] [PubMed]
25. Knoops, K.; Kikkert, M.; Worm, S.H.V.D.; Zevenhoven-Dobbe, J.C.; van der Meer, Y.; Koster, A.J.; Mommaas, A.M.; Snijder, E.J. SARS-Coronavirus Replication Is Supported by a Reticulovesicular Network of Modified Endoplasmic Reticulum. *PLoS Biology* **2008**, *6*, e226. [CrossRef]
26. Wolff, G.; Limpens, R.W.A.L.; Zevenhoven-Dobbe, J.C.; Laugks, U.; Zheng, S.; de Jong, A.W.M.; Koning, R.I.; Agard, D.A.; Grünewald, K.; Koster, A.J.; et al. A Molecular Pore Spans the Double Membrane of the Coronavirus Replication Organelle. *Science* **2020**, *369*, 1395–1398. [CrossRef]
27. Cencic, R.; Desforges, M.; Hall, D.R.; Kozakov, D.; Du, Y.; Min, J.; Dingledine, R.; Fu, H.; Vajda, S.; Talbot, P.J.; et al. Blocking eIF4E-eIF4G Interaction as a Strategy To Impair Coronavirus Replication. *J. Virol.* **2011**, *85*, 6381–6389. [CrossRef]
28. Xue, M.; Fu, F.; Ma, Y.; Zhang, X.; Li, L.; Feng, L.; Liu, P. The PERK Arm of the Unfolded Protein Response Negatively Regulates Transmissible Gastroenteritis Virus Replication by Suppressing Protein Translation and Promoting Type I Interferon Production. *J. Virol.* **2018**, *92*, e00431-18. [CrossRef]
29. Dolliver, S.M.; Kleer, M.; Bui-Marinos, M.P.; Ying, S.; Corcoran, J.A.; Khaperskyy, D.A. Nsp1 Proteins of Human Coronaviruses HCoV-OC43 and SARS-CoV2 Inhibit Stress Granule Formation. *PLoS Pathog.* **2022**, *18*, e1011041. [CrossRef]
30. Moffat, J.; Grueneberg, D.A.; Yang, X.; Kim, S.Y.; Kloepfer, A.M.; Hinkle, G.; Piqani, B.; Eisenhaure, T.M.; Luo, B.; Grenier, J.K.; et al. A Lentiviral RNAi Library for Human and Mouse Genes Applied to an Arrayed Viral High-Content Screen. *Cell* **2006**, *124*, 1283–1298. [CrossRef]
31. Sanjana, N.E.; Shalem, O.; Zhang, F. Improved Vectors and Genome-Wide Libraries for CRISPR Screening. *Nat. Methods* **2014**, *11*, 783–784. [CrossRef]
32. Ying, S.; Khaperskyy, D.A. UV Damage Induces G3BP1-Dependent Stress Granule Formation That Is Not Driven by Translation Arrest via mTOR Inhibition. *J. Cell Sci.* **2020**, *133*, jcs248310. [CrossRef]
33. Oda, J.M.; den Hartigh, A.B.; Jackson, S.M.; Tronco, A.R.; Fink, S.L. The Unfolded Protein Response Components IRE1α and XBP1 Promote Human Coronavirus Infection. *mBio* **2023**, *14*, e0054023. [CrossRef]
34. Xue, M.; Feng, L. The Role of Unfolded Protein Response in Coronavirus Infection and Its Implications for Drug Design. *Front. Microbiol.* **2021**, *12*, 808593. [CrossRef]
35. Sidrauski, C.; Acosta-Alvear, D.; Khoutorsky, A.; Vedantham, P.; Hearn, B.R.; Li, H.; Gamache, K.; Gallagher, C.M.; Ang, K.K.-H.; Wilson, C.; et al. Pharmacological Brake-Release of mRNA Translation Enhances Cognitive Memory. *eLife* **2013**, *2*, e00498. [CrossRef]
36. Sidrauski, C.; Tsai, J.C.; Kampmann, M.; Hearn, B.R.; Vedantham, P.; Jaishankar, P.; Sokabe, M.; Mendez, A.S.; Newton, B.W.; Tang, E.L.; et al. Pharmacological Dimerization and Activation of the Exchange Factor eIF2B Antagonizes the Integrated Stress Response. *eLife* **2015**, *4*, e07314. [CrossRef]
37. Clavarino, G.; Cláudio, N.; Couderc, T.; Dalet, A.; Judith, D.; Camosseto, V.; Schmidt, E.K.; Wenger, T.; Lecuit, M.; Gatti, E.; et al. Induction of GADD34 Is Necessary for dsRNA-Dependent Interferon-β Production and Participates in the Control of Chikungunya Virus Infection. *PLoS Pathog.* **2012**, *8*, e1002708. [CrossRef]
38. Haneda, M.; Xiao, H.; Hasegawa, T.; Kimura, Y.; Nakashima, I.; Isobe, K. Regulation of mouse GADD34 gene transcription after DNA damaging agent methylmethane sulfonate. *Gene* **2004**, *336*, 139–146. [CrossRef]
39. Gambardella, G.; Staiano, L.; Moretti, M.N.; De Cegli, R.; Fagnocchi, L.; Di Tullio, G.; Polletti, S.; Braccia, C.; Armirotti, A.; Zippo, A.; et al. GADD34 is a modulator of autophagy during starvation. *Sci. Adv.* **2020**, *6*, eabb0205. [CrossRef]
40. Walter, P.; Ron, D. The Unfolded Protein Response: From Stress Pathway to Homeostatic Regulation. *Science* **2011**, *334*, 1081–1086. [CrossRef]
41. Liao, Y.; Fung, T.S.; Huang, M.; Fang, S.G.; Zhong, Y.; Liu, D.X. Upregulation of CHOP/GADD153 during Coronavirus Infectious Bronchitis Virus Infection Modulates Apoptosis by Restricting Activation of the Extracellular Signal-Regulated Kinase Pathway. *J. Virol.* **2013**, *87*, 8124–8134. [CrossRef]
42. Krähling, V.; Stein, D.A.; Spiegel, M.; Weber, F.; Mühlberger, E. Severe Acute Respiratory Syndrome Coronavirus Triggers Apoptosis via Protein Kinase R but Is Resistant to Its Antiviral Activity. *J. Virol.* **2009**, *83*, 2298–2309. [CrossRef]
43. Zhou, Y.; Zhang, Y.; Dong, W.; Gan, S.; Du, J.; Zhou, X.; Fang, W.; Wang, X.; Song, H. Porcine Epidemic Diarrhea Virus Activates PERK-ROS Axis to Benefit Its Replication in Vero E6 Cells. *Vet. Res.* **2023**, *54*, 9. [CrossRef]
44. Keramidas, P.; Papachristou, E.; Papi, R.M.; Mantsou, A.; Choli-Papadopoulou, T. Inhibition of PERK Kinase, an Orchestrator of the Unfolded Protein Response (UPR), Significantly Reduces Apoptosis and Inflammation of Lung Epithelial Cells Triggered by SARS-CoV-2 ORF3a Protein. *Biomedicines* **2023**, *11*, 1585. [CrossRef]
45. Gao, B.; Gong, X.; Fang, S.; Weng, W.; Wang, H.; Chu, H.; Sun, Y.; Meng, C.; Tan, L.; Song, C.; et al. Inhibition of Anti-Viral Stress Granule Formation by Coronavirus Endoribonuclease Nsp15 Ensures Efficient Virus Replication. *PLoS Pathog.* **2021**, *17*, e1008690. [CrossRef]

46. Nguyen, L.C.; Renner, D.M.; Silva, D.; Yang, D.; Parenti, N.A.; Medina, K.M.; Nicolaescu, V.; Gula, H.; Drayman, N.; Valdespino, A.; et al. SARS-CoV-2 Diverges from Other Betacoronaviruses in Only Partially Activating the IRE1α/XBP1 Endoplasmic Reticulum Stress Pathway in Human Lung-Derived Cells. *mBio* **2022**, *13*, e0241522. [CrossRef]
47. Favreau, D.J.; Desforges, M.; St-Jean, J.R.; Talbot, P.J. A Human Coronavirus OC43 Variant Harboring Persistence-Associated Mutations in the S Glycoprotein Differentially Induces the Unfolded Protein Response in Human Neurons as Compared to Wild-Type Virus. *Virology* **2009**, *395*, 255–267. [CrossRef]
48. Song, P.; Yang, S.; Hua, H.; Zhang, H.; Kong, Q.; Wang, J.; Luo, T.; Jiang, Y. The Regulatory Protein GADD34 Inhibits TRAIL-Induced Apoptosis via TRAF6/ERK-Dependent Stabilization of Myeloid Cell Leukemia 1 in Liver Cancer Cells. *J. Biol. Chem.* **2019**, *294*, 5945–5955. [CrossRef]
49. Farook, J.M.; Shields, J.; Tawfik, A.; Markand, S.; Sen, T.; Smith, S.B.; Brann, D.; Dhandapani, K.M.; Sen, N. GADD34 Induces Cell Death through Inactivation of Akt Following Traumatic Brain Injury. *Cell Death Dis.* **2013**, *4*, e754. [CrossRef]
50. Shi, W.; Sun, C.; He, B.; Xiong, W.; Shi, X.; Yao, D.; Cao, X. GADD34-PP1c Recruited by Smad7 Dephosphorylates TGFbeta Type I Receptor. *J. Cell Biol.* **2004**, *164*, 291–300. [CrossRef]
51. Ku, H.-C.; Cheng, C.-F. Master Regulator Activating Transcription Factor 3 (ATF3) in Metabolic Homeostasis and Cancer. *Front. Endocrinol.* **2020**, *11*, 556. [CrossRef]
52. Rohini, M.; Haritha Menon, A.; Selvamurugan, N. Role of Activating Transcription Factor 3 and Its Interacting Proteins under Physiological and Pathological Conditions. *Int. J. Biol. Macromol.* **2018**, *120*, 310–317. [CrossRef]
53. Dalskov, L.; Gad, H.H.; Hartmann, R. Viral Recognition and the Antiviral Interferon Response. *EMBO J.* **2023**, *42*, e112907. [CrossRef]
54. Duncan, J.K.S.; Xu, D.; Licursi, M.; Joyce, M.A.; Saffran, H.A.; Liu, K.; Gohda, J.; Tyrrell, D.L.; Kawaguchi, Y.; Hirasawa, K. Interferon Regulatory Factor 3 Mediates Effective Antiviral Responses to Human Coronavirus 229E and OC43 Infection. *Front. Immunol.* **2023**, *14*, 930086. [CrossRef] [PubMed]
55. Loo, S.-L.; Wark, P.A.B.; Esneau, C.; Nichol, K.S.; Hsu, A.C.-Y.; Bartlett, N.W. Human Coronaviruses 229E and OC43 Replicate and Induce Distinct Antiviral Responses in Differentiated Primary Human Bronchial Epithelial Cells. *Am. J. Physiol. Lung Cell Mol. Physiol.* **2020**, *319*, L926–L931. [CrossRef]
56. Kalvakolanu, D.V.; Bandyopadhyay, S.K.; Harter, M.L.; Sen, G.C. Inhibition of interferon-inducible gene expression by adenovirus E1A proteins: Block in transcriptional complex formation. *Proc. Natl. Acad. Sci. USA* **1991**, *88*, 7459–7463. [CrossRef]
57. Chahal, J.S.; Qi, J.; Flint, S.J. The human adenovirus type 5 E1B 55 kDa protein obstructs inhibition of viral replication by type I interferon in normal human cells. *PLoS Pathog.* **2012**, *8*, e1002853. [CrossRef]

viruses

MDPI

Article

Shiftless Is a Novel Member of the Ribosome Stress Surveillance Machinery That Has Evolved to Play a Role in Innate Immunity and Cancer Surveillance

Jamie A. Kelly † and Jonathan D. Dinman *

Department of Cell Biology and Molecular Genetics, University of Maryland, College Park, MD 20742, USA; twz3@cdc.gov

* Correspondence: dinman@umd.edu

† Current address: Viral Special Pathogens Branch, Division of High-Consequence Pathogens and Pathology, National Center for Emerging and Zoonotic Infectious Diseases, Centers for Disease Control and Prevention, Atlanta, GA 30329, USA.

Abstract: A longstanding paradox in molecular biology has centered on the question of how very long proteins are synthesized, despite numerous measurements indicating that ribosomes spontaneously shift reading frame at rates that should preclude their ability completely translate their mRNAs. Shiftless (SFL; C19orf66) was originally identified as an interferon responsive gene encoding an antiviral protein, indicating that it is part of the innate immune response. This activity is due to its ability to bind ribosomes that have been programmed by viral sequence elements to shift reading frame. Curiously, Shiftless is constitutively expressed at low levels in mammalian cells. This study examines the effects of altering Shiftless homeostasis, revealing how it may be used by higher eukaryotes to identify and remove spontaneously frameshifted ribosomes, resolving the apparent limitation on protein length. Data also indicate that Shiftless plays a novel role in the ribosome-associated quality control program. A model is proposed wherein SFL recognizes and arrests frameshifted ribosomes, and depending on SFL protein concentrations, either leads to removal of frameshifted ribosomes while leaving mRNAs intact, or to mRNA degradation. We propose that SFL be added to the growing pantheon of proteins involved in surveilling translational fidelity and controlling gene expression in higher eukaryotes.

Keywords: ribosome; frameshifting; translation; fidelity; shiftless; RQC; virus; innate immunity; stress; surveillance

1. Introduction

Protein synthesis is the most energy intensive of all cellular processes [1,2]. Accordingly, protein synthesis is controlled and monitored at many levels in order to maximize cellular energy utilization by rapidly identifying and responding to translational errors [3]. One of the core requirements for efficient protein synthesis is that ribosomes faithfully maintain translational reading frame. As evidence of this, the intrinsic rate of spontaneous ribosomal frameshifting (sRF) has been measured to be very low, on the order of one event per 5000 codons or ~2×10^{-4} [4–8]. Although this should not present a problem for simple organisms whose coding sequences fall below this limit, it poses a conundrum for higher eukaryotes which utilize larger proteins. For example, at the rate of 2×10^{-4} frameshifts/codon, the probability that any single ribosome would be able to fully translate an mRNA of 3500 codons without shifting reading frame is approximately 50% (Figure S1). Decreasing the rate of random frameshifting to 10^{-4} only raises this threshold to ~7000 codons and doubling the rate to 4×10^{-4} lowers this threshold to ~1700 codons. Given that the human genome encodes at least 23 proteins greater than 5000 codons, some of which also have a broad tissue distribution [9], the energetic costs incurred by their translation would be prohibitive. Taking this analysis further, it should be impossible to fully synthesize very large proteins such as Titan, whose mRNA encodes 38,138 amino acid residues [10]. Additionally, accumulation of truncated proteins resulting from spontaneous frameshifts can have trans-dominant negative and/or deleterious gain-of-function effects on cell growth and homeostasis [11]. In short, how mRNAs encoding very long proteins can be efficiently synthesized is a longstanding unsolved paradox of molecular biology.

Despite strong limitations on spontaneous ribosomal frameshifting, many RNA viruses evade this by programming ribosomes to shift into alternate reading frames, often at frequencies of three orders of magnitude greater than the background rate, through a process called Programmed Ribosomal Frameshifting (PRF) [12–15]. It is generally accepted that RNA viruses use PRF to maximize their gene expression programs without altering their genomes [16]. PRF has also been documented to occur on mRNAs of cellular origin [17–19], although this field remains controversial due to issues recently raised regarding dual luciferase reporter systems that had been in wide use for approximately two decades [20]. Regardless, PRF is directed to occur at specific locations along mRNAs, typically by a combination of specific 'slippery' sequences and structured RNA elements [12,21]. Experimentally, altering viral PRF frequencies has negative consequences on viral particle biogenesis, thus rendering PRF a focus for development of antiviral therapeutics [22–24]. In the evolutionary battle between viruses and their host cells, this reliance of viruses for high and precise rates of PRF has been targeted by higher eukaryotes in the form of a protein called Shiftless (a.k.a. C19orf66). Shiftless (SFL) recognizes and binds to ribosomes that are in the process of shifting reading frame, arresting them at the PRF signals [25]. While much remains unknown about how SFL works, it appears to lock frameshifting ribosomes in place on the mRNA and recruits eRF3 (and presumably its partner eRF1) to terminate translation by frameshifted ribosomes. Ongoing research efforts are revealing that SFL overexpression inhibits the replication of a growing number of viruses [25–27]. Not surprisingly, SFL transcription is upregulated by interferons, i.e., SFL is an interferon responsive gene (IRG), identifying it as a participant in the innate immune response [27].

Curiously, an immunoblot analysis indicated that SFL appears to be constitutively expressed in mammalian cells, albeit at much lower levels than during viral infection [28]. The current study addresses the reason for constitutive SFL expression. In so doing, it illuminates how SFL may be used by higher eukaryotes to overcome the apparent limitation on protein length. Genetic evidence is also presented connecting SFL with the molecular apparatus involved in translational surveillance known as ribosome stress surveillance [29]. A model is proposed wherein SFL recognizes and arrests frameshifted ribosomes, and depending on SFL protein concentrations, either leads to removal of frameshifted ribosomes while leaving mRNAs intact, or to mRNA degradation. As such, we propose that SFL be added to the growing pantheon of proteins involved in surveilling translational fidelity and controlling gene expression in higher eukaryotes.

2. Materials and Methods

2.1. Cell Culture and Plasmid Transfections

Human embryonic kidney (HEK293T/17) (CRL-11268), HeLa (CCL-2), and U87 MG (HTB-14) cells were purchased from the American Type Culture Collection (Manassas, VA, USA). HEK 293T cells were maintained in Dulbecco's modified Eagle's medium (Corning Life Sciences 10-013-CV Durham, NC, USA) supplemented with 10% fetal bovine serum (Life Technologies 26140079 Carlsbad, CA, USA) 1% GlutaMAX (35050061), 1% nonessential amino acids (Gibco 11140050), 1% HEPES buffer (Cytiva Life Sciences SH30237.01) and 1x Penicillin/Streptomycin (Gibco 15140122). HeLa and U87 MG cells were maintained in Dulbecco's modified Eagle's medium supplemented with 10% fetal bovine serum, 1% GlutaMAX, and 1x Penicillin/Streptomycin. Both cell lines were incubated at 37 °C in 5% CO_2. The cells were transfected with a total of 500 ng (dual luciferase assay) or 1 μg (bifluorescence and qRT PCR assays) plasmid DNA 24 h after seeding using Lipofectamine 3000 (Invitrogen L3000015) per the manufacturer's protocol.

2.2. *shRNA Knockdowns*

A panel of short hairpin shRNA plasmids targeting SFL and commercially validated shRNAs targeting Pelota, SMG1, ASCC3, Hbs1L, and ZNF598 were purchased from Millipore Sigma. shRNA targets are listed in Table 1. Five different SFL shRNAs (SFL shA–E) were assayed and knockdowns were conducted by transfecting 250 ng (dual luciferase assay) or 500 ng (bifluorescence assay) shRNA plasmid DNA into HEK293T cells 24 h after seeding using Lipofectamine 3000 per the manufacturer's protocol. RT-PCR analysis identified two shRNAs, shRNA A and shRNA B, that promoted the strongest knockdown of SFL, and quantitatively validated by qRT PCR (Figure S2B).

Table 1. shRNA target sequences used in *this work*.

shRNA Target	Target Sequence (5′ to 3′)	Cat#
SFL shA	GTGTATCCAACACGGATCCTC	TRCN0000434142
SFL shB	CCAAGAACTAAGTAACGATCT	TRCN0000161786
SFL shC	AGCAACCCTCACATTAGCAGT	TRCN0000420530
SFL shD	GAAGTTTCATGGGAAGGTATC	TRCN0000164654
SFL shE	GAAGTTCTGTGGGACACATTG	TRCN0000418723
ASCC3	TGAGGAGCGAACTGGATATTT	TRCN0000296023
Hbs1L	GCGATCTATTGACAAACCTTT	TRCN0000353597
PELO	GCAGTGAAGACCGACAACAAA	TRCN0000163394
SMG1	GCACTGTAACTACGGCTACAA	TRCN0000037413
ZNF598	CCAACCCTCTAAAGTTGGGAA	TRCN0000222610

2.3. *Generation of CRISPR Knockout Cell Lines*

An SFL$^{-/-}$ HEK293T cell line was generated using CRSIPR-Cas9 as described in Ran et al. [30] HEK923T cells were transfected with spCas9(BB)-2A-Puro plasmid containing a gRNA specific to C19orf66 (5′ CGTGTATCCAACACGGATCC 3′) designed to result in a 1 base deletion in the ORF using Lipofectamine3000 per manufacturer's protocol. Cells were selected for the presence of Cas9 by incubating with 1.0 µg/mL puromycin. Clonal cell lines were obtained by seeding cells at low density in a 10 cm tissue culture dish and then selecting well-isolated colonies for expansion. Clonal lines were screened for mutations by PCR-amplifying C19orf66 from genomic DNA and assessing for mutations using the Surveyor mutation detection kit (Integrated DNA Technologies 706020) per manufacturer's protocol. Positive hits from the Surveyor screen were sequence-verified (Figure S2C) and knockout of SFL was validated via qRT PCR (Figure S2B).

2.4. *Growth Curve of SFL$^{-/-}$ HEK293T Cells*

HEK WT and SFL$^{-/-}$ cells were seeded at a density of 10^4 cells per well of a 12-well plate. Cells were trypsinized and counted at 24 h intervals for a total of 168 h. Cell doubling times were determined using the equation:

$$DT = \frac{T * \ln(2)}{\ln\left(\frac{X_e}{X_b}\right)}$$

where T = time in hours, X_e = endpoint cell count, X_b = beginning cell count.

These data are shown in Figure S2D.

2.5. *Preparation of Reporter Plasmids*

Dual luciferase reporters containing -1 PRF signals of SARS-CoV, SARS-CoV-2, and human CCR5 listed in Table 2 were made by digesting pJD2257 with *Sal* I and *Bam* HI, gel-purifying digest products, and ligating a DNA oligonucleotide insert (IDT) containing the

-1 PRF signal of interest into the plasmid using T4 DNA ligase (NEB). pJD2257 is derived from pSGDLuc [20] into which *Sal* I and *Bam* HI sites were inserted so as to preclude possible distortions of luciferase reporter readouts due to insertion of sequences of viral or cellular origin. The spontaneous -1 frameshift (-1 sFS) reporter plasmid was made using site-directed mutagenesis of pJD2257. Site directed mutagenesis primers (Table 3) were synthesized and purified by IDT.

Table 2. Plasmids used in this study.

Plasmid Name	Description
pJD2257	Modified pSGDluc (dual luciferase with inteins) with 0-frame control sequence insert
pJD2256	Modified pSGDluc (dual luciferase with inteins) with HIV-1 -1 PRF sequence insert
pJD2258	Modified pSGDluc (dual luciferase with inteins) with CCR5 -1 PRF sequence insert
pJD2359	Modified pSGDluc (dual luciferase with inteins) with SARS-CoV -1 PRF sequence insert
pJD2514	Modified pSGDluc (dual luciferase with inteins) with SARS-CoV-2 -1 PRF sequence insert
pJD2450	Modified pSGDluc (dual luciferase with inteins) with spontaneous -1 FS sequence insert
pJD2262	Dual fluorescent construct with 0-frame control sequence insert
pJD2261	Dual fluorescent construct with HIV-1 -1 PRF sequence insert
pJD2281	Dual fluorescent construct with CCR5 -1 PRF sequence insert
pJD2529	Dual fluorescent construct with SARS-CoV-2 sequence insert
pJD2350	Dual fluorescent construct with OAZ1 +1 PRF sequence insert
pJD2455	Dual fluorescent construct with VEEV stop codon readthrough sequence insert
pJD2260	pMAX GFP monocistronic reporter
pJD2612	C19orf66 ORF in pCMV-Myc
pJD175f	P2luci (first-generation dual luciferase) with 0-frame control sequence insert
pJD827	P2luci with CCR5 -1 PRF sequence insert

Table 3. Oligonucleotides.

Oligo Name	Sequence (5′ to 3′)
H_GAPDH_F	GGATGATGTTCTGGAGAGCC
H_GAPDH_R	CATCACCATCTTCCAGGAGC
GFP_qPCR_F	GGCTACGGCTTCTACCACTT
GFP_qPCR_R	CTCGTACTTCTCGATGCGGG
Firefly Luciferase qPCR Fwd	TCGCCTCTCTGATTAACGCC
Firefly Luciferase qPCR Rev	ATTACACCCGAGGGGGATGA
C19orf66_qPCR	Biorad PrimePCR qHsaCED0003572
C19orf66_RT_PCR_F	CTCAGGAAGGTGTGGAGCTG
C19orff_RT_PCR_R	GCCACTGCTAATGTGAGGGT
Pelo_qPCR_F	GACCGACAACAAACTGCTCCTG

Table 3. *Cont.*

Oligo Name	Sequence (5′ to 3′)
Pelo_qPCR_R	AGCCACAGTAGGGTCACAAAGG
SMG1_qPCR_F	ATGCTGGTGAGCTTCGGCAGTA
SMG1_qPCR_R	CGCACATACACTTCAGGGTGGT
CRISPR validation primer Fwd	AAATCTGGCTTCTGAACCTCCT
CRISPR validation primer Rev	GTGGGAGACAAAGTGGACTGAG
SFL gRNA Top	CACCG GGATCCGTGTTGGATACACG
SFL gRNA Bottom	AAAC CGTGTATCCAACACGGATCC C
CRISPR validation primer Fwd	AAATCTGGCTTCTGAACCTCCT
CRISPR validation primer Rev	GTGGGAGACAAAGTGGACTGAG
-1 sFS_SDM_F	AAAGAGGCTGCGGCAAAAGC
-1 sFS_SDM_R	GCGGCTGCTTCGGTCGAC

A bifluorescent reporter was initially constructed by swapping AcGFP and mCherry coding sequences into pJD2257, replacing the dual luciferase cassettes to construct pJD2262. A panel of bifluorescent reporters containing human and viral translational control element inserts (Table 2) were made by digesting pJD2261 with *Sal* I and *Bam* HI, gel-purifying digest products, and ligating a DNA oligonucleotide insert (IDT) containing sequences of interest into the plasmid using T4 DNA ligase (NEB). Products were transformed into DH5α Escherichia coli cells (NEB) and spread onto LB agar plates containing 50 mg/mL carbenicillin. Positive clones were verified by DNA sequencing (Genewiz).

2.6. Dual Luciferase Assays of -1 PRF

The frameshifting efficiency of luciferase reporter plasmids in cultured cells was assayed as previously described using a dual luciferase reporter assay system kit (Promega Madison, WI, USA) [31,32]. Then, 24 h after transfection, cells were washed with 1x PBS then lysed with 1x passive lysis buffer (E194A, Promega). Reporter activity was calculated by measuring the luminescence of firefly or Renilla luciferase in 50 μL of cell lysate. Assays were conducted in triplicate in 96-well plates and quantified using a GloMax microplate luminometer (Promega). Percent frameshift was calculated by averaging the three Firefly or Renilla luciferase technical replicate reads per sample then forming a ratio of firefly to Renilla luminescence per sample. Each sample ratio was compared to a 0-frame control set to 100%. At least three biological replicates with three technical replicates each were assayed for each sample. Statistical analyses were conducted using one-way analysis of variance using Prism 9 software (GraphPad).

2.7. Bifluorescence Assays of -1 PRF

Frameshift efficiency of bifluorescent reporters in cultured cells were assayed as described previously [33]. HEK293T or HeLa cells were seeded at density of 10^5 cells per well of a 12-well plate in appropriate growth media. After a 24 h incubation, cells were transfected with 500 ng bifluorescence reporter plasmid and 500 ng of either a SFL overexpression plasmid, shRNA plasmid targeting SFL, or shRNA scramble non-targeting control. Additional media was added to cells 24 h post transfection and incubated for an additional 24 h. Cells were collected by scraping into Dulbecco's phosphate-buffered saline (Corning 21-031-CV), pelleted by centrifugation then lysed in 150 uL Triton lysis buffer (1% Triton X-100, 150 mM NaCl, 50 mM Tris pH8, 1x Halt protease inhibitor cocktail (Thermo Scientific Waltham, MA, USA)). Cell lysates were clarified by centrifugation and assayed in a clear-bottom black-walled 96-well plate (Grenier Bio-One Atlanta, GA, USA). Fluorescence was quantified using a GloMax microplate luminometer (Promega Madison, WI, USA) with the "green" optical kit (Excitation 525 nm, Emission 580–640 nm) for mCherry and the "blue" optical kit (excitation 490 nm, emission 510–570 nm) for AcGFP. Reporter activity and -1 PRF efficiency was corrected for AcGFP bleed over into the mCherry channel by subtracting background fluorescence from mock transfected cells. Reporter activity was calculated by measuring the fluorescence of 150 µL clarified cell lysate and subtracting background fluorescence from mock transfected cells. -1 PRF efficiency was calculated as previously described [33]. Statistical analyses were conducted using one-way analysis of variance using Prism 9 software (GraphPad).

2.8. qRT PCR and RT PCR Methods

To quantitatively measure mRNA abundances (qRT-PCR), total RNA was extracted from cells grown to 70–80% confluency using the Total RNA miniprep kit (NEB T2010S) according to the manufacturer's protocol. cDNA synthesis was performed using the iScript gDNA Clear cDNA synthesis kit (BioRad 1725034) according to the manufacturer's protocol using 500 ng total RNA. qPCR was performed with 100 ng of total cDNA, 250 nM primers in a final volume of 10 µL using BioRad CFX96 and SSOAdvanced SYBR green master mix (BioRad 1725270). mRNA abundances for SFL overexpression and $SFL^{-/-}$ conditions were normalized to WT HEK293T conditions. Primers for this assay are listed in Table 3. qPCR conditions were 4 min at 95 °C followed by 40 cycles at 95 °C for 5 s and 60 °C for 30 s followed by a melt curve from 65 °C to 95 °C. Experiments were repeated with at least three independent biological replicates.

To monitor the presence of specific mRNAs (RT-PCR), RNA was extracted from HEK293T, HeLa, and U87MG cells grown to 70–80% confluency and an overnight *E. coli* culture using the Total RNA miniprep kit (NEB T2010S) according to the manufacturer's protocol. cDNA synthesis was performed using the iScript gDNA Clear cDNA synthesis kit (BioRad 1725034) according to the manufacturer's protocol. RT PCR was performed using 1 µL of cDNA, and 500 nM primers in a final volume of 35 µL using Q5 Hotstart Mastermix (NEB M0494S). Primers for this assay are listed in Table 3. PCR conditions were: 98 °C for 30 s followed by 40 cycles at 98 °C for 10 s, 71 °C for 15 s and 72 °C for 20 s, then 72 °C final extension for 2 min.

To assess splicing of reporter plasmid transcripts, RNA was extracted from HEK293T cells transfected with first-generation dual luciferase (p2luci), second-generation dual luciferase (pSGD), or bifluorescent (BiFl) reporters containing 0-frame control, CCR5 -1 PRF, or HIV-1 -1 PRF inserts. cDNA synthesis was performed as stated above. RT PCR was performed using 1 uL of cDNA or 100 pg of reporter plasmid as an input control, and 500 nM primers in a final volume of 20 uL using DreamTaq Green PCR Mastermix (Thermofisher K1081). Primers for this assay are listed in Table 3. PCR conditions were 95 °C for 2 min followed by 40 cycles at 95 °C for 30 s, 64.4 °C (p2luci), 57 °C (pSGD) or 59.8 °C (BiFl) for 30 s, and 72 °C for 3 min, then 72 °C final extension for 5 min.

2.9. Bioinformatics Analysis of SFL Expression in Cancer Cells

The expression of SFL was analyzed in common cancer tissues and normal tissues deposited in the TGCA database using Genome Browser [34] (http://genome.ucsc.edu (accessed on 4 February 2022)). Average transcripts per million (TPM) for each normal and cancerous cell type were displayed on a heatmap using GraphPad Prism 9 software. Survival corresponding to SFL expression levels among patients with various cancers was analyzed using the UALCAN platform [35].

3. Results

3.1. SFL Is Constitutively Expressed in Human-Derived Cells

Previous studies demonstrated that SFL is constitutively expressed at low levels in human-derived cell lines and can be upregulated by the type I interferon response [25,28]. To confirm that it is indeed constitutively expressed, we used RT-PCR to probe for SFL expression in RNAs extracted from three human derived cell lines, HeLa, HEK293T, and U87 MG cells. RNA from *E. coli* was used as a negative control. Our results show that both SFL and the shorter splice isoform, SFLS [25], are constitutively expressed in the human-derived cell lines but not in bacteria (Figure S2A). In order to manipulate SFL levels in the cell, we used an overexpression plasmid, or short hairpin RNA (shRNAs) targeting SFL. Overexpressing SFL resulted in a 1.8-fold increase in SFL mRNA levels (p = 0.005), while SFL-targeting shRNAs shA and shB reduced mRNA 3.25 (p = 0.01) and 3.42-fold (p = 0.0098), respectively (Figure S2B).

3.2. First-Generation Dual Luciferase Reporters, Not the CCR5 -1 PRF Element, Promote Off-Target mRNA Splicing

The first generation of dual luciferase plasmids used to assess frameshift efficiency, P2luci, were subject to multiple issues including off-target mRNA splicing and inefficient cleavage of reporter proteins resulting in inaccurate values of frameshifting. Though subsequent reporter generations largely resolved these issues, questions arose regarding if some frameshift elements were subject to off-target mRNA splicing. In particular, the validity of the human -1 PRF signal in CCR5 came into question after one report suggested apparent frameshifting of this sequence was a product of splicing and not a ribosomal frameshift [36]. If some reported -1 PRF signals are the result of off-target mRNA splicing and not a ribosomal frameshift event due to properties of the reporter system used or those intrinsic to the -1 PRF element itself, it is imperative to identify true frameshift elements from splicing artifacts. To ascertain whether the CCR5 -1 PRF reporter promoted off target mRNA splicing in the context of the second-generation dual luciferase and bifluorescent reporters, mRNAs were extracted from cells expressing these reporters, and primers complementary to the 5′ end of the 5′ reporters, and 3′ ends of the 3′ reporters were used to amplify the full-length mRNAs (Figure S3). Controls included the same reporters alone or containing the HIV-1 -1 PRF signal. The PCR products were separated by gel electrophoresis. Plasmid controls revealed bands of the expected lengths, i.e., ~2500 nt for the dual luciferase reporters and ~1500 nt for the bifluorescent reporters. Small molecular weight products were observed in the first generation (p2luci) 0-frame control and CCR5 -1 PRF reporter mRNA samples, consistent with the propensity of this reporter to undergo off-target splicing events [20]. These PCR products were absent from the second generation (pSGD) and bifluorescent (BiFl) reporter samples. Additional PCR products of ~1200 and ~800 bp were observed with all the BiFl samples. In addition, no "reporter crash", i.e., 3–4 order of magnitude decreases in raw reporter protein activities indicative of off-target splicing, were observed in CCR5 samples as compared to HIV-1 samples from either the pSGD or BiFl data (Supplemental Data Set S1). These findings indicate that off-target splicing was due to sequences in the first-generation p2Luci reporters rather than the CCR5 -1 PRF signal as previously reported [36].

3.3. SFL Overexpression or Knockdown Results in 2-Fold Reciprocal Effects on -1 PRF

The ability, accuracy, and reproducibility of the bifluorescent reporter system was assessed by cloning validated PRF signals representing a range of frameshifting efficiencies and comparing the results with the same elements cloned into the dual luciferase reporter vector. The parallel analysis revealed that reporter systems yielded comparable PRF efficiencies, but that the bifluorescent system generated more reproducible data, particularly at higher PRF efficiencies (Figure S4). The bifluorescent reporter assay was used to monitor -1 frameshift efficiency in HEK293T cells over- or under-expressing SFL. Disruption of SFL homeostasis resulted in reciprocal two-fold changes to -1 PRF efficiencies in a panel of human and viral-derived -1 PRF signals (Figure 1). Consistent with previous studies, SFL overexpression resulted in ~2-fold decreases in -1 PRF efficiency promoted by the -1 PRF signals derived from SARS-CoV, HIV-1 and CCR5 (Figure 1). Specifically, SARS-CoV -1 PRF decreased from ~35% ± 5.5% in WT HEKs to ~18.9% ± 2.2% upon SFL overexpression (SFL OE) (p = 0.0057) (Figure 1A). SFL overexpression decreased HIV-1 frameshift efficiency from 4.5% ± 0.45% in WT to 2.89% ± 0.36% in (p = 0.157) (Figure 1B), and frameshifting promoted by the CCR5 -1 PRF signal decreased from 1.76% ± 0.13% in WT to 0.95% ± 0.21%

($p = 0.0298$) (Figure 1C). While significant decreases in -1 PRF due to SFL overexpression were observed using the bifluorescent reporter system, SFL overexpression did not yield significant decreases in -1 PRF using dual-luciferase reporters (Figure S5). In each individual experiment SFL overexpression did decrease -1 PRF as monitored by the dual luciferase reporters, but the variability in % -1 PRF between experimental replicates masked this effect.

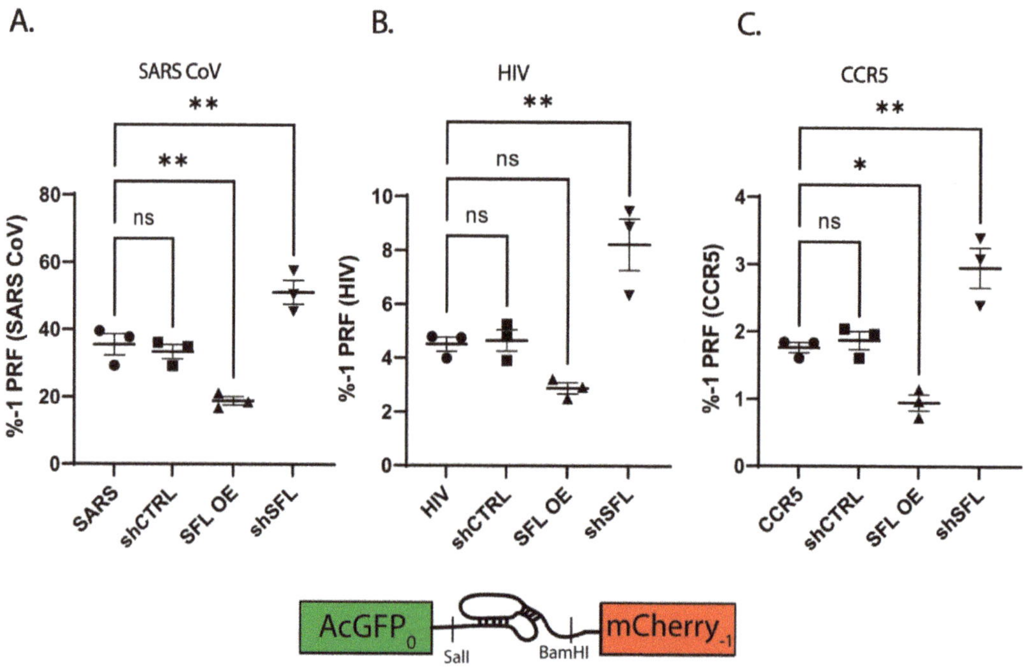

Figure 1. SFL overexpression or knockdown alters -1 PRF. Frameshift efficiency of three translational recoding elements measured using dual fluorescence reporters in HEK293T cells over or under-expressing SFL. (**A**) SARS-CoV -1 PRF signal, (**B**) HIV-1 -1 PRF signal, (**C**) CCR5 -1 PRF signal. Error bars denote standard error of the mean (SEM). * p-Value < 0.05, ** p-Value $\leq$ 0.01, ns, not significant.

Conversely, knockdown of expression using an shRNA targeting SFL (shSFL) resulted in a roughly two-fold increase in apparent frameshifting. SARS-CoV promoted -1 PRF increased from ~35% in WT HEKs to 51.2% ± 6.08% in SFL shRNA knockdown (shSFL) ($p = 0.0081$). HIV-1 stimulated -1 PRF increased from 4.5% in WT to 8.24% ± 1.65% in shSFL ($p = 0.003$), and CCR5 frameshifting increased from 1.76% in WT to 2.96% ± 0.51% ($p = 0.0036$). Similar results were observed using dual luciferase reporters (Figure S5).

3.4. SFL Is Not Limited to -1 PRF Signals

Although SFL was originally identified as a -1 PRF specific frameshift inhibitor [25], a recent study demonstrated SFL can also modulate programmed termination codon readthrough and that its activity is not limited to any specific type of translational recoding element [37]. To further probe the range of possible SFL substrates, SFL was over- or under-expressed in the context of cells expressing either a +1 PRF signal from ornithine decarboxylase antizyme 1 (OAZ1) [38] or a termination codon readthrough (TCR) signal from Venezuelan equine encephalitis virus (VEEV) [39–41]. Similar to the findings with -1 PRF signals, reciprocal two-fold changes were observed with these recoding signals. Specifically, SFL overexpression in OAZ1 decreased +1 PRF from 20.9% ± 1.85% to 12% ± 1.35% ($p < 0.0001$) and shSFL increased apparent frameshifting to 35.9% ± 1.78% ($p < 0.0001$), respectively (Figure 2A). SFL overexpression in VEEV decreased TCR from 1.25% ± 0.24%

to 0.52% ± 0.05% (p = 0.0005) and shSFL increased TCR to 1.89% ± 0.22% (p = 0.0011) (Figure 2B).

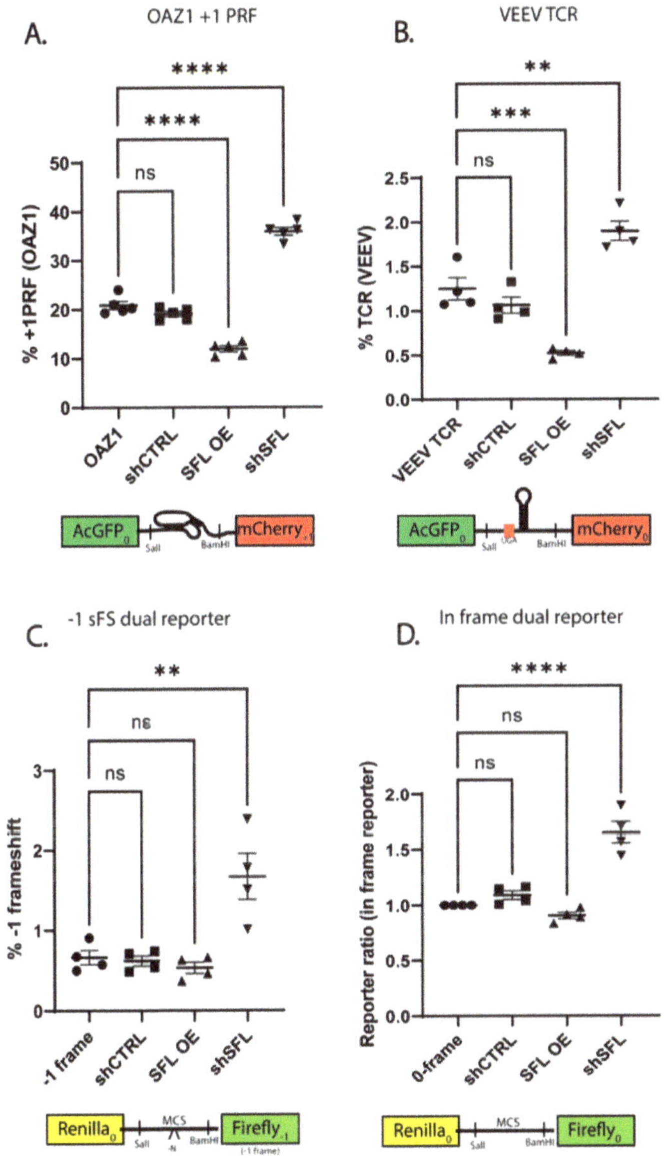

Figure 2. SFL is not limited to -1 PRF signals. Translational recoding efficiency and reporter ratio measured using dual fluorescence reporters in HEK293T cells over or under-expressing SFL. (**A**) OAZ1 +1 ribosomal frameshift signal, (**B**) VEEV termination codon readthrough signal, (**C**) -1 sFS dual luciferase reporter, (**D**) 0-frame dual luciferase reporter. Error bars denote SEM. ** p-Value ≤ 0.01, *** p-Value ≤ 0.001, **** p-Value ≤ 0.0001, ns not significant.

To determine whether these effects are limited to *bona fide* recoding signals, the effects of SFL over- or under-expression were assayed using a spontaneous -1 frameshift (-1sFS) reporter in which the downstream (firefly luciferase) reporter is in the -1 frame relative to the upstream reporter. SFL overexpression did not significantly change -1 sFS, however SFL knockdown resulted in a 2.5-fold increase, from 0.66% ± 0.18% to 1.67% ± 0.57% (p = 0.0018) (Figure 2C). Following this observation, we examined the effects of SFL over- and under-expression on the 0-frame reporter, i.e., where both reporters are in-frame with one another. Although SFL overexpression did not significantly alter the 3′ to 5′ reporter ratio, knocking down SFL expression resulted in a 1.65-fold increase in reporter ratio ($p < 0.0001$) (Figure 2D). This suggests that in-frame mRNAs may also be substrates for SFL.

3.5. Disrupting SFL Homeostasis Reduces Reporter Protein Activity

The experiments described above examined ratios of ratios of two reporter proteins expressed in cells either over- or under-expressing SFL as compared to mock transfected cells. Since calculation of recoding efficiency can reduce information content by averaging ratios of ratios, an issue that particularly impacts dual luciferase-based assays due to their greater variability, measuring actual reporter gene activities presents a more accurate view of the effects of SFL on gene expression. Because SFL over- or under-expression influenced these ratios independent of the type, or even the presence of a recoding element, we examined actual reporter gene activities to determine the extent to which SFL directly influences reporter output. In WT HEK cells, expression of both AcGFP and mCherry from the bicistronic dual-fluorescence reporter was ~14,000 and ~1000 fluorescence units respectively per 150 uL of cell lysate. Over-expression of SFL promoted a ~2-fold decrease in both AcGFP and mCherry activities using this reporter (Figure 3A). Gene deletion of SFL (in isogenic SFL$^{-/-}$ HEK cells) resulted in a 1.6-fold decrease in expression of the 5′ AcGFP reporter with no significant change in mCherry expression. Similar inhibitory trends of SFL over- or under-expression of gene expression were observed with 0-frame dual luciferase reporters (Figure 3B) and the -1 sFS dual luciferase reporter (Figure 3C). Over-expression of SFL promoted >10-fold decreases in both Renilla and firefly luciferase activities using the 0-frame reporter. SFL$^{-/-}$ reduced reporter activities to a lesser extent, i.e., a ~6.8-fold in Renilla and ~5-fold in firefly (Figure 3B). We note that the baseline firefly luciferase activity is almost three orders of magnitude less than Renilla activity in the -1 sFS reporter construct, reflecting the baseline rate of spontaneous -1 frameshifting (Figure 3C). These observations suggested that SFL may normally monitor in-frame messages for spontaneous frameshifting activity. Consistent with this, the activity of a monocistronic GFP reporter was similarly impacted by SFL overexpression (>10-fold decrease) or in an isogenic SFL$^{-/-}$ HEK cell line (~30% decrease) (Figure 3D).

3.6. Disruption of SFL Homeostasis Decreases mRNA Steady State Abundances

Given the negative impacts of increased or decreased SFL expression on reporter protein activities, we examined the effects on mRNA steady state abundances in isogenic WT HEK cells or HEK cells overexpressing SFL by qRT-PCR using 100 ng of total cDNA derived from cellular RNAs. SFL overexpression decreased the steady state abundances of mRNAs expressed from the GFP monocistronic reporter by 77% ± 5.2% (p = 0.0002), Renilla luciferase in the 0-frame dual luciferase reporter by 82% ± 13.9% (p = 0.0003), and Renilla luciferase in the SARS-CoV-2 -1 PRF reporter by 75.6% ± 26.1% (p = 0.0001) compared to WT HEKs. Importantly, SFL overexpression also decreased the abundance of the endogenous GAPDH mRNA by 44% ± 17.9% (p = 0.0021) (Figure 4).

In parallel, comparison of mRNA steady state abundances in isogenic WT HEK cells versus SFL$^{-/-}$ HEK cells revealed that the lack of SFL also reduced the abundance of all four mRNAs, although to lesser extents. Specifically, in SFL$^{-/-}$ cells, GFP mRNA was reduced by 25.6% ± 17.8% (p = 0.046), 0-frame Renilla mRNA levels by 53.7% ± 15% (p = 0.0001), SARS-CoV-2 Renilla mRNA levels by 72.4% ± 6% (p = 0.0002), and GAPDH levels were decreased by 40.9% + 14.3% (p = 0.0034) (Figure 4).

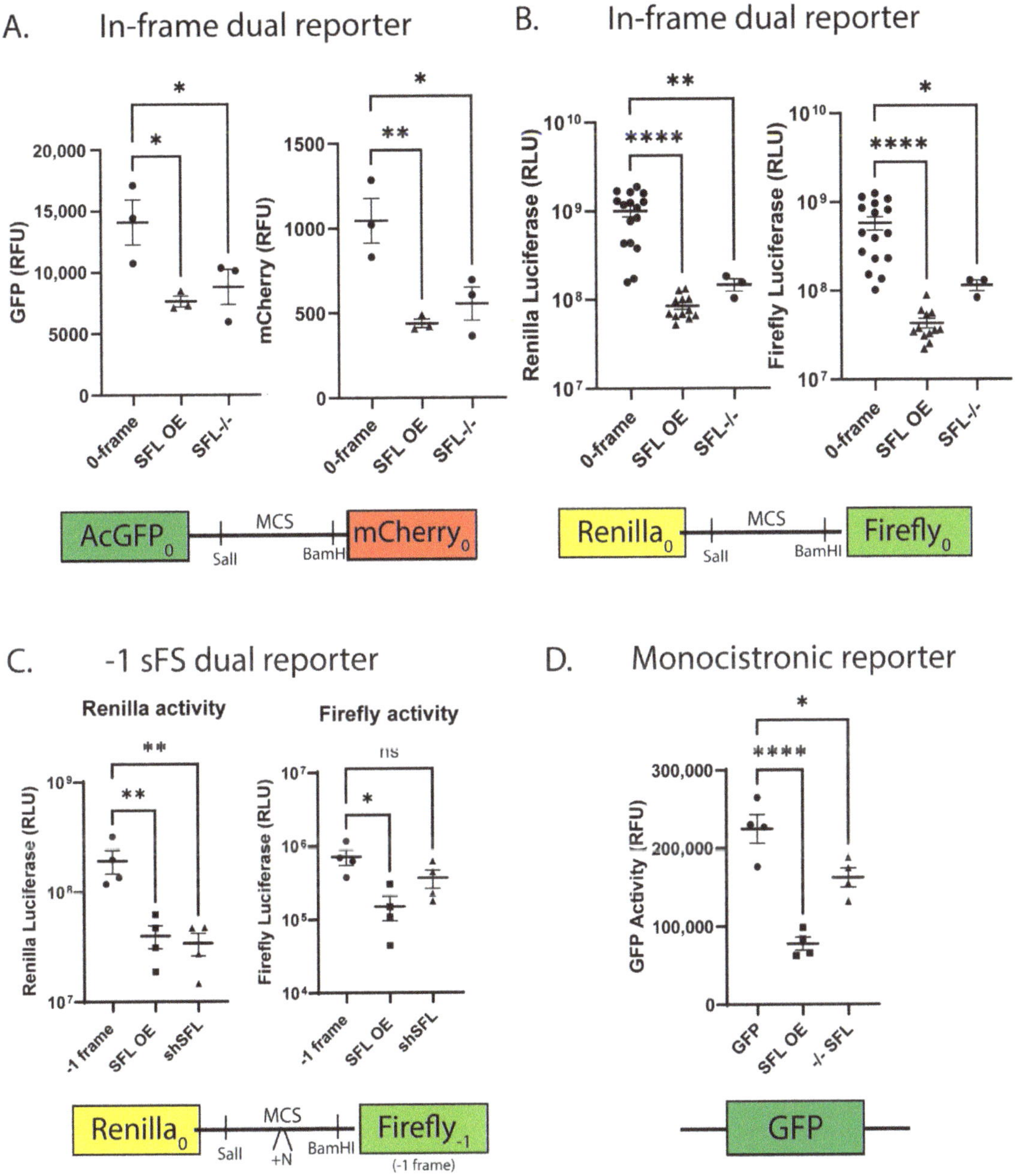

Figure 3. Disruption of SFL homeostasis inhibits reporter protein activity. Activity of reporters in HEK293T cells over or under-expressing SFL (**A**) 0-frame dual fluorescence reporter, (**B**) 0-frame dual luciferase reporter, (**C**) -1 sFS reporter, and (**D**) monocistronic GFP reporter. Error bars denote SEM. * p-Value < 0.05, ** p-Value $\leq$ 0.01, **** p-Value $\leq$ 0.0001, ns not significant.

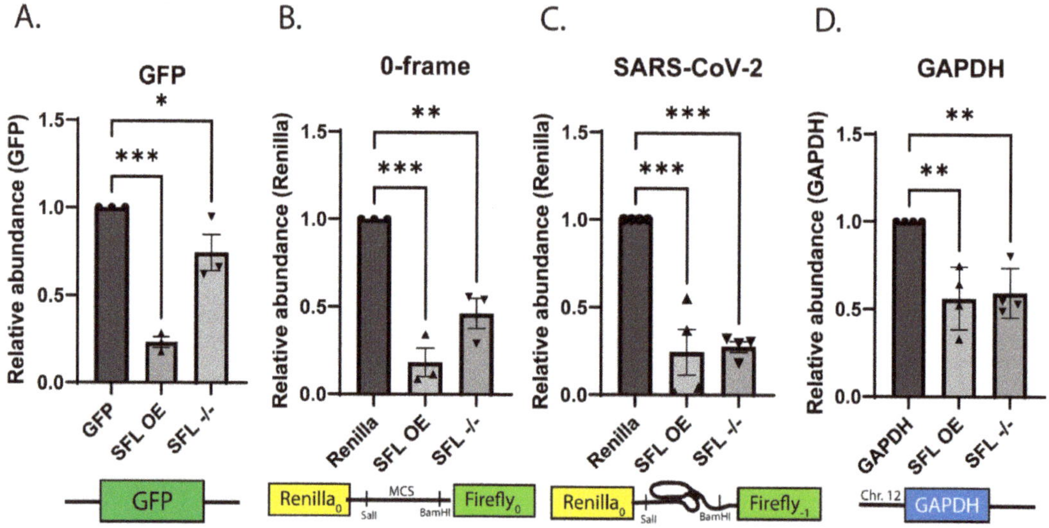

Figure 4. Disruption of SFL homeostasis decreases mRNA steady-state abundances relative to wild-type conditions. Relative mRNA abundance in $SFL^{+/+}$ HEK293T or $SFL^{-/-}$ HEK293T cells. Relative abundance of (**A**) GFP mRNA from a monocistronic reporter, Renilla luciferase mRNA from the (**B**) 0-frame control or (**C**) SARS-CoV-2 dual luciferase reporter, and (**D**) GAPDH mRNA. Error bars denote SEM. * p-Value < 0.05, ** p-Value ≤ 0.01, *** p-Value ≤ 0.001, ns not significant.

3.7. Genetic Analysis of SFL in the Context of Ribosome-Associated Quality Control (RQC)

RQC is initiated by disome formation when a trailing ribosome encounters a paused downstream ribosome [42,43]. SFL is known to bind tightly to hyper-rotated, frameshifted ribosomes, where it presumably causes them to pause [25,44]. This suggests that defects in SFL expression may affect the kinetics of pausing of frameshifted ribosomes, thus affecting RQC activation. Paused ribosomes direct recruitment of numerous factors including the GIGYF2-4EBP complex, ZNF598, and EDF1 [45]. GIGYF2-4EBP inhibits initiation by new ribosomes. Easily resolved pauses result in recruitment of the tRNA-eEF1A-GTP ternary complex, resulting in resumption of translation elongation. In contrast, persistent collisions stimulate eS10 and uS10 ubiquitination by ZNF598, which in turn stimulates ASCC3 recruitment to the disome [46–48]. This complex then recruits the no-go mRNA machinery which includes Pelota and Hbs1L.

The effects of SFL homeostasis disruption on RQC was tested by knocking down the expression of ZNF598, ASCC3, Hbs1L, or Pelota in HEK cells either overexpressing SFL, or in the $SFL^{-/-}$ background using the in-frame dual luciferase reporter. shRNA knockdown of any of the RQC factors inhibited expression of the upstream Firefly reporter to approximately the same extent as SFL overexpression, but not as much as in $SFL^{-/-}$ cells (Figure 5A). The combination of ZNF598 knockdown plus SFL overexpression was comparable to either of these conditions alone. In contrast, ZNF598 knockdown in $SFL^{-/-}$ cells appeared to have a synergistic effect, reducing Renilla expression to an even greater extent than either of the two conditions alone. Similar patterns were observed upon shRNA knockdown of ASCC3 (Figure 5B), and Hbs1L (Figure 5C). In contrast, synergy between Pelota and SFL overexpression was not observed (Figure 5D). Rather, all conditions inhibited Renilla expression to similar extents. To probe potential positional effects, the effects of these conditions on expression of the in-frame downstream firefly luciferase reporter were also evaluated. These analyses revealed similar results, i.e., the absence of SFL had synergistic effects in combination with shRNA knockdown of ZNF598, ASCC3, and Hbs1L, while SFL overexpression was dominant to Pelota knockdown (Figure 5).

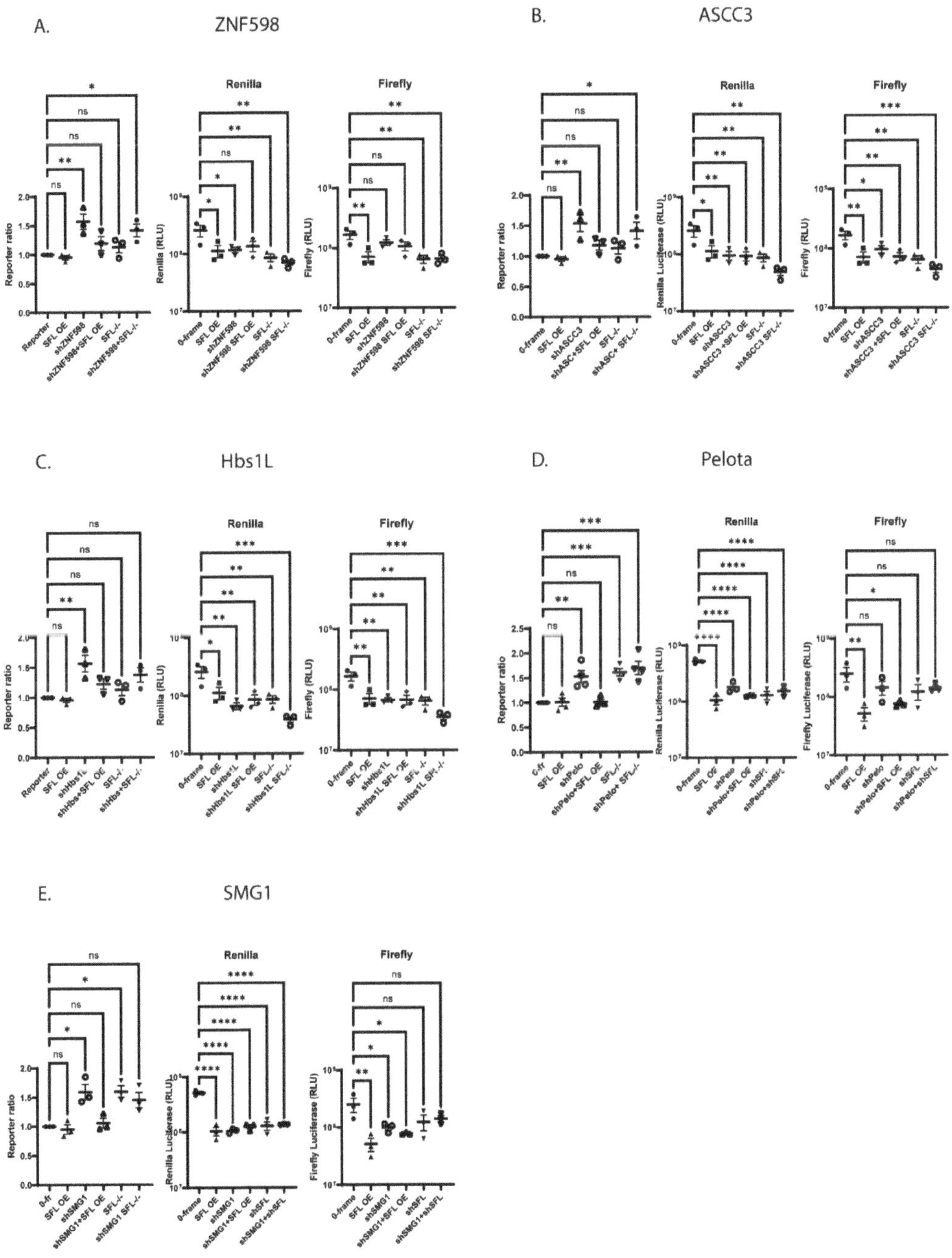

Figure 5. SFL and RQC. In-frame control reporter protein ratios and luciferase activity with shRNA knockdown of RQC proteins in WT HEK293T or SFL$^{-/-}$ HEK293T. (**A**) ZNF598, (**B**) ASCC3, (**C**) Hbs1L, (**D**) Pelota, or (**E**) SMG1 shRNA knockdown. Error bars denote SEM. * p-Value < 0.05, ** p-Value $\leq$ 0.01, *** p-Value $\leq$ 0.001, **** p-Value $\leq$ 0.0001, ns not significant.

Nonsense-mediated mRNA decay (NMD) represents a second arm of the RQC apparatus. Importantly, ribosomes that encounter a termination codon in the wrong context have been observed to slip back and forth on mRNAs [49], suggesting that they too may recruit SFL. Knockdown of the NMD factor SMG1 also inhibited expression of the upstream Renilla reporter; there was no synergy with either SFL overexpression or lack of expression (Figure 5E). In contrast, expression of the downstream firefly reporter was more strongly affected by SFL overexpression, and lack of SMG1 in combination with SFL overexpression more closely resembled SMG1 knockdown than SFL overexpression alone, suggesting that SMG1 may be epistatic to SFL.

The effects of knocking down RQC factors in the context of the HIV-1 -1 PRF were also examined. Lack of ZNF598 enhanced expression of the downstream firefly reporter relative to upstream Renilla luciferase reporter, resulting in a net increase in -1 PRF efficiency (Figure S6A). SFL overexpression resulted in a small but statistically insignificant decrease in the downstream reporter relative to the upstream one, resulting in a net decrease in -1 PRF efficiency (again, not statistically significant). The same patterns were observed upon shRNA knockdown of ASCC3, Hbs1L, Pelota, or SMG1 (Figure S6B–D).

3.8. SFL Expression Is Significantly Reduced in Many Common Cancers and Correlates with Worse Clinical Outcomes

Given observations above, changes in SFL expression are expected to globally affect both the quality and quantity of protein synthesis. Although the transient alteration of SFL expression has therapeutically valuable antiviral effects, sustained defects in SFL expression should be pathological. To search for evidence of pathological effects, the Cancer Genome Atlas (https://www.cancer.gov/tcga (accessed on 28 February 2022)) and Genome Browser [34] databases were analyzed for changes in SFL expression across a panel of the most common cancer types [50] and their corresponding normal tissues. SFL expression was found to be decreased in all cancers examined by more than 30%, with the exception of cutaneous melanoma (SKCM) (Figure 6A, Table S1). In addition, ZNF598 expression was also decreased in most cancer types except mesothelioma (MESO) (Figure 6B). Expression of other members of RQC including Pelota (Figure 6C), SMG1 (Figure 6D) did not correlate across cancer types, nor were patterns observed in the expression of reference genes (GAPDH, HNRPL, PCBP1, SNW1, and RER1) identified to have stable expression patterns in human cancer and matching normal cell types (Figure 6F–I) [51,52]. Strikingly, lower levels of SFL expression correlated with reduced survival in lung mesothelioma, bladder, and skin cancers (Figure S5).

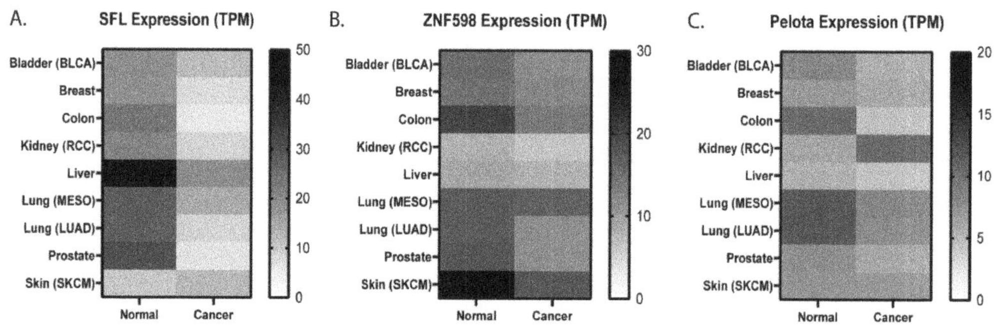

Figure 6. *Cont.*

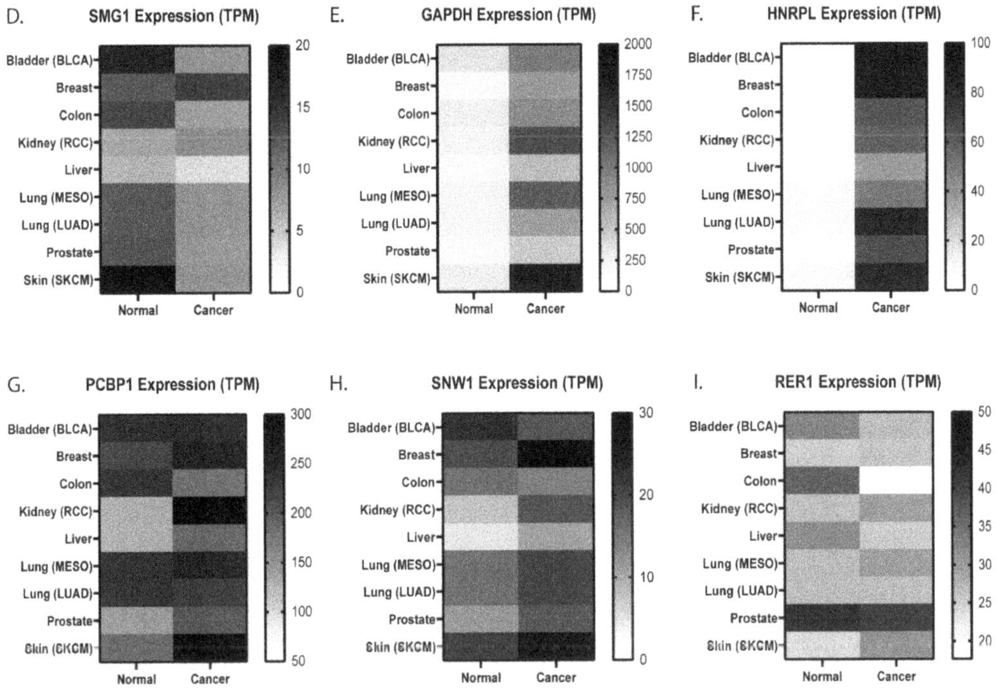

Figure 6. Gene expression pattern changes in cancer. Average TPM of RQC components or control genes in common cancers and corresponding normal tissue. (**A**) SFL, (**B**) ZNF598, (**C**) Pelota, (**D**) SMG1, (**E**) GAPDH, (**F**) HNRPL, (**G**) PCBP1, (**H**) SNW1, and (**I**) RER1. Shading intensity indicates the degree of SFL expression from high (dark) to low (light).

4. Discussion

Over 30 years ago, multiple independent measurements of spontaneous ribosomal frameshifting generated the following paradox: long mRNAs encoding very large proteins (e.g., titin, alpha 5 laminin and BRCA2), are statistically untranslatable. While researching prior literature on SFL, we were struck by data indicating that this interferon-inducible antiviral protein appeared to be constitutively expressed, albeit at low levels, across a broad range of cell lines (see Figure 1D in [28]). Combining these two observations generated the hypothesis that constitutive low level SFL expression may enable the translational apparatus to identify and remove spontaneously frameshifted ribosomes while leaving mRNAs intact to be fully translated by trailing ribosomes. Corollary to this is the hypothesis that IFN induction of high levels of SFL may have evolved as an arm of the innate immune system to identify and disable viral mRNAs, many of which are known to program ribosomes to shift reading frame by >3 orders of magnitude.

The data shown in Figures 3 and 4 are most informative: either too much or too little SFL results in decreased mRNA abundance (Figure 4), resulting in decreased protein expression (Figure 3). This held true regardless of the reporter used, or whether it encoded a synthetic bicistronic reporter or a natural monocistronic mRNA. This is a critical observation; it suggests that SFL functions to surveil cellular mRNAs for spontaneously frameshifted ribosomes. The ability of SFL to co-immunoprecipitate with Poly(A) binding protein cytoplasmic 1 (PABPC1p), La-motif related protein 1 (LARP1p), MOV10p, and UPF1p [28,53] further suggested a connection with the molecular machinery that is involved in translational surveillance in general. The reporter protein expression data shown in Figure 5 supports this, showing that disrupting SFL homeostasis in combination with knockdown of mRNAs encoding various elements of the RQC apparatus all promoted ~10-fold decreases in reporter protein activities. In this same figure, the ratios of downstream firefly to upstream Renilla luciferases generated from the in-frame reporter demonstrated that SFL overexpression tended to be dominant to shRNA knockdown of the RQC apparatus (Figure 5). In contrast, the effects of shRNA knockdown of SFL, which resulted in ~1.5-fold increases in reporter protein ratios, tended to remain the same or greater when combined shRNA knockdown of various RQC factors (Figure 5).

The synergistic effects of SFL knockdown shRNA knockdown of ZNF598, ASCC3, and Hbs1L (Figure 5) suggests SFL is recruited early in the process, stabilizing frameshifted ribosomes on mRNAs, thus enhancing disome formation, consistent with the early recruitment of these factors consequent to disome formation (reviewed in [54,55]). The dominance of SFL overexpression to Pelota knockdown also supports this view. We propose a model beginning with a frameshifted ribosome stalling on an mRNA (Figure 7A). Frameshifted ribosomes assume a hyper-rotated conformation [56], bringing the ribosomal proteins uL5 and eS31 in close enough proximity for SFL to interact with them and be recruited (see [44]). In this model, SFL recruits the release factors eRF1/3, terminating translation and initiating releasing the ribosome from the mRNA. Thus, SFL may function to save mRNAs from spontaneous frameshift events, therefore solving the problem of how long mRNAs are fully translated. Importantly, SFL concentration appears to matter. Current models suggest that the context and duration of ribosome collisions guide downstream consequences with increasing severity, ranging from rapid resolution of the problem which saves the mRNA for translation by successive ribosomes, all the way to complete translational shutdown and triggering of apoptotic pathways [57]. Accordingly, in the absence of SFL, frameshifted ribosomes are free to continue in the new reading frame where they will eventually either encounter a premature termination codon or continue translate into the 3′ UTR, in the former case activating NMD, and non-stop mRNA decay (NSD) in the latter (Figure 7B). This model is supported by the observation that the absence of SFL promoted a greater decrease on mRNA abundance when frameshifting rates are elevated over background by a -1 PRF signal (compare Figure 4A,B,D, with Figure 4C). It also follows that SFL constitutive under-expression would have globally negative impacts gene expression and cellular homeostasis, thus accounting for the link between low levels of SFL in many cancers (Figure 6) and the correlation between SFL expression and cancer survival rates (Figure S7). Conversely, overexpression (i.e., high concentrations) of SFL (Figure 7C) would stall frameshifted ribosomes for longer periods of time on the mRNA, providing more time for disome formation and recruitment of RQC factors, resulting in much larger decreases in mRNA abundance and gene expression. This is reflected by the data shown in Figures 4 and 5. We suggest that this is the basis for the antiviral activity caused by SFL overexpression consequent to IFN induction (reviewed in [27]). By this model, IFN induction of SFL expression would represent an early step in the integrated stress response (reviewed in [58]) (Figure 7D). Elucidating when SFL is recruited during the RQC process, and the role played by ribosome collisions awaits deeper biophysical analyses.

Interestingly, SFL knockdown or knockout resulted in an ~2-fold increase in PRF (Figures 1 and 2). The same observation was made when the space/time between elongating ribosomes was increased by decreasing translation initiation rates [59]. This two-fold effect was observed, regardless of the extent of initiation deficiency, and a model was suggested involving collisions between ribosomes stalled at a -1 PRF signal and the trailing ribosome [59]. Since then, the importance of ribosome collisions and RQC have been discovered (reviewed in [54,55,60,61]). We suggest that when ribosomes collide at a PRF signal, the leading one can either shift or not shift (the frequency of which is determined by the particular PRF signal), but the trailing ribosome cannot shift because the PRF stimulating mRNA structural element has been denatured by the transit of the leading ribosome and does not have enough time to reform before the sequence is covered by the lagging ribosome. In other words, only half of the ribosomes that would normally encounter a PRF signal can actually shift. This would explain the two-fold increase in PRF efficiency by either increasing the time between successive ribosomes or by the absence of SFL. Conversely, overexpression of SFL resulted in a two-fold decrease in PRF. How this works is complicated by the dynamics of mRNA decay, but points to the efficient clearance of frameshifted ribosomes. Illuminating the precise molecular mechanisms underlying how altering SFL expression results in the apparent two-fold changes in translational recoding promises to reveal novel insights into the biophysical interactions that occur between this class of cis-acting RNA elements and elongating ribosomes within the context of RQC.

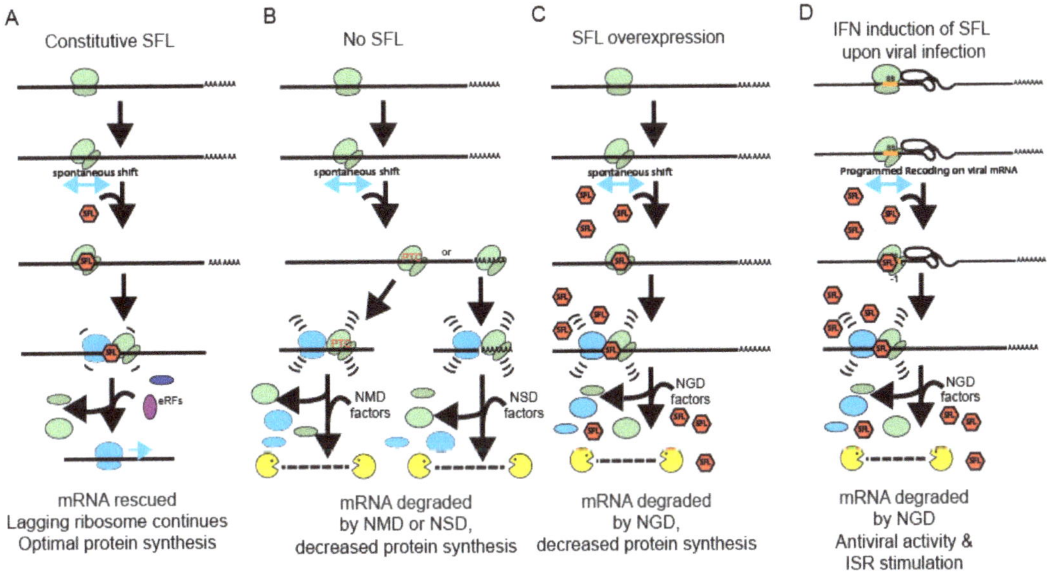

Figure 7. Model. (**A**) In constitutive low level SFL expression, SFL recognizes spontaneously frameshifted ribosomes, and locks them onto the mRNA. Disomes do not persist for long, as SFL recruits release factors and the frameshifted (leading) ribosome is removed from the mRNA. The lagging ribosome is free to continue translating the mRNA. (**B**) In the absence of SFL, frameshifted ribosomes are free to continue translating in the −1 or +1 frame. In most cases, they will encounter a PTC, activating mRNA degradation by NMD. In rare cases, they will continue translating past the in-frame termination codon and into the polyA tail, where they will activate NSD. (**C**) At high concentrations, SFL saturates the frameshifted ribosome, generating a long disome pause, activating NGD and decreasing protein expression. (**D**) Viral infection results in IFN induction of SFL expression, generating high concentrations of SFL. The resulting long pauses at -1 PRF signals results in disome stabilization, activating NGD and the ISR.

We wish to address two final issues raised by the data presented here. First, the data presented in Figure 2C show that the rate of spontaneous ribosomal frameshifting is ~0.7×10^{-3}, significantly larger than the 2×10^{-4} described in the prior literature. We suggest two possible explanations. One is that this is an inherent feature of bicistronic reporters. However, the activities of both the upstream and downstream reporter proteins in the spontaneous frameshift reporter are consistently an order of magnitude lower than those of the in-frame reporter, consistent with the former being a substrate for NMD (compare Figure 3B,C). Indeed, dual luciferase bicistronic reporters were first introduced into the translational recoding field to replace monocistronic reporters to control for apparent changes in recoding rates consequent to defects in NMD. This was because recoding reporters, which by their very nature are nonsense-containing mRNAs, are compared to in-frame control reporters (which are not NMD substrates) to generate recoding efficiency. Because of this discrepancy, the use of monocistronic reporters to measure recoding rates resulted in the mistaken conclusion that NMD factors influenced -1 PRF rates [62,63]. This raises the possibility that the 2×10^{-4} frequency of spontaneous frameshifting as originally measured using monocistronic reporters represents an underestimation of the frequency of these events. If substantiated, it would further emphasize the importance of SFL in surveilling mRNAs for spontaneous frameshifts. The second issue regards the data presented in Figure 1C and Figure S5C showing that sequence derived from the human CCR5 mRNA promoted ~2% -1 PRF. A previous publication disputed the ability of this sequence to promote -1 PRF due to artifactual splicing issues with the first generation of dual luciferase reporters [36]. In the current work, the CCR5 sequence was cloned into the second generation dual luciferase reporter used by those authors (pSGDLuc) [20,36], the only difference being the insertion of *Bam* HI and *Sal* I restriction sites into the multiple cloning site of the vector (Figure S5C). We also note that similar results were observed using the bifluorescent reporters (Figure 1C). RT PCR analyses suggest reporter splicing does indeed occur in the first-generation reporters but is not exclusive to the CCR5 sequence and can be seen in the 0-frame control reporter as well (Figure S3). The second-generation dual luciferase and bifluorescent reporters, however, lack such splicing artifacts regardless of frameshift sequence. Additionally, we did not observe the "reporter crash" seen in the prior study, consistent with absence of splicing artifacts in these reporter systems. Thus, at least in our hands, the CCR5-derived sequence appears to function as an efficient -1 PRF signal. The raw and collated data from these experiments are presented as Excel spreadsheets for the convenience of the reader (Supplemental Datasets S1 and S2). We leave it to the scientific community to resolve this matter.

Supplementary Materials: The following supporting information can be downloaded at: https://www.mdpi.com/article/10.3390/v15122296/s1, Figure S1: Probability plot of the likelihood of a spontaneous ribosomal frameshift as a function of codon length. Figure S2: Modifying SFL expression in the cell. Figure S3: Assessment of reporter splicing. Figure S4: Comparison of translational recoding measured with dual luciferase and bifluorescent reporters. Figure S5: SFL overexpression or knockdown alters -1 PRF. Figure S6: Frameshift efficiency of the HIV-1 -1 PRF signal in HEK293T cells over or under-expressing SFL in combination with shRNA knockdown of (A) ZNF598, (B) ASCC3, (C) Hbs1L, (D) Pelota, & SMG1. Figure S7: Lower SFL expression decreases cancer patient survival. Supplemental Data Set S1: Shiftless reporter data-all. Supplemental Data Set S2: Shiftless figures data.

Author Contributions: J.A.K. and J.D.D. conceptualization; J.A.K. and J.D.D. data curation; J.A.K. formal analysis; J.A.K. and J.D.D. validation; J.A.K. and J.D.D. investigation; J.A.K. and J.D.D. methodology; J.A.K. and J.D.D. writing-review and editing; J.D.D. resources; J.D.D. funding acquisition; J.D.D. supervision; J.A.K. writing original draft; J.D.D. project administration. All authors have read and agreed to the published version of the manuscript.

Funding: This work was supported by Defense Threat Reduction Agency Grant HDTRA1-13-1-0005; NIGMS, National Institutes of Health Grant R01 GM117177; and a University of Maryland Coronavirus Research Program Seed Grant to J.D.D. J.A.K. was supported by NIH Institutional Training Grant 2T32AI051967-06A1.

Institutional Review Board Statement: Not applicable.

Informed Consent Statement: Not applicable.

Data Availability Statement: Data are available as Supplemental Data Sets S1 and S2.

Acknowledgments: We would like to acknowledge Jonathan Jacobs with assistance with Figure S1, and Joseph Briggs for developing and testing the bifluorescent reporters. J.D.D. would like to thank the Translational Control community for providing him with an intellectually stimulating, exciting and mostly friendly collegial environment over the past thirty-plus years.

Conflicts of Interest: The authors declare no conflict of interest.

References

1. Warner, J.R. The Economics of Ribosome Biosynthesis in Yeast. *Trends Biochem.* **1999**, *24*, 437–440. [CrossRef]
2. Rudra, D.; Warner, J.R. What Better Measure than Ribosome Synthesis? *Genes Dev.* **2004**, *18*, 2431–2436. [CrossRef]
3. Calamita, P.; Gatti, G.; Miluzio, A.; Scagliola, A.; Biffo, S. Translating the Game: Ribosomes as Active Players. *Front. Genet.* **2018**, *9*, 418681. [CrossRef]
4. Kurland, C.G. Translational Accuracy In Vitro. *Cell* **1982**, *28*, 201–202. [CrossRef]
5. Belcourt, M.F.; Farabaugh, P.J. Ribosomal Frameshifting in the Yeast Retrotransposon Ty: TRNAs Induce Slippage on a 7 Nucleotide Minimal Site. *Cell* **1990**, *62*, 339–352. [CrossRef]
6. Dinman, J.D.; Icho, T.; Wickner, R.B. A -1 Ribosomal Frameshift in a Double-Stranded RNA Virus Forms a Gag-Pol Fusion Protein. *Proc. Natl. Acad. Sci. USA* **1991**, *88*, 174–178. [CrossRef]
7. Parker, J. Errors and Alternatives in Reading the Universal Genetic-Code. *Microbiol. Rev.* **1989**, *53*, 273–298. [CrossRef]
8. Kurland, C.G. Translational Accuracy and the Fitness of Bacteria. *Annu. Rev. Genet.* **1992**, *26*, 29–50. [CrossRef]
9. Gu, J.; Yan, H.; Huang, Y.; Zhu, Y.; Sun, H.; Qiu, Y.; Chen, S. Heightened Protein-Translation Activities in Mammalian Cells and the Disease/Treatment Implications. *Natl. Sci. Rev.* **2020**, *7*, 1851–1855. [CrossRef]
10. Bang, M.L.; Centner, T.; Fornoff, F.; Geach, A.J.; Gotthardt, M.; McNabb, M.; Witt, C.C.; Labeit, D.; Gregorio, C.C.; Granzier, H.; et al. The Complete Gene Sequence of Titin, Expression of an Unusual ≈700-KDa Titin Isoform, and Its Interaction with Obscurin Identify a Novel Z-Line to I-Band Linking System. *Circ. Res.* **2001**, *89*, 1065–1072. [CrossRef]
11. Chang, Y.F.; Imam, J.S.; Wilkinson, M.F. The Nonsense-Mediated Decay RNA Surveillance Pathway. *Annu. Rev. Biochem.* **2007**, *76*, 51–74. [CrossRef]
12. Dever, T.E.; Dinman, J.D.; Green, R. Translation Elongation and Recoding in Eukaryotes. *Cold Spring Harb. Perspect. Biol.* **2018**, *10*, a032649. [CrossRef] [PubMed]
13. Hill, C.H.; Brierley, I. Structural and Functional Insights into Viral Programmed Ribosomal Frameshifting. *Annu. Rev. Virol.* **2023**, *10*, 217–242. [CrossRef]
14. Atkins, J.F.; Loughran, G.; Bhatt, P.R.; Firth, A.E.; Baranov, P.V. Ribosomal Frameshifting and Transcriptional Slippage: From Genetic Steganography and Cryptography to Adventitious Use. *Nucleic Acids Res.* **2016**, *44*, 7007–7078. [CrossRef]
15. Penn, W.D.; Mukhopadhyay, S. Abracadabra, One Becomes Two: The Importance of Context in Viral -1 Programmed Ribosomal Frameshifting. *MBio* **2022**, *13*, e02468-21. [CrossRef] [PubMed]
16. Dinman, J.D.; Wickner, R.B. Ribosomal Frameshifting Efficiency and Gag/Gag-Pol Ratio Are Critical for Yeast M1 Double-Stranded RNA Virus Propagation. *J. Virol.* **1992**, *66*, 3669–3676. [CrossRef] [PubMed]
17. Dinman, J.D. Programmed Ribosomal Frameshifting Goes Beyond Viruses: Organisms from All Three Kingdoms Use Frameshifting to Regulate Gene Expression, Perhaps Signaling a Paradigm Shift. *Microbe* **2006**, *1*, 521–527. [PubMed]
18. Belew, A.T.; Meskauskas, A.; Musalgaonkar, S.; Advani, V.M.; Sulima, S.O.; Kasprzak, W.; Shapiro, B.A.; Dinman, J.D. Ribosomal Frameshifting in the CCR5 MRNA Is Regulated by MiRNAs and the NMD Pathway. *Nature* **2014**, *512*, 265–269. [CrossRef]
19. Riegger, R.J.; Caliskan, N. Thinking Outside the Frame: Impacting Genomes Capacity by Programmed Ribosomal Frameshifting. *Front. Mol. Biosci.* **2022**, *9*, 842261. [CrossRef] [PubMed]
20. Loughran, G.; Howard, M.T.; Firth, A.E.; Atkins, J.F. Avoidance of Reporter Assay Distortions from Fused Dual Reporters. *RNA* **2017**, *23*, 1285–1289. [CrossRef]
21. Dinman, J.D. Translational Recoding Signals: Expanding the Synthetic Biology Toolbox. *J. Biol. Chem.* **2019**, *294*, 7537–7545. [CrossRef]
22. Kelly, J.; Woodside, M.; Dinman, J. Programmed -1 Ribosomal Frameshifting in Coronaviruses: A Therapeutic Target. *Virology* **2021**, *554*, 75–82. [CrossRef]
23. Dinman, J.D.; Ruiz-Echevarria, M.J.; Peltz, S.W. Translating Old Drugs into New Treatments: Ribosomal Frameshifting as a Target for Antiviral Agents. *Trends Biotechnol.* **1998**, *16*, 190–196. [CrossRef] [PubMed]
24. Hung, M.; Patel, P.; Davis, S.; Green, S.R. Importance of Ribosomal Frameshifting for Human Immumodeficiency Virus Type 1 Assembly and Replication. *J. Virol.* **1998**, *72*, 4819–4824. [CrossRef] [PubMed]
25. Wang, X.; Xuan, Y.; Han, Y.; Ding, X.; Ye, K.; Yang, F.; Gao, P.; Goff, S.P.; Gao, G. Regulation of HIV-1 Gag-Pol Expression by Shiftless, an Inhibitor of Programmed-1 Ribosomal Frameshifting. *Cell* **2019**, *176*, 625–635.e14. [CrossRef] [PubMed]
26. Hanners, N.W.; Mar, K.B.; Boys, I.N.; Eitson, J.L.; Cruz-Rivera, P.C.D.L.; Richardson, R.B.; Fan, W.; Wight-Carter, M.; Schoggins, J.W. Shiftless Inhibits Flavivirus Replication In Vitro and Is Neuroprotective in a Mouse Model of Zika Virus Pathogenesis. *Proc. Natl. Acad. Sci. USA* **2021**, *118*, e2111266118. [CrossRef]
27. Rodriguez, W.; Muller, M. Shiftless, a Critical Piece of the Innate Immune Response to Viral Infection. *Viruses* **2022**, *14*, 1338. [CrossRef]
28. Balinsky, C.A.; Schmeisser, H.; Wells, A.I.; Ganesan, S.; Jin, T.; Singh, K.; Zoon, K.C. IRAV (FLJ11286), an Interferon-Stimulated Gene with Antiviral Activity against Dengue Virus, Interacts with MOV10. *J. Virol.* **2017**, *91*, e01606-16. [CrossRef]
29. Snieckute, G.; Genzor, A.V.; Vind, A.C.; Ryder, L.; Stoneley, M.; Chamois, S.; Dreos, R.; Nordgaard, C.; Sass, F.; Blasius, M.; et al. Ribosome Stalling Is a Signal for Metabolic Regulation by the Ribotoxic Stress Response. *Cell Metab.* **2022**, *34*, 2036–2046.e8. [CrossRef]
30. Ran, F.A.; Hsu, P.D.; Wright, J.; Agarwala, V.; Scott, D.A.; Zhang, F. Genome Engineering Using the CRISPR-Cas9 System. *Nat. Protoc.* **2013**, *8*, 2281–2308. [CrossRef]

31. Harger, J.W.; Dinman, J.D. An In Vivo Dual-Luciferase Assay System for Studying Translational Recoding in the Yeast Saccharomyces Cerevisiae. *RNA* **2003**, *9*, 1019–1024. [CrossRef]
32. Jacobs, J.L.; Dinman, J.D. Systematic Analysis of Bicistronic Reporter Assay Data. *Nucleic Acids Res.* **2004**, *32*, e160. [CrossRef] [PubMed]
33. Munshi, S.; Neupane, K.; Ileperuma, S.M.; Halma, M.T.J.; Kelly, J.A.; Halpern, C.F.; Dinman, J.D.; Loerch, S.; Woodside, M.T. Identifying Inhibitors of −1 Programmed Ribosomal Frameshifting in a Broad Spectrum of Coronaviruses. *Viruses* **2022**, *14*, 177. [CrossRef]
34. Lee, B.T.; Barber, G.P.; Benet-Pagès, A.; Casper, J.; Clawson, H.; Dlekhans, M.; Fischer, C.; Gonzalez, J.N.; Hinrichs, A.S.; Lee, C.M.; et al. The UCSC Genome Browser Database: 2022 Update. *Nucleic Acids Res.* **2022**, *50*, D1115–D1122. [CrossRef]
35. Chandrashekar, D.S.; Bashel, B.; Balasubramanya, S.A.H.; Creighton, C.J.; Ponce-Rodriguez, I.; Chakravarthi, B.V.S.K.; Varambally, S. UALCAN: A Portal for Facilitating Tumor Subgroup Gene Expression and Survival Analyses. *Neoplasia* **2017**, *19*, 649–658. [CrossRef] [PubMed]
36. Khan, Y.A.; Loughran, G.; Steckelberg, A.-L.; Brown, K.; Baranov, P.V.; Kieft, J.S.; Firth, A.E.; Atkins, J.F. Evaluating the Evidence for −1 Frameshifting in Immune-Functioning C-C Chemokine Receptor (CCR5)—The HIV1 Co-Receptor. *Nature* **2019**, *604*, E16–E23. [CrossRef] [PubMed]
37. Napthine, S.; Hill, C.H.; Nugent, H.C.M.; Brierley, I. Modulation of Viral Programmed Ribosomal Frameshifting and Stop Codon Readthrough by the Host Restriction Factor Shiftless. *Viruses* **2021**, *13*, 1230. [CrossRef] [PubMed]
38. Matsufuji, S.; Matsufuji, T.; Miyazaki, Y.; Murakami, Y.; Atkins, J.F.; Gesteland, R.F.; Hayashi, S. Autoregulatory Frameshifting in Decoding Mammalian Ornithine Decarboxylase Antizyme. *Cell* **1995**, *80*, 51–60. [CrossRef]
39. Firth, A.E.; Wills, N.M.; Gesteland, R.F.; Atkins, J.F. Stimulation of Stop Codon Readthrough: Frequent Presence of an Extended 3′ RNA Structural Element. *Nucleic Acids Res.* **2011**, *39*, 6679–6691. [CrossRef] [PubMed]
40. Strauss, E.G.; Rice, C.M.; Strauss, J.H. Sequence Coding for the Alphavirus Nonstructural Proteins Is Interrupted by an Opal Termination Codon. *Proc. Natl. Acad. Sci. USA* **1983**, *80*, 5271–5275. [CrossRef]
41. Kendra, J.A.; de la Fuente, C.; Brahms, A.; Woodson, C.; Bell, T.M.; Chen, B.; Khan, Y.A.; Jacobs, J.L.; Kehn-Hall, K.; Dinman, J.D. Ablation of Programmed −1 Ribosomal Frameshifting in Venezuelan Equine Encephalitis Virus Results in Attenuated Neuropathogenicity. *J. Virol.* **2017**, *91*, e01766-16. [CrossRef] [PubMed]
42. Wu, C.C.-C.; Peterson, A.; Zinshteyn, B.; Regot, S.; Green, R. Ribosome Collisions Trigger General Stress Responses to Regulate Cell Fate. *Cell* **2020**, *182*, 404–416.e14. [CrossRef]
43. Meydan, S.; Guydosh, N.R. Disome and Trisome Profiling Reveal Genome-Wide Targets of Ribosome Quality Control. *Mol. Cell* **2020**, *79*, 588–602.e6. [CrossRef] [PubMed]
44. Dinman, J.D. Scaring Ribosomes Shiftless. *Biochemistry* **2019**, *58*, 1831–1832. [CrossRef]
45. Hickey, K.L.; Dickson, K.; Cogan, J.Z.; Replogle, J.M.; Schoof, M.; D'Orazio, K.N.; Sinha, N.K.; Hussmann, J.A.; Jost, M.; Frost, A.; et al. GIGYF2 and 4EHP Inhibit Translation Initiation of Defective Messenger RNAs to Assist Ribosome-Associated Quality Control. *Mol. Cell* **2020**, *79*, 950–962.e6. [CrossRef] [PubMed]
46. Matsuo, Y.; Ikeuchi, K.; Saeki, Y.; Iwasaki, S.; Schmidt, C.; Udagawa, T.; Sato, F.; Tsuchiya, H.; Becker, T.; Tanaka, K.; et al. Ubiquitination of Stalled Ribosome Triggers Ribosome-Associated Quality Control. *Nat. Commun.* **2017**, *8*, 159. [CrossRef]
47. Garzia, A.; Jafarnejad, S.M.; Meyer, C.; Chapat, C.; Gogakos, T.; Morozov, P.; Amiri, M.; Shapiro, M.; Molina, H.; Tuschl, T.; et al. The E3 Ubiquitin Ligase and RNA-Binding Protein ZNF598 Orchestrates Ribosome Quality Control of Premature Polyadenylated MRNAs. *Nat. Commun.* **2017**, *8*, 16056. [CrossRef]
48. D'Orazio, K.N.; Green, R. Ribosome States Signal RNA Quality Control. *Mol. Cell* **2021**, *81*, 1372–1383. [CrossRef]
49. Amrani, N.; Ganesan, R.; Kervestin, S.; Mangus, D.A.; Ghosh, S.; Jacobson, A. A Faux 3′-UTR Promotes Aberrant Termination and Triggers Nonsense-Mediated MRNA Decay. *Nature* **2004**, *432*, 112–118. [CrossRef]
50. Howlader, N.; Noone, A.; Krapcho, M.; Miller, D.; Brest, A.; Yu, M.; Ruhl, J.; Tatalovich, Z.; Mariotto, A.; Lewis, D.R.; et al. *SEER Cancer Statistics Review, 1975–2017*; National Cancer Institute: Bethesda, MD, USA, 2020. Available online: https://seer.cancer.gov/statistics/interactive.html (accessed on 28 February 2022).
51. Rácz, G.A.; Nagy, N.; Tóvári, J.; Apáti, Á.; Vértessy, B.G. Identification of New Reference Genes with Stable Expression Patterns for Gene Expression Studies Using Human Cancer and Normal Cell Lines. *Sci. Rep.* **2021**, *11*, 19459. [CrossRef]
52. Jo, J.; Choi, S.; Oh, J.; Lee, S.G.; Choi, S.Y.; Kim, K.K.; Park, C. Conventionally Used Reference Genes Are Not Outstanding for Normalization of Gene Expression in Human Cancer Research. *BMC Bioinform.* **2019**, *20*, 13–21. [CrossRef] [PubMed]
53. Suzuki, Y.; Chin, W.X.; Han, Q.; Ichiyama, K.; Lee, C.H.; Eyo, Z.W.; Ebina, H.; Takahashi, H.; Takahashi, C.; Tan, B.H.; et al. Characterization of RyDEN (C19orf66) as an Interferon-Stimulated Cellular Inhibitor against Dengue Virus Replication. *PLoS Pathog.* **2016**, *12*, e1005357. [CrossRef]
54. Parker, M.D.; Karbstein, K. Quality Control Ensures Fidelity in Ribosome Assembly and Cellular Health. *J. Cell Biol.* **2023**, *222*, e202209115. [CrossRef] [PubMed]
55. Kim, K.Q.; Zaher, H.S. Canary in a Coal Mine: Collided Ribosomes as Sensors of Cellular Conditions. *Trends Biochem. Sci.* **2022**, *47*, 82–97. [CrossRef] [PubMed]
56. Chen, J.; Petrov, A.; Tsai, A.; O'Leary, S.E.; Puglisi, J.D. Coordinated Conformational and Compositional Dynamics Drive Ribosome Translocation. *Nat. Struct. Mol. Biol.* **2013**, *20*, 718–727. [CrossRef]
57. Olson, A.N.; Dinman, J.D. Two Ribosomes Are Better than One. Sometimes. *Mol. Cell* **2020**, *79*, 541–543. [CrossRef]

58. Costa-Mattioli, M.; Walter, P. The Integrated Stress Response: From Mechanism to Disease. *Science* **2020**, *368*, eaat5314. [CrossRef]
59. Lopinski, J.D.; Dinman, J.D.; Bruenn, J.A. Kinetics of Ribosomal Pausing during Programmed -1 Translational Frameshifting. *Mol. Cell Biol.* **2000**, *20*, 1095–1103. [CrossRef]
60. Schuller, A.P.; Green, R. Roadblocks and Resolutions in Eukaryotic Translation. *Nat. Rev. Mol. Cell Biol.* **2018**, *19*, 526–541. [CrossRef] [PubMed]
61. De, S.; Mühlemann, O. A Comprehensive Coverage Insurance for Cells: Revealing Links between Ribosome Collisions, Stress Responses and MRNA Surveillance. *RNA Biol.* **2022**, *19*, 609–621. [CrossRef]
62. Grentzmann, G.; Ingram, J.A.; Kelly, P.J.; Gesteland, R.F.; Atkins, J.F. A Dual-Luciferase Reporter System for Studying Recoding Signals. *RNA* **1998**, *4*, 479–486. [PubMed]
63. Harger, J.W.; Dinman, J.D. Evidence against a Direct Role for the Upf Proteins in Frameshfiting or Nonsense Codon Readthrough. *RNA* **2004**, *10*, 1721–1729. [CrossRef] [PubMed]

viruses

Review

Host-like RNA Elements Regulate Virus Translation

Debjit Khan * and Paul L. Fox *

Department of Cardiovascular and Metabolic Sciences, Lerner Research Institute, Cleveland Clinic, Cleveland, OH 44195, USA
* Correspondence: debjit.khan@gmail.com (D.K.); foxp@ccf.org (P.L.F.); Tel.: +1-216-444-8053 (P.L.F.)

Abstract: Viruses are obligate, intracellular parasites that co-opt host cell machineries for propagation. Critical among these machineries are those that translate RNA into protein and their mechanisms of control. Most regulatory mechanisms effectuate their activity by targeting sequence or structural features at the RNA termini, i.e., at the 5′ or 3′ ends, including the untranslated regions (UTRs). Translation of most eukaryotic mRNAs is initiated by 5′ cap-dependent scanning. In contrast, many viruses initiate translation at internal RNA regions at internal ribosome entry sites (IRESs). Eukaryotic mRNAs often contain upstream open reading frames (uORFs) that permit condition-dependent control of downstream major ORFs. To offset genome compression and increase coding capacity, some viruses take advantage of out-of-frame overlapping uORFs (oORFs). Lacking the essential machinery of protein synthesis, for example, ribosomes and other translation factors, all viruses utilize the host apparatus to generate virus protein. In addition, some viruses exhibit RNA elements that bind host regulatory factors that are not essential components of the translation machinery. SARS-CoV-2 is a paradigm example of a virus taking advantage of multiple features of eukaryotic host translation control: the virus mimics the established human GAIT regulatory element and co-opts four host aminoacyl tRNA synthetases to form a stimulatory binding complex. Utilizing discontinuous transcription, the elements are present and identical in all SARS-CoV-2 subgenomic RNAs (and the genomic RNA). Thus, the virus exhibits a post-transcriptional regulon that improves upon analogous eukaryotic regulons, in which a family of functionally related mRNA targets contain elements that are structurally similar but lacking sequence identity. This "thrifty" virus strategy can be exploited against the virus since targeting the element can suppress the expression of all subgenomic RNAs as well as the genomic RNA. Other 3′ end viral elements include 3′-cap-independent translation elements (3′-CITEs) and 3′-tRNA-like structures. Elucidation of virus translation control elements, their binding proteins, and their mechanisms can lead to novel therapeutic approaches to reduce virus replication and pathogenicity.

Keywords: internal ribosome entry sites; RNA element; SARS-CoV-2; translation control; untranslated region; upstream open reading frames; virus; virus-host interaction

1. Introduction

The flow of genetic information in biological systems is tightly regulated at all levels, that is, at replication, transcription, and translation. Post-transcriptional gene regulation allows rapid biological responses when challenged by cellular signals and environmental stimuli. A principal component of post-transcriptional gene regulation is translational control [1–3]. mRNA translation occurs in four stages: initiation, elongation, termination, and ribosome recycling. Translation initiation is generally the rate-limiting step in protein synthesis and is exquisitely regulated by a variety of mechanisms [1,2,4]. In eukaryotes, translation of most mRNAs is initiated via cap-dependent scanning [5,6]. Viruses are obligate, intracellular parasites that co-opt host cell machineries for propagation. Eukaryotic viruses subvert host translation machinery by employing diverse and elaborate RNA-centered strategies, often mimicking host features [7]. Utilizing such features that draw upon shared resources, viral RNAs can compete with, and even outpace, host mRNAs for selfish benefits. In this review, we present the RNA elements (Figure 1) and their protein partners that are key to shared virus and host translational control mechanisms underlying the pathogenesis of viral diseases and are targetable with a synthesis of understanding of the molecular interactions.

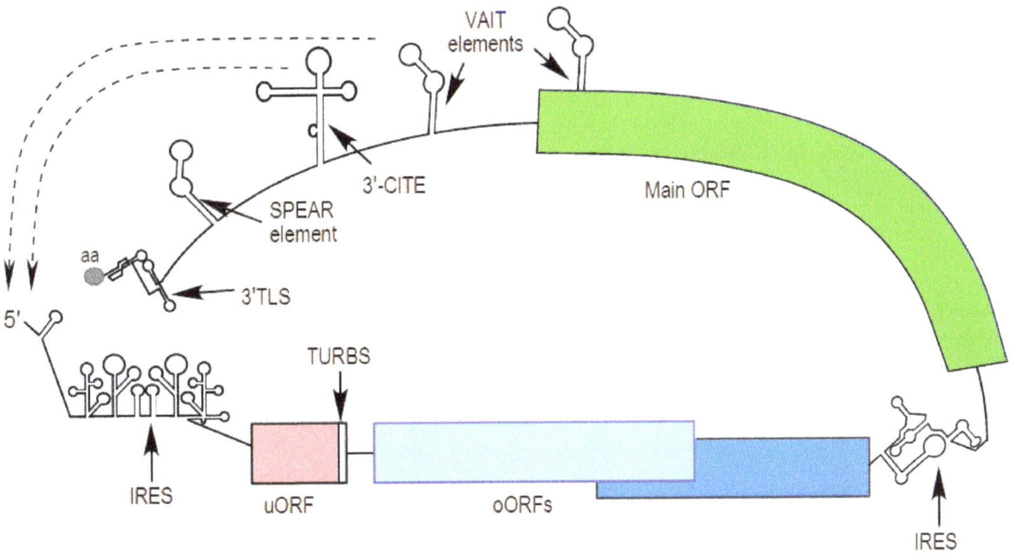

Figure 1. Translation in viruses is regulated by themes shared with hosts. Various RNA elements shown here are IRES (Internal Ribosome Entry Site), uORF (upstream Open Reading Frame), oORF (overlapping Open Reading Frame), TURBS (Termination Upstream Ribosome Binding Site), VAIT (Virus Activated Inhibitor of Translation element), 3′-CITE (3′-Cap Independent Translation Enhancer), SPEAR (Sarbecoviral Pan-End Activating RNA element), and 3′TLS (3′ tRNA-Like Structure) that can be charged with amino acid (aa). IRESs and 3′-CITEs are diverse, as discussed in the text. 3′-CITEs and VAIT activate and inhibit virus translation in a 5′-end dependent manner, as shown with dashed lines.

2. Virus 5′ End Elements

An archetype of eukaryotic translation regulation was initially described about 50 years ago in mouse liver extracts, where ferritin mRNA was induced by iron [8]. In the absence of iron, iron responsive elements (IRE) containing 25- to 30-nt stem-loop secondary structures in the ferritin and eALAS mRNA 5′-untranslated regions (UTRs) were shown to recruit iron regulatory protein 1 (IRP1) and block scanning of the 43S pre-initiation complex, and thus translation [9–11]. In the presence of iron, the IRP1 does not bind the IRE, thus permitting scanning and translation and representing a paradigm of 5′-end-directed translation regulation.

2.1. Internal Ribosome Entry Sites

Parallel to the landmark studies on the eukaryote IRE, pioneering studies were reported on viral translation regulation through internal ribosome entry sites (IRESs) (Figure 1). Increasing secondary structure complexity and content in 5′ UTRs of eukaryotic mRNAs, as well as viral RNAs, generally decreases translation efficiency [12]. In contrast, IRESs are RNA elements, defined by structure, sequence, or both, that recruit ribosomes at internal regions of mRNAs to initiate translation. IRESs were first discovered in picornaviruses such as poliovirus (PV) and encephalomyocarditis virus (EMCV) [13,14]. Soon afterwards, human immunodeficiency virus (HIV), hepatitis C virus (HCV), and foot and mouth disease virus (FMDV) IRESs were reported [15–17], and the inventory of viruses utilizing IRESs continues to expand [18]. Although the viral IRESs contain a diversity of sequences, many have similar secondary structures and initiate translation through similar mechanisms. Importantly, viruses and host cells often share translation control mechanisms, and cellular IRESs were soon reported in mRNAs encoding proteins required in physiological and pathological stress responses, e.g., apoptosis, hypoxia, and starvation [19–24]. Generally, cellular IRESs are not well conserved in either sequence or structure and are difficult to predict from sequence ab initio, and require rigorous experimental verification [25–29]. Notwithstanding, high-throughput methods such as mRNA display have revealed the genome-wide presence of cap-independent translation-enhancing elements [30], and nearly 10% of mammalian mRNAs contain RNA elements that potentially function as IRESs [31]. The scope and challenges in cellular IRES annotation and validation have been described in detail [32]. Mis-annotation of 5′UTRs, cryptic promoter activity, presence of 5′ introns and alternate splice sites, confusion of cap-independence with 5′end-independence, and unnatural experimental platforms have led to contested interpretation of cellular IRESs [33,34], and an essential workflow for characterizing cellular IRESs has been postulated [35]. Many viral IRESs function completely independent of the 5′-terminal cap structure, while cap-dependence of some viral IRESs and many cellular IRESs is more nuanced and context-dependent [32,36–41]. In addition to common translation initiation factors (eIFs), IRESs often recruit IRES-specific RNA-binding proteins, i.e., IRES trans-acting factors (ITAFs), for translation initiation (reviewed in [20,42–44]).

Many viruses, including picornaviruses, deploy a variety of strategies directing global shutdown of host cap-dependent translation while the IRESs promote translation of viral RNAs, thereby conferring a competitive advantage to the virus [7,45,46]. For example, PV infection cleaves eIF4G, preventing end-to-end circularization and translation of host mRNAs, while the IRES utilizes the cleaved carboxy-terminal fragment of eIF4G that binds eIF3 for translation initiation. In PV and EMCV infections, eIF4E is sequestered by hypophosphorylated 4E-BP, reducing cellular cap-dependent translation, without affecting IRES-dependent viral translation [47]. In both cases, viral infection essentially frees up the cellular translation machinery for self-serving protein synthesis.

2.2. Class 1 and 2 IRESs

Viral IRESs can be grouped into four classes by structure and the requirement of eIFs and ITAFs. In general, Class 1 and 2 IRESs do not bind the 40S subunit directly and require all canonical translation initiation factors, except the cap-binding protein eIF4E [48,49]. They generally present as complex secondary structures comprised of multiple domains, each containing stem-loops, bulges, and junctions. These structures are recognized by ITAFs, facilitating eIF binding and 43S pre-initiation complex recruitment. Long-range RNA–RNA interactions are also involved in the function of these IRESs [50,51]. Despite their similarity in length and multi-domain RNA structure, Class 1 and Class 2 IRESs are not identical in their secondary structures, ITAF requirements, or mechanism of initiation (reviewed in [39,48]); however, ribosomal subunit scanning after internal recruitment of the pre-initiation complex may not be considered a distinguishing factor between the two classes. For example, initiation on Class 1 IRESs can occur at the IRES-adjacent upstream codon as well as downstream at the polyprotein initiation codon [52–54]. Similarly, initiation on Class 2 IRESs can occur at the IRES-adjacent codon but also occurs downstream in all FMDV isolates [55,56], and in Theiler's murine encephalomyelitis virus (TMEV), downstream initiation yields the L* protein by translation of an overlapping ORF [57,58]. Class 1 and 2 IRESs are partially or completely refractory to translation inhibition by eIF2α phosphorylation [59,60], suggesting they do not require canonical Met-tRNAi^{Met} delivery by the ternary complex. In general, eIF2-less initiation can be facilitated by eIF2A [61], eIF2D (Ligatin) [62], eIF5B [63], or by the concerted action of MCT-1 and DENR [64]. Multiple ITAFs have been identified for the Class 1 and 2 IRESs; the specific ITAF required is variable [42,43,46]. The Class 2 EMCV IRES binds the ITAF polypyrimidine tract-binding protein (PTB), thus altering RNA conformation [65,66]. The Class 1 PV IRES binds both PTB and poly(rC)-binding protein 2 (PCBP2) as ITAFs [65,67]. Domain V in PV IRES binds cellular glycyl-tRNA synthetase (GARS), promoting association of the IRES in the mRNA binding site of the ribosome [68]. Cellular Unr protein is required for human rhinovirus (HRV) IRES activity and acts synergistically with PTB, while such synergy is absent for PV IRES, highlighting differential ITAF requirements within the same IRES class [69].

2.3. Class 3 IRESs

Class 3 IRESs, exemplified by the hepatitis C virus (HCV) of the family Flaviviridae, are present at the 5′ end of viral RNAs and contain multiple stem-loop structures organized around helical junctions and a pseudoknot with additional requirements of ITAFs, e.g., PCBP2 and La [70–72]. Class 3 IRESs require a small set of canonical eIFs and a notable ability for eIF2-less initiation [63,73], but can operate in either eIF2-dependent or eIF2-independent modes depending on availability [62,64,74–76]. Incoming 40S subunits take a "land and initiate" approach; the stem-loop structures directly capture the ribosome at the AUG start codon, thereby bypassing scanning [77–79]. The 40S ribosomal subunit and eIF3 bind and are stabilized by domains II and III of the IRES RNA [73,80]. The interaction of eIF3 and 40S with the IRES is structurally distinct from the cap-dependent scanning mode of translation initiation [78,79,81–84]. HCV IRES subdomain IIIb as well as a stem-loop in the specialized translation initiation element in the cellular *c-JUN* transcript 5′UTR share structural similarity and bind eIF3d [85]. Nonetheless, recruitment of the 60S subunit leads to 80S ribosome formation and subsequent elongation [77,86,87]. An extensive catalog of HCV IRES-40S interactions has been described [88]. This study also demonstrates how eIF5B reorients initiator tRNA prior to ribosome subunit joining, promoting eIF2-independent initiation.

2.4. Class 4 IRESs

Class 4 IRESs have been identified to date only in the Dicistroviridae family of picornaviruses, e.g., cricket paralysis virus (CrPV). The single stranded viral RNA genome contains two open reading frames (ORFs) separated by an intergenic region (IGR) that harbors an IRES [89,90] (Figure 1). Class 4 IRESs are considered "factorless", that is, without the requirement of eIFs and ITAFs, yet display strong translation initiation potencies in multiple cell types [91–94]. These IRESs are folded into a compact, ordered three-dimensional structure consisting of multiple pseudoknots and stem-loops that directly interact with the 40S ribosomal subunit and permit assembly of the 80S ribosome [95–97]. Moreover, the IRES structure can directly recruit an 80S ribosome in vitro [98]. Remarkably, the IRES initiates translation from a non-AUG start codon, bypassing the requirement for the eIF2/Met-tRNAi/GTP ternary complex [97,99,100]. An RNA pseudoknot in the Class 4 IRES mimics a cellular tRNA that occupies the ribosome P-site and directs a tRNA-less elongation step, i.e., pseudotranslocation, poising translation to start from the second elongation step [91,101–105]. eIF2α phosphorylation, a hallmark of viral infections and other stresses, can increase translation efficiency of these IRESs, thereby flipping this antiviral response mechanism to the benefit of the virus [106].

2.5. Hybrid IRESs and Other IRESs

In many viruses, IRESs operate outside the bounds of easy classification into the schemes outlined above. Uniquely, the hepatitis A virus IRES depends on eIF4E, requiring its association with eIF4G to unwind RNA [107,108]. A "Class 5" IRES was identified in *Kobuvirus*, *Salivirus*, and *Paraturdivirus* genera of Picornaviridae [109]. Various domains in this IRES class, as typified in Aichivirus, are shared with both Class 1 and Class 2 IRESs, with a unique requirement of cellular DHX29 protein as an ITAF [109]. IRESs that are not classified within the four common types are present in pegiviruses, pestiviruses, and hepaciviruses [110]. Retroviral IRES function, including HIV-1 IRES, is debated as viral transcripts are capped [38,111,112]. In addition, a specialized translation mechanism of HIV-1 mRNAs sustained during eIF4E repression can be directed by a hypermethylated cap instead of being cap-independent, requiring interaction of the primer binding site in viral 5′UTR with cellular nuclear RNA helicase A (RHA)/DHX9 [113]. The IRES in the iflavirus *Ectropis obliqua* picorna-like virus (EoPV) is similar in several functional and structural aspects to Class 2 IRESs in cardioviruses and aphthoviruses; however, unlike Class 2 IRESs, it lacks a GNRA motif, has a functionally redundant RAAA motif, and utilizes a non-homologous initiation codon [114]. The dicistrovirus CrPV 5′UTR-IRES assembles a functional initiation complex utilizing an upstream start codon within the IRES [115]. In another dicistrovirus, the Halastavi árva virus (HalV), the 5′UTR IRES utilizes an unusual retrograde scanning of the 43S complex for initiation [36].

In summary, IRES structures are diverse and employ equally diverse mechanisms with limited common principles that serve as a framework for features specific to a single IRES or IRES class. Focused reviews contain further information on viral IRESs [39,46,48,49,116]. Beyond IRESs, viral and cellular transcripts can share specialized themes in cap-dependent translation initiation as well, e.g., cis-acting elements in the 5′UTR of transcripts from *Mononegavirales* family members such as vesicular stomatitis virus (VSV), measles virus, and rabies virus, as well as cellular transcripts such as *DDR2*, interact uniquely with RPL40 (ribosomal protein 40, large subunit) for cap-dependent translation initiation, suggesting these viruses might have usurped an endogenous translation pathway [117]. It remains to be seen if other alternative pathways of cap-dependent cellular translation initiation, e.g., eIF3d-dependent [118] and eIF4EHP-dependent [119], are utilized by multiple viruses.

3. Upstream Open Reading Frames

3.1. Translation Control by Upstream Open Reading Frames

A series of landmark studies revealed translation repression of *GCN4* mRNA in brewer's yeast (*Saccharomyces cerevisiae*) through initiation events at small ORFs in its 5′UTR [120–123]. Upstream open reading frames (uORFs) are *cis*-regulatory RNA elements prevalent in eukaryotic mRNAs that regulate translation initiation of downstream coding sequences (CDSs) (Figure 1). The pre-initiation complex (PIC) scans the 5′ UTR for a start codon until it recognizes an upstream AUG (uAUG), thereby permitting assembly of 80S ribosomes at the uORF. uORFs serve as "roadblocks", translation elongation following uORF recognition prevents PIC scanning to the downstream AUG of the main ORF. The Kozak sequence context of an uAUG, as in a canonical CDS, is a critical determinant of translation initiation of an uORF (reviewed in [124]). A weak context at the uAUG leads to "leaky scanning" by 43S PIC and initiation at the main ORF, while a strong context results in translation elongation at the uORF, necessitating re-initiation events by 40S subunits at the downstream AUG of the main ORF. Non-canonical initiation codons, e.g., CUG, GUG, and UUG, can also direct upstream initiation at uORFs [125,126]. uORFs attenuate main ORF translation by multiple mechanisms: by the "first dibs" scheme of translation initiation, by ribosome dissociation at the uORF stop codon, by stalling-mediated mRNA decay, and by encoding regulatory peptides [4,127–129]. uORF-mediated repression is widespread across the mammalian transcriptome [130–132] and is associated with diseases and stress responses (reviewed in [133–136]). For example, the integrated stress response depletes pre-initiation complexes to specifically de-repress translation from uORF-containing mRNAs [137–139]. A PIC reinitiates at a main ORF initiation site only when an uORF does not overlap with the main ORF. In contrast, an upstream AUG preceding the main ORF start codon can be in an overlapping ORF, with 2 further subclasses: out-of-frame overlapping uORFs (oORFs), the stop codons of which are downstream of main ORF AUGs and in different reading frames, and N-terminal extensions, which are basically overlapping uORFs in-frame with the main ORF [140–144] (Figure 1). The downstream CDS can be engaged by the re-initiating PIC only when the translated uORFs do not overlap with CDSs [145], or in special cases through a TURBS (termination upstream ribosome binding site) RNA element at the end of the preceding ORF (reviewed in [145,146], see below) (Figure 1). In addition to leaky scanning, bi-directional scanning of the PIC between closely spaced start codons, as in oORFs, can influence start site selection [147,148].

3.2. Virus Utilization of uORFs and uAUGs

Viruses take advantage of oORFs to offset genome compression and increase coding capacity [149]. Likewise, utilization of uORFs and uAUGs is commonplace in DNA viruses as well as in positive- and negative-sense RNA viruses [150,151]. Examples representing multiple mechanisms and viruses are summarized here.

Hepatitis B virus (HBV) polymerase ORF is preceded by an overlapping core ORF and by a more upstream AUG in the viral pre-genomic RNA [152]. Translation from the uAUG represses translation of the core ORF while allowing reinitiation of translation at the polymerase ORF [153,154]. Translation of the Human Cytomegalovirus (HCMV) gpUL4 ORF is regulated by an uORF [155]. Kaposi's sarcoma-associated herpesvirus (KSHV) ORF35 and ORF36 are translated from a polycistronic transcript regulated by a pair of uORFs. The second uORF overlaps with ORF35 and allows translation of ORF36 by a reinitiation mechanism, which is essential for viral propagation [156,157]. An uORF in the ebolavirus (EBOV) L gene suppresses translation of the L ORF under normal conditions and increases it under stress to maintain optimal polymerase activity, and uORF mutations attenuate EBOV growth [158]. Human immunodeficiency virus type 1 (HIV1) *tat* ORF is followed by *rev* and *nef* ORFs, and translation of upstream *tat* strongly represses Rev production [159]. A minimal uORF, containing only a start and a stop codon within the HIV1 *vpu* leader, permits efficient translation initiation at the downstream *env* start codon in viral vpu-env bicistronic mRNAs [160]. In arteriviruses, an uORF is present in the 5′-leader of the genomic RNA (gRNA), as well as in all sub-genomic RNAs (sgRNAs). In related coronaviruses, the uORF maps downstream of the genomic leader and is present only in the gRNA [161]. Bovine coronaviruses are an exception in which the uORF is utilized in all sgRNAs during persistent infection, causing translation attenuation [162]. Simian immunodeficiency virus (SIV) Rev and Env proteins are regulated by up to five uAUGs in multiple splice variants [163,164]. uORF translation enhances ribosomal shunting in prototype foamy virus genomic RNA [165] and in rice tungro virus [166], in a way that enables the mRNA or viral RNA to escape translation inhibition.

Termination-reinitiation events in uORF/main ORF pairs are regulated by TURBS elements extensively characterized in viruses of the Caliciviridae family and in influenza B virus from the Orthomyxoviridae family [167–169]. In these cases, the uORF stop codon and downstream ORF start codon are in close proximity, out-of-frame, or overlap in consecutive nucleotides. The viral TURBS regulates reinitiation through base-pairing with 18S rRNA and tethering of the post-termination 40S subunit (reviewed in [145,146]). It has been suggested that uORFs in the 5′UTR regulate translation of cellular *SLAMF1* mRNA utilizing a TURBS-like element, highlighting shared strategies in eukaryotic translational control [170]. Additionally, functional polypeptides encoded by viral uORFs can regulate virulence and tropism. For example, a conserved uORF overlaps the viral polyprotein ORF in enteroviruses, including echovirus 7 and poliovirus 1, encoding a transmembrane protein that facilitates virus growth in gut epithelial cells [53].

4. 3′ End Elements

4.1. GAIT and VAIT Elements

The Gamma-interferon-Activated Inhibitor of Translation (GAIT) RNA element was first identified in the 3′UTR of human ceruloplasmin mRNA as an inducible, translation repressive element [171]. Subsequent studies observed similar structurally conserved GAIT elements in *VEGFA*, *DAPK*, and *ZIPK* mRNAs. To date, translation-repressive GAIT elements have been reported in multiple inflammation-related human mRNAs [171–174]. The GAIT RNA element is characterized by a split stem-loop secondary structure, with generally conserved A and U residues (absent in the *ZIPK* GAIT element [172]) in an asymmetric bulge separating the stems. The GAIT element recruits a gamma-interferon-inducible, heterotetrameric complex in which the direct RNA-interacting constituent is the bifunctional glutamyl-prolyl tRNA synthetase EPRS1 [175,176]. EPRS1 contains two catalytic domains joined by a linker containing three repeats of an atypical RNA-binding domain termed the WHEP domain, also found in human tryptophanyl, histidyl, glycyl, and methionyl tRNA synthetases. Three other constituents of the human GAIT complex are ribosomal protein L13a, heterogenous ribonucleoprotein Q (or NSAP1), and GAPDH. Eukaryotic translation occurs in a closed loop where the 5′ and 3′ ends of translated RNAs are generally bridged by eIF4G and PABP interactions [177–179]. L13a is central to the repressive function of the GAIT complex, as it blocks eIF4G from recruiting the pre-initiation complex [180].

The GAIT system is a classical archetype of a post-transcriptional regulon in which families of functionally related mRNAs are co-regulated by specific RNA-binding proteins that target similar sequences or structural elements [181,182]. Mimicking host RNA structures, GAIT-like RNA elements have been described in the RNA viruses RSV (respiratory syncytial virus) and the pig alphacoronavirus TGEV (transmissible gastroenteritis virus); both repress viral RNA translation in response to viral cues [183,184]. The cellular GAIT element, the RSV GAIT-like element binds L13a and has been termed a VAIT or Virus-Activated Inhibitor of Translation element [184] (Figure 1). Two coding-region VAITs have been reported in SARS-CoV-2, one in ORF1a of the genomic RNA and another in the coding region of Spike subgenomic (sgRNA) [185]. Unlike the VAITs, the inhibitory TGEV GAIT-like element does not bind L13a but binds two aminoacyl-tRNA synthetases (aaRSs), EPRS1 and RARS1, showcasing L13a-independent translation repression [183], and suggesting similar RNA structural motifs can recruit distinct but functionally similar machineries.

4.2. *Sarbecoviral Pan-End Activating RNA Element*

We reported a novel 39-nucleotide GAIT-like RNA element in the 3′-end of SARS-CoV-2 genomic and sgRNAs [186] (Figure 1). Discontinuous transcription of the viral gRNA generates an ensemble of nested 3′-co-terminal sgRNAs that contain 5′-leader and 3′-end sequences identical to each other and to the genomic RNA (gRNA) (Figure 2). Thus, the element defines a unique post-transcriptional regulon distinct from non-viral systems, in which the elements are structurally related but not identical in sequence [187]. The RNA sequence is conserved in viruses of the subgenus Sarbecovirus of the genus Betacoronavirus, including SARS-CoV-1, suggesting an invariant function. Host-derived insulin and IFN-γ, agents associated with COVID-19 severity and outcome, singly or additively with SARS-CoV-2 spike subunit 1 that contains the host ACE2-receptor binding domain, increase sgRNA expression and translation only when the cis-element is intact [186]. Therefore, the element was termed the Sarbecoviral Pan-End Activating RNA (SPEAR) element. Strikingly, EPRS1 binds SPEAR as part of an unconventional, inducible Tetra-Aminoacyl-tRNA synthetase Sarbecoviral RNA-Interacting (TASRI) complex, containing three other aaRSs: arginyl-, lysyl- and methionyl-tRNA synthetases (RARS1, KARS1, and MARS1, respectively). Of note, SPEAR, such as the repressive TGEV GAIT-like element, does not bind L13a, L13a-interacting RSV, or SARS-CoV-2 VAITs [184,186]. Binding of RARS1 and EPRS1 to the repressive TGEV element [183] suggests the translation-enhancing function of SPEAR might depend on the other binding partners, i.e., KARS1 or MARS1, or both. Potential internal initiation at ORF10, a 3′-end co-terminal feature newly acquired in SARS-CoV-2, appears to play a critical role in SPEAR-mediated induction of viral sgRNA expression [186]. The SPEAR element and TASRI complex together form a host system hijacked by SARS-CoV-2 to direct an outcome, i.e., translation enhancement, opposite to the original host function of translation inhibition.

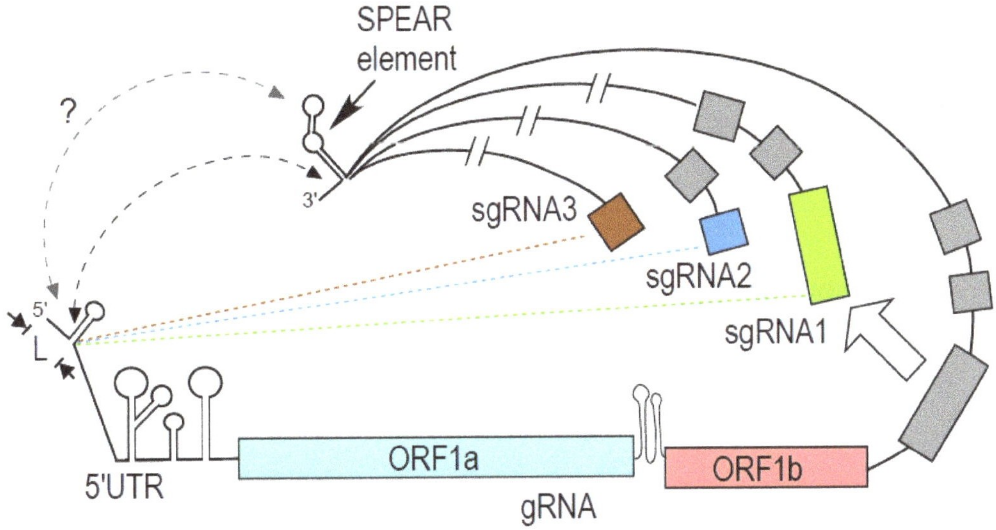

Figure 2. A unique regulon in viruses that employ discontinuous transcription can regulate genomic and sub-genomic RNAs. As an example, a simplified schematic of the SPEAR regulon of SARS-CoV-2 is shown. L: 5′-leader (within solid black arrows) that resides in the 5′UTR, gRNA: genomic RNA, sgRNA: subgenomic RNA. Frameshift element between ORF1a and ORF1b is shown in grey. Green, brown and blue dashed lines represent sgRNA-generating events between identical transcription regulatory sequences contained in 5′-leader and in nucleotides 5′ to ORFs encoded in sgRNA1 (green), sgRNA2 (blue), sgRNA3 (brown) and so on. The genomic terminus, common to gRNA and sgRNAs, reportedly pairs with the 5′-leader (darker gray dashed line). 5′-3′ communication through the SPEAR element is yet unknown (lighter gray dashed line).

Surprisingly, the SPEAR element at the 3′-end of gRNA supports ribosomal frameshifting in reporter assays, but only in the presence of the genomic 5′UTR [186], indicative of a 5′-end-dependent role of the SPEAR element. Circularization of eukaryotic mRNAs by interactions of proteins bound to both termini have been proposed to augment translation efficiency, possibly by facilitating ribosome transfer to the initiation site following a round of elongation [177]. In an early example of mRNA circularization influencing translation control, rather than translation efficiency, translational silencing by the 3′UTR GAIT element was shown to require the elements of mRNA circularization, including eIF4G [188,189]. In an alternative control mechanism that takes advantage of the closed loop, the 3′-Cap-Independent Translation Elements (CITEs, described in detail in Section 4.3), structurally diverse elements in the 3′UTRs of many plant viral genomes facilitate translation by recruiting initiation factors, ribosomes, and ITAFs to the 3′UTR, which exhibit trans-terminal effects on initiation at the 5′-end [116,190–195]. Indeed, SARS-CoV-2 initiates a global shutdown of translation, possibly necessitating a CITE-like function of SPEAR [196]. A proposed closed-loop mechanism placing the SPEAR element in proximity of the start codon across all viral sgRNAs awaits verification [197]. The 5′UTR of SARS-CoV-2 gRNA exhibits cap-independent translation activity; however, the smaller 5′-leaders of sgRNAs are eIF4E-dependent [198]. Since the SPEAR element is present in all SARS-CoV-2 RNAs and might facilitate 5′-3′ end communication, the difference in translation-initiation modes between 5′UTR of gRNA and 5′-leaders of sgRNAs can further indicate diverse roles of SPEAR in the 5′-3′ communications of gRNA compared to sgRNAs (Figure 2).

A critical feature of the discontinuous transcription program utilized by SARS-CoV-2 is the generation of an ensemble of nested 3′-co-terminal sgRNAs containing 5′ leader and 3′ end sequences identical to each other and to the genomic sequence [186]. Thus, the system forms a "post-transcriptional regulon" analogous to those found in eukaryotes, in which families of functionally related mRNAs are co-regulated by specific RNA-binding proteins that target similar sequences or structural elements [181]. However, the viral system is distinct in that the RNAs regulated are contained within the same genetic unit, versus scattered among chromosomes in eukaryotic systems. Also, the element is identical in the RNAs of the viral system, compared to the similarity of sequence or structure in eukaryotic systems. The "thrifty" virus mechanism that compacts the genome, i.e., reusing the genomic termini in every sgRNA, can be turned against the virus as a therapeutic strategy. Thus, interfering with the SPEAR-TASRI complex interaction is an attractive antiviral strategy to target the entire SPEAR regulon. Importantly, interception of the SPEAR element/TASRI complex interaction with peptide-conjugated phosphorodiamidate morpholine oligonucleotides (PPMOs) antisense to the SPEAR element markedly reduces SARS-CoV-2 replication in Caco-2-hACE2 cells [186].

4.3. 3′-Cap-Independent Translation Elements

The trans-terminal-acting 3′-CITEs are structurally diverse elements in seven identified classes [39] (Figure 1). The 5′ region of the viral RNA is essential for the function of certain 3′-CITEs through 5′-3′ communication, even if the cap is not [199–201] (Figure 1). These elements can compete with cellular mRNAs for the eIFs; e.g., barley yellow dwarf virus translation element (BTE) binds eIF4G [202–204] and panicum mosaic virus-like translation enhancers (PTEs) bind eIF4E while base-paired with the 5′UTR [205,206]. Crystal structures have shown that a guanosine is extruded from a G-rich stretch in the PTE of the pea-enation mosaic virus (PEMV2), providing a structural basis for eIF4E recognition [207]. Similar studies in the saguaro cactus virus (SCV) revealed a flipped-out G residue ready to dock into the 5′ cap-binding pocket of eIF4E, highlighting 5′ cap mimicry by a 3′-CITE [208]. 3′-CITEs in T-shaped structures (TSSs) in TCV and cardamine chlorotic fleck virus RNAs can recycle post-termination ribosomes or ribosomal subunits directly to the 5′-end [209,210]. The 55-nt Cucurbit aphid-borne yellows virus-Xinjiang-like translation element (CXTE) functions in eIF4E-depleted lysate and is enhanced by the viral 5′UTR; however, detailed molecular mechanisms remain unknown [211]. Interfamilial recombination of 3′-CITEs between viruses has been observed, and they can act in combination with other elements, e.g., the melon necrotic spot virus has a CXTE and an I-shaped structure (ISS) [211], and the PTE and an upstream TSS in PEMV2 [212]. Furthermore, 3′-CITEs are conformationally dynamic [213], e.g., the TCV TSS is disassembled through viral RNA-dependent RNA polymerase binding, a requirement for replication [214]. Structural rearrangement is also proposed for the ORF10-dependent function of the SPEAR element in SARS-CoV-2 sgRNA 3′-end, comparable to 3′-CITEs [186,197]. The TSSs in TCV and PEMV2 are predicted to form internal tRNA-like structures (TLS), ubiquitous host molecules linking the amino acid and RNA worlds [192].

4.4. 3′-tRNA-Like Structures

The plant RNA virus Tobamovirus and Tymovirus genomes contain 3′-terminal TLSs that control their replication [215–217] (Figure 1). Like the aaRS-interacting GAIT and TGEV elements, these 3′-TLSs recruit specific aaRSs and are aminoacylated with histidine (e.g., tobacco mosaic virus, TMV), tyrosine (e.g., brome mosaic virus, BMV), or valine (e.g., turnip yellow mosaic virus, TYMV) to stabilize viral RNAs, enhance 5′-cap-dependent translation by communicating with 5′UTR elements in the closed-loop model of translation initiation, and possibly recruit eIFs and ribosomes [215,218,219]. The TLS in TYMV interacts with an upstream pseudoknot domain, possibly constituting a longer, modular structural element that increases the stability of this RNA substrate for valylation [220]. It remains contentious whether the TLS delivers valine as the first amino acid for viral protein synthesis [221,222]. The BMV TLS binds tyrosyl-tRNA synthetase in a conformation that markedly differs from a tRNA [223], further illustrating the dynamics of viral RNA structures. The TLS of the *Hordeivirus* genus in Virgaviridae, although structurally distinct from BMV TLS, can be tyrsylated, and, more importantly, a chimeric *hordeivirus*-BMV TLS can be similarly aminoacylated, illustrating modularity in TLSs [224]. Likewise, tRNA mimicry is seen in 5′-localized structures in viral RNAs, e.g., poliovirus IRES domain V resembles glycine tRNA and binds glycyl-tRNA synthetase for translation initiation [68].

5. Conclusions

Viruses employ a diverse array of host-directed translational strategies for efficient expression of viral proteins. Lacking the essential machinery of protein synthesis, for example, ribosomes and other translation factors, all viruses take advantage of the host apparatus to generate virus protein. Layered above the absolute requirement for the translation machinery, viruses have adopted certain features of host translation control to the benefit of the virus. In some cases, the mechanism itself is imitated; for example, virus utilization of IRESs and uORFs is used to increase coding capacity while reducing genome size. Some viruses have co-opted eukaryotic sequence or structural elements generally employed for condition-dependent regulation, often translation repression. Viruses can incorporate similar elements to recruit essential constituents of the eukaryote translation apparatus, generally to enhance translation initiation. Other virus elements recruit host regulatory factors that are not components of the translation machinery. The utilization of multiple aaRSs as trans-acting factors interacting with a variety of virus elements forms an unexpected molecular leitmotif. SARS-CoV-2 is a paradigm example of a virus conjoining several distinct features of eukaryotic host translation control: the virus mimics an established regulatory element and co-opts four aaRSs to form a stimulatory binding complex. Utilizing discontinuous transcription, identical elements are present in all SARS-CoV-2 sgRNAs (and the gRNA). Thus, the virus post-transcriptional regulon improves upon comparable regulons in eukaryotes, in which related but different elements are present in mRNA targets scattered among the chromosomes. The same strategy exploited by the virus can be pitted against the virus, as targeting the element can simultaneously inhibit the expression of all sgRNAs. In-depth elucidation of virus translation control elements, their binding proteins, and their mechanisms can lead to novel therapeutic approaches to reduce virus replication and pathogenicity.

Author Contributions: Conceptualization, D.K. and P.L.F.; writing—original draft preparation, D.K.; writing—review and editing, D.K. and P.L.F.; funding acquisition, P.L.F. All authors have read and agreed to the published version of the manuscript.

Funding: This research was funded by N.I.H. grants R01 DK130377 and R01 NS124547 (to P.L.F.).

Institutional Review Board Statement: The study did not require ethical approval.

Informed Consent Statement: The study did not require informed consent.

Data Availability Statement: There are no original data in the paper.

Conflicts of Interest: The authors declare no conflicts of interest.

References

1. Jackson, R.J.; Hellen, C.U.; Pestova, T.V. The mechanism of eukaryotic translation initiation and principles of its regulation. *Nat. Rev. Mol. Cell Biol.* **2010**, *11*, 113–127. [CrossRef] [PubMed]
2. Sonenberg, N.; Hinnebusch, A.G. Regulation of translation initiation in eukaryotes: Mechanisms and biological targets. *Cell* **2009**, *136*, 731–745. [CrossRef] [PubMed]
3. Gebauer, F.; Hentze, M.W. Molecular mechanisms of translational control. *Nat. Rev. Mol. Cell Biol.* **2004**, *5*, 827–835. [CrossRef] [PubMed]
4. Hinnebusch, A.G.; Ivanov, I.P.; Sonenberg, N. Translational control by 5′-untranslated regions of eukaryotic mRNAs. *Science* **2016**, *352*, 1413–1416. [CrossRef] [PubMed]
5. Hinnebusch, A.G. The scanning mechanism of eukaryotic translation initiation. *Annu. Rev. Biochem.* **2014**, *83*, 779–812. [CrossRef] [PubMed]
6. Shirokikh, N.E.; Preiss, T. Translation initiation by cap-dependent ribosome recruitment: Recent insights and open questions. *Wiley Interdiscip. Rev. RNA* **2018**, *9*, e1473. [CrossRef] [PubMed]

7. Stern-Ginossar, N.; Thompson, S.R.; Mathews, M.B.; Mohr, I. Translational control in virus-infected cells. *Cold Spring Harb. Perspect. Biol.* **2019**, *11*, a033001. [CrossRef] [PubMed]
8. Zahringer, J.; Baliga, B.S.; Munro, H.N. Novel mechanism for translational control in regulation of ferritin synthesis by iron. *Proc. Natl. Acad. Sci. USA* **1976**, *73*, 857–861. [CrossRef]
9. Hentze, M.W.; Caughman, S.W.; Rouault, T.A.; Barriocanal, J.G.; Dancis, A.; Harford, J.B.; Klausner, R.D. Identification of the iron-responsive element for the translational regulation of human ferritin mRNA. *Science* **1987**, *238*, 1570–1573. [CrossRef]
10. Hentze, M.W.; Rouault, T.A.; Caughman, S.W.; Dancis, A.; Harford, J.B.; Klausner, R.D. A cis-acting element is necessary and sufficient for translational regulation of human ferritin expression in response to iron. *Proc. Natl. Acad. Sci. USA* **1987**, *84*, 6730–6734. [CrossRef]
11. Gray, N.K.; Hentze, M.W. Iron regulatory protein prevents binding of the 43S translation pre-initiation complex to ferritin and eALAS mRNAs. *EMBO J.* **1994**, *13*, 3882–3891. [CrossRef] [PubMed]
12. Pelletier, J.; Sonenberg, N. Insertion mutagenesis to increase secondary structure within the 5′ noncoding region of a eukaryotic mRNA reduces translational efficiency. *Cell* **1985**, *40*, 515–526. [CrossRef]
13. Jang, S.K.; Krausslich, H.G.; Nicklin, M.J.; Duke, G.M.; Palmenberg, A.C.; Wimmer, E. A segment of the 5′ nontranslated region of encephalomyocarditis virus RNA directs internal entry of ribosomes during in vitro translation. *J. Virol.* **1988**, *62*, 2636–2643. [CrossRef] [PubMed]
14. Pelletier, J.; Sonenberg, N. Internal initiation of translation of eukaryotic mRNA directed by a sequence derived from poliovirus RNA. *Nature* **1988**, *334*, 320–325. [CrossRef] [PubMed]
15. Belsham, G.J.; Brangwyn, J.K. A region of the 5′ noncoding region of foot-and-mouth disease virus RNA directs efficient internal initiation of protein synthesis within cells: Involvement with the role of L protease in translational control. *J. Virol.* **1990**, *64*, 5389–5395. [CrossRef]
16. Tsukiyama-Kohara, K.; Iizuka, N.; Kohara, M.; Nomoto, A. Internal ribosome entry site within hepatitis C virus RNA. *J. Virol.* **1992**, *66*, 1476–1483. [CrossRef] [PubMed]
17. Buck, C.B.; Shen, X.; Egan, M.A.; Pierson, T.C.; Walker, C.M.; Siliciano, R.F. The human immunodeficiency virus type 1 gag gene encodes an internal ribosome entry site. *J. Virol.* **2001**, *75*, 181–191. [CrossRef]
18. Zhao, J.; Li, Y.; Wang, C.; Zhang, H.; Zhang, H.; Jiang, B.; Guo, X.; Song, X. IRESbase: A comprehensive database of experimentally validated internal ribosome entry sites. *Genom. Proteom. Bioinform.* **2020**, *18*, 129–139. [CrossRef]
19. Komar, A.A.; Hatzoglou, M. Cellular IRES-mediated translation: The war of ITAFs in pathophysiological states. *Cell Cycle* **2011**, *10*, 229–240. [CrossRef]
20. King, H.A.; Cobbold, L.C.; Willis, A.E. The role of IRES trans-acting factors in regulating translation initiation. *Biochem. Soc. Trans.* **2010**, *38*, 1581–1586. [CrossRef]
21. Spriggs, K.A.; Bushell, M.; Willis, A.E. Translational regulation of gene expression during conditions of cell stress. *Mol. Cell* **2010**, *40*, 228–237. [CrossRef] [PubMed]
22. Yoon, A.; Peng, G.; Brandenburger, Y.; Zollo, O.; Xu, W.; Rego, E.; Ruggero, D. Impaired control of IRES-mediated translation in X-linked dyskeratosis congenita. *Science* **2006**, *312*, 902–906. [CrossRef] [PubMed]
23. Leprivier, G.; Rotblat, B.; Khan, D.; Jan, E.; Sorensen, P.H. Stress-mediated translational control in cancer cells. *Biochim. Biophys. Acta* **2015**, *1849*, 845–860. [CrossRef] [PubMed]
24. Kovalski, J.R.; Kuzuoglu-Ozturk, D.; Ruggero, D. Protein synthesis control in cancer: Selectivity and therapeutic targeting. *EMBO J.* **2022**, *41*, e109823. [CrossRef] [PubMed]
25. Martinez-Salas, E.; Pineiro, D.; Fernandez, N. Alternative mechanisms to initiate translation in eukaryotic mRNAs. *Comp. Funct. Genom.* **2012**, *2012*, 391546. [CrossRef] [PubMed]
26. Dotu, I.; Lozano, G.; Clote, P.; Martinez-Salas, E. Using RNA inverse folding to identify IRES-like structural subdomains. *RNA Biol.* **2013**, *10*, 1842–1852. [CrossRef] [PubMed]
27. Van Eden, M.E.; Byrd, M.P.; Sherrill, K.W.; Lloyd, R.E. Demonstrating internal ribosome entry sites in eukaryotic mRNAs using stringent RNA test procedures. *RNA* **2004**, *10*, 720–730. [CrossRef] [PubMed]
28. Kolekar, P.; Pataskar, A.; Kulkarni-Kale, U.; Pal, J.; Kulkarni, A. IRESPred: Web server for prediction of cellular and viral internal ribosome entry site (IRES). *Sci. Rep.* **2016**, *6*, 27436. [CrossRef]
29. Gritsenko, A.A.; Weingarten-Gabbay, S.; Elias-Kirma, S.; Nir, R.; de Ridder, D.; Segal, E. Sequence features of viral and human Internal Ribosome Entry Sites predictive of their activity. *PLoS Comput. Biol.* **2017**, *13*, e1005734. [CrossRef]
30. Wellensiek, B.P.; Larsen, A.C.; Stephens, B.; Kukurba, K.; Waern, K.; Briones, N.; Liu, L.; Snyder, M.; Jacobs, B.L.; Kumar, S.; et al. Genome-wide profiling of human cap-independent translation-enhancing elements. *Nat. Methods* **2013**, *10*, 747–750. [CrossRef]
31. Weingarten-Gabbay, S.; Elias-Kirma, S.; Nir, R.; Gritsenko, A.A.; Stern-Ginossar, N.; Yakhini, Z.; Weinberger, A.; Segal, E. Comparative genetics. Systematic discovery of cap-independent translation sequences in human and viral genomes. *Science* **2016**, *351*, aad4939. [CrossRef] [PubMed]
32. Lozano, G.; Francisco-Velilla, R.; Martinez-Salas, E. Deconstructing internal ribosome entry site elements: An update of structural motifs and functional divergences. *Open Biol.* **2018**, *8*, 180155. [CrossRef]
33. Xue, S.; Tian, S.; Fujii, K.; Kladwang, W.; Das, R.; Barna, M. RNA regulons in Hox 5′ UTRs confer ribosome specificity to gene regulation. *Nature* **2015**, *517*, 33–38. [CrossRef] [PubMed]

34. Akirtava, C.; May, G.E.; McManus, C.J. False-positive IRESes from *Hoxa9* and other genes resulting from errors in mammalian 5′ UTR annotations. *Proc. Natl. Acad. Sci. USA* **2022**, *119*, e2122170119. [CrossRef] [PubMed]
35. Terenin, I.M.; Smirnova, V.V.; Andreev, D.E.; Dmitriev, S.E.; Shatsky, I.N. A researcher's guide to the galaxy of IRESs. *Cell. Mol. Life Sci.* **2017**, *74*, 1431–1455. [CrossRef] [PubMed]
36. Abaeva, I.S.; Pestova, T.V.; Hellen, C.U. Attachment of ribosomal complexes and retrograde scanning during initiation on the Halastavi arva virus IRES. *Nucleic Acids Res.* **2016**, *44*, 2362–2377. [CrossRef] [PubMed]
37. Thompson, S.R. Tricks an IRES uses to enslave ribosomes. *Trends Microbiol.* **2012**, *20*, 558–566. [CrossRef] [PubMed]
38. Smirnova, V.V.; Terenin, I.M.; Khutornenko, A.A.; Andreev, D.E.; Dmitriev, S.E.; Shatsky, I.N. Does HIV-1 mRNA 5′-untranslated region bear an internal ribosome entry site? *Biochimie* **2016**, *121*, 228–237. [CrossRef]
39. Jaafar, Z.A.; Kieft, J.S. Viral RNA structure-based strategies to manipulate translation. *Nat. Rev. Microbiol.* **2019**, *17*, 110–123. [CrossRef]
40. Jackson, R.J. The current status of vertebrate cellular mRNA IRESs. *Cold Spring Harb. Perspect. Biol.* **2013**, *5*, a011569. [CrossRef]
41. Gilbert, W.V. Alternative ways to think about cellular internal ribosome entry. *J. Biol. Chem.* **2010**, *285*, 29033–29038. [CrossRef] [PubMed]
42. Lopez-Ulloa, B.; Fuentes, Y.; Pizarro-Ortega, M.S.; Lopez-Lastra, M. RNA-binding proteins as regulators of internal initiation of viral mRNA translation. *Viruses* **2022**, *14*, 188. [CrossRef] [PubMed]
43. Lee, K.M.; Chen, C.J.; Shih, S.R. Regulation mechanisms of viral IRES-driven translation. *Trends Microbiol.* **2017**, *25*, 546–561. [CrossRef] [PubMed]
44. Godet, A.C.; David, F.; Hantelys, F.; Tatin, F.; Lacazette, E.; Garmy-Susini, B.; Prats, A.C. IRES trans-acting factors, key actors of the stress response. *Int. J. Mol. Sci.* **2019**, *20*, 924. [CrossRef] [PubMed]
45. Walsh, D.; Mathews, M.B.; Mohr, I. Tinkering with translation: Protein synthesis in virus-infected cells. *Cold Spring Harb. Perspect. Biol.* **2013**, *5*, a012351. [CrossRef]
46. Martinez-Salas, E.; Francisco-Velilla, R.; Fernandez-Chamorro, J.; Lozano, G.; Diaz-Toledano, R. Picornavirus IRES elements: RNA structure and host protein interactions. *Virus Res.* **2015**, *206*, 62–73. [CrossRef]
47. Gingras, A.C.; Svitkin, Y.; Belsham, G.J.; Pause, A.; Sonenberg, N. Activation of the translational suppressor 4E-BP1 following infection with encephalomyocarditis virus and poliovirus. *Proc. Natl. Acad. Sci. USA* **1996**, *93*, 5578–5583. [CrossRef]
48. Lozano, G.; Martinez-Salas, E. Structural insights into viral IRES-dependent translation mechanisms. *Curr. Opin. Virol.* **2015**, *12*, 113–120. [CrossRef]
49. Mailliot, J.; Martin, F. Viral internal ribosomal entry sites: Four classes for one goal. *Wiley Interdiscip. Rev. RNA* **2018**, *9*, e1458. [CrossRef]
50. Fernandez-Miragall, O.; Ramos, R.; Ramajo, J.; Martinez-Salas, E. Evidence of reciprocal tertiary interactions between conserved motifs involved in organizing RNA structure essential for internal initiation of translation. *RNA* **2006**, *12*, 223–234. [CrossRef]
51. Fernandez, N.; Fernandez-Miragall, O.; Ramajo, J.; Garcia-Sacristan, A.; Bellora, N.; Eyras, E.; Briones, C.; Martinez-Salas, E. Structural basis for the biological relevance of the invariant apical stem in IRES-mediated translation. *Nucleic Acids Res.* **2011**, *39*, 8572–8585. [CrossRef] [PubMed]
52. Pestova, T.V.; Hellen, C.U.; Wimmer, E. A conserved AUG triplet in the 5′ nontranslated region of poliovirus can function as an initiation codon in vitro and in vivo. *Virology* **1994**, *204*, 729–737. [CrossRef] [PubMed]
53. Lulla, V.; Dinan, A.M.; Hosmillo, M.; Chaudhry, Y.; Sherry, L.; Irigoyen, N.; Nayak, K.M.; Stonehouse, N.J.; Zilbauer, M.; Goodfellow, I.; et al. An upstream protein-coding region in enteroviruses modulates virus infection in gut epithelial cells. *Nat. Microbiol.* **2019**, *4*, 280–292. [CrossRef] [PubMed]
54. Guo, H.; Li, Y.; Liu, G.; Jiang, Y.; Shen, S.; Bi, R.; Huang, H.; Cheng, T.; Wang, C.; Wei, W. A second open reading frame in human enterovirus determines viral replication in intestinal epithelial cells. *Nat. Commun.* **2019**, *10*, 4066. [CrossRef] [PubMed]
55. Clarke, B.E.; Sangar, D.V.; Burroughs, J.N.; Newton, S.E.; Carroll, A.R.; Rowlands, D.J. Two initiation sites for foot-and-mouth disease virus polyprotein in vivo. *J. Gen. Virol.* **1985**, *66 Pt 12*, 2615–2626. [CrossRef] [PubMed]
56. Andreev, D.E.; Fernandez-Miragall, O.; Ramajo, J.; Dmitriev, S.E.; Terenin, I.M.; Martinez-Salas, E.; Shatsky, I.N. Differential factor requirement to assemble translation initiation complexes at the alternative start codons of foot-and-mouth disease virus RNA. *RNA* **2007**, *13*, 1366–1374. [CrossRef]
57. Yamasaki, K.; Weihl, C.C.; Roos, R.P. Alternative translation initiation of Theiler's murine encephalomyelitis virus. *J. Virol.* **1999**, *73*, 8519–8526. [CrossRef]
58. van Eyll, O.; Michiels, T. Non-AUG-initiated internal translation of the L* protein of Theiler's virus and importance of this protein for viral persistence. *J. Virol.* **2002**, *76*, 10665–10673. [CrossRef]
59. Meurs, E.F.; Watanabe, Y.; Kadereit, S.; Barber, G.N.; Katze, M.G.; Chong, K.; Williams, B.R.; Hovanessian, A.G. Constitutive expression of human double-stranded RNA-activated p68 kinase in murine cells mediates phosphorylation of eukaryotic initiation factor 2 and partial resistance to encephalomyocarditis virus growth. *J. Virol.* **1992**, *66*, 5805–5814. [CrossRef]
60. White, J.P.; Reineke, L.C.; Lloyd, R.E. Poliovirus switches to an eIF2-independent mode of translation during infection. *J. Virol.* **2011**, *85*, 8884–8893. [CrossRef]
61. Ventoso, I.; Sanz, M.A.; Molina, S.; Berlanga, J.J.; Carrasco, L.; Esteban, M. Translational resistance of late alphavirus mRNA to eIF2alpha phosphorylation: A strategy to overcome the antiviral effect of protein kinase PKR. *Genes Dev.* **2006**, *20*, 87–100. [CrossRef] [PubMed]

62. Dmitriev, S.E.; Terenin, I.M.; Andreev, D.E.; Ivanov, P.A.; Dunaevsky, J.E.; Merrick, W.C.; Shatsky, I.N. GTP-independent tRNA delivery to the ribosomal P-site by a novel eukaryotic translation factor. *J. Biol. Chem.* **2010**, *285*, 26779–26787. [CrossRef] [PubMed]
63. Terenin, I.M.; Dmitriev, S.E.; Andreev, D.E.; Shatsky, I.N. Eukaryotic translation initiation machinery can operate in a bacterial-like mode without eIF2. *Nat. Struct. Mol. Biol.* **2008**, *15*, 836–841. [CrossRef] [PubMed]
64. Pestova, T.V.; de Breyne, S.; Pisarev, A.V.; Abaeva, I.S.; Hellen, C.U. eIF2-dependent and eIF2-independent modes of initiation on the CSFV IRES: A common role of domain II. *EMBO J.* **2008**, *27*, 1060–1072. [CrossRef] [PubMed]
65. Kafasla, P.; Morgner, N.; Poyry, T.A.; Curry, S.; Robinson, C.V.; Jackson, R.J. Polypyrimidine tract binding protein stabilizes the encephalomyocarditis virus IRES structure via binding multiple sites in a unique orientation. *Mol. Cell* **2009**, *34*, 556–568. [CrossRef] [PubMed]
66. Borovjagin, A.; Pestova, T.; Shatsky, I. Pyrimidine tract binding protein strongly stimulates in vitro encephalomyocarditis virus RNA translation at the level of preinitiation complex formation. *FEBS Lett.* **1994**, *351*, 299–302. [CrossRef] [PubMed]
67. Sweeney, T.R.; Abaeva, I.S.; Pestova, T.V.; Hellen, C.U. The mechanism of translation initiation on Type 1 picornavirus IRESs. *EMBO J.* **2014**, *33*, 76–92. [CrossRef] [PubMed]
68. Andreev, D.E.; Hirnet, J.; Terenin, I.M.; Dmitriev, S.E.; Niepmann, M.; Shatsky, I.N. Glycyl-tRNA synthetase specifically binds to the poliovirus IRES to activate translation initiation. *Nucleic Acids Res.* **2012**, *40*, 5602–5614. [CrossRef]
69. Hunt, S.L.; Hsuan, J.J.; Totty, N.; Jackson, R.J. unr, a cellular cytoplasmic RNA-binding protein with five cold-shock domains, is required for internal initiation of translation of human rhinovirus RNA. *Genes Dev.* **1999**, *13*, 437–448. [CrossRef]
70. Isoyama, T.; Kamoshita, N.; Yasui, K.; Iwai, A.; Shiroki, K.; Toyoda, H.; Yamada, A.; Takasaki, Y.; Nomoto, A. Lower concentration of La protein required for internal ribosome entry on hepatitis C virus RNA than on poliovirus RNA. *J. Gen. Virol.* **1999**, *80 Pt 9*, 2319–2327. [CrossRef]
71. Wang, L.; Jeng, K.S.; Lai, M.M. Poly(C)-binding protein 2 interacts with sequences required for viral replication in the hepatitis C virus (HCV) 5′ untranslated region and directs HCV RNA replication through circularizing the viral genome. *J. Virol.* **2011**, *85*, 7954–7964. [CrossRef] [PubMed]
72. Kumar, A.; Ray, U.; Das, S. Human La protein interaction with GCAC near the initiator AUG enhances hepatitis C Virus RNA replication by promoting linkage between 5′ and 3′ untranslated regions. *J. Virol.* **2013**, *87*, 6713–6726. [CrossRef] [PubMed]
73. Pestova, T.V.; Shatsky, I.N.; Fletcher, S.P.; Jackson, R.J.; Hellen, C.U. A prokaryotic-like mode of cytoplasmic eukaryotic ribosome binding to the initiation codon during internal translation initiation of hepatitis C and classical swine fever virus RNAs. *Genes Dev.* **1998**, *12*, 67–83. [CrossRef] [PubMed]
74. Kim, J.H.; Park, S.M.; Park, J.H.; Keum, S.J.; Jang, S.K. eIF2A mediates translation of hepatitis C viral mRNA under stress conditions. *EMBO J.* **2011**, *30*, 2454–2464. [CrossRef] [PubMed]
75. Skabkin, M.A.; Skabkina, O.V.; Dhote, V.; Komar, A.A.; Hellen, C.U.; Pestova, T.V. Activities of Ligatin and MCT-1/DENR in eukaryotic translation initiation and ribosomal recycling. *Genes Dev.* **2010**, *24*, 1787–1801. [CrossRef]
76. de Breyne, S.; Yu, Y.; Pestova, T.V.; Hellen, C.U. Factor requirements for translation initiation on the Simian picornavirus internal ribosomal entry site. *RNA* **2008**, *14*, 367–380. [CrossRef] [PubMed]
77. Khawaja, A.; Vopalensky, V.; Pospisek, M. Understanding the potential of hepatitis C virus internal ribosome entry site domains to modulate translation initiation via their structure and function. *Wiley Interdiscip. Rev. RNA* **2015**, *6*, 211–224. [CrossRef]
78. Filbin, M.E.; Vollmar, B.S.; Shi, D.; Gonen, T.; Kieft, J.S. HCV IRES manipulates the ribosome to promote the switch from translation initiation to elongation. *Nat. Struct. Mol. Biol.* **2013**, *20*, 150–158. [CrossRef]
79. Spahn, C.M.; Kieft, J.S.; Grassucci, R.A.; Penczek, P.A.; Zhou, K.; Doudna, J.A.; Frank, J. Hepatitis C virus IRES RNA-induced changes in the conformation of the 40s ribosomal subunit. *Science* **2001**, *291*, 1959–1962. [CrossRef]
80. Kieft, J.S.; Zhou, K.; Jubin, R.; Doudna, J.A. Mechanism of ribosome recruitment by hepatitis C IRES RNA. *RNA* **2001**, *7*, 194–206. [CrossRef]
81. Hashem, Y.; des Georges, A.; Dhote, V.; Langlois, R.; Liao, H.Y.; Grassucci, R.A.; Pestova, T.V.; Hellen, C.U.; Frank, J. Hepatitis-C-virus-like internal ribosome entry sites displace eIF3 to gain access to the 40S subunit. *Nature* **2013**, *503*, 539–543. [CrossRef] [PubMed]
82. Jaafar, Z.A.; Oguro, A.; Nakamura, Y.; Kieft, J.S. Translation initiation by the hepatitis C virus IRES requires eIF1A and ribosomal complex remodeling. *eLife* **2016**, *5*, e21198. [CrossRef] [PubMed]
83. Filbin, M.E.; Kieft, J.S. HCV IRES domain IIb affects the configuration of coding RNA in the 40S subunit's decoding groove. *RNA* **2011**, *17*, 1258–1273. [CrossRef] [PubMed]
84. Locker, N.; Easton, L.E.; Lukavsky, P.J. HCV and CSFV IRES domain II mediate eIF2 release during 80S ribosome assembly. *EMBO J.* **2007**, *26*, 795–805. [CrossRef] [PubMed]
85. Walker, M.J.; Shortridge, M.D.; Albin, D.D.; Cominsky, L.Y.; Varani, G. Structure of the RNA specialized translation initiation element that recruits eIF3 to the 5′-UTR of c-Jun. *J. Mol. Biol.* **2020**, *432*, 1841–1855. [CrossRef]
86. Yamamoto, H.; Unbehaun, A.; Loerke, J.; Behrmann, E.; Collier, M.; Burger, J.; Mielke, T.; Spahn, C.M. Structure of the mammalian 80S initiation complex with initiation factor 5B on HCV-IRES RNA. *Nat. Struct. Mol. Biol.* **2014**, *21*, 721–727. [CrossRef] [PubMed]
87. Quade, N.; Boehringer, D.; Leibundgut, M.; van den Heuvel, J.; Ban, N. Cryo-EM structure of Hepatitis C virus IRES bound to the human ribosome at 3.9-A resolution. *Nat. Commun.* **2015**, *6*, 7646. [CrossRef] [PubMed]

88. Brown, Z.P.; Abaeva, I.S.; De, S.; Hellen, C.U.T.; Pestova, T.V.; Frank, J. Molecular architecture of 40S translation initiation complexes on the hepatitis C virus IRES. *EMBO J.* **2022**, *41*, e110581. [CrossRef]
89. Wilson, J.E.; Powell, M.J.; Hoover, S.E.; Sarnow, P. Naturally occurring dicistronic cricket paralysis virus RNA is regulated by two internal ribosome entry sites. *Mol. Cell. Biol.* **2000**, *20*, 4990–4999. [CrossRef]
90. Khong, A.; Bonderoff, J.M.; Spriggs, R.V.; Tammpere, E.; Kerr, C.H.; Jackson, T.J.; Willis, A.E.; Jan, E. Temporal regulation of distinct internal ribosome entry sites of the Dicistroviridae cricket paralysis virus. *Viruses* **2016**, *8*, 25. [CrossRef]
91. Jan, E.; Sarnow, P. Factorless ribosome assembly on the internal ribosome entry site of cricket paralysis virus. *J. Mol. Biol.* **2002**, *324*, 889–902. [CrossRef] [PubMed]
92. Kerr, C.H.; Jan, E. Commandeering the ribosome: Lessons learned from dicistroviruses about translation. *J. Virol.* **2016**, *90*, 5538–5540. [CrossRef] [PubMed]
93. Hertz, M.I.; Thompson, S.R. Mechanism of translation initiation by Dicistroviridae IGR IRESs. *Virology* **2011**, *411*, 355–361. [CrossRef] [PubMed]
94. Colussi, T.M.; Costantino, D.A.; Zhu, J.; Donohue, J.P.; Korostelev, A.A.; Jaafar, Z.A.; Plank, T.D.; Noller, H.F.; Kieft, J.S. Initiation of translation in bacteria by a structured eukaryotic IRES RNA. *Nature* **2015**, *519*, 110–113. [CrossRef] [PubMed]
95. Pfingsten, J.S.; Costantino, D.A.; Kieft, J.S. Structural basis for ribosome recruitment and manipulation by a viral IRES RNA. *Science* **2006**, *314*, 1450–1454. [CrossRef] [PubMed]
96. Schuler, M.; Connell, S.R.; Lescoute, A.; Giesebrecht, J.; Dabrowski, M.; Schroeer, B.; Mielke, T.; Penczek, P.A.; Westhof, E.; Spahn, C.M. Structure of the ribosome-bound cricket paralysis virus IRES RNA. *Nat. Struct. Mol. Biol.* **2006**, *13*, 1092–1096. [CrossRef] [PubMed]
97. Spahn, C.M.; Jan, E.; Mulder, A.; Grassucci, R.A.; Sarnow, P.; Frank, J. Cryo-EM visualization of a viral internal ribosome entry site bound to human ribosomes: The IRES functions as an RNA-based translation factor. *Cell* **2004**, *118*, 465–475. [CrossRef]
98. Petrov, A.; Grosely, R.; Chen, J.; O'Leary, S.E.; Puglisi, J.D. Multiple parallel pathways of translation initiation on the CrPV IRES. *Mol. Cell* **2016**, *62*, 92–103. [CrossRef]
99. Costantino, D.A.; Pfingsten, J.S.; Rambo, R.P.; Kieft, J.S. tRNA-mRNA mimicry drives translation initiation from a viral IRES. *Nat. Struct. Mol. Biol.* **2008**, *15*, 57–64. [CrossRef]
100. Muhs, M.; Hilal, T.; Mielke, T.; Skabkin, M.A.; Sanbonmatsu, K.Y.; Pestova, T.V.; Spahn, C.M. Cryo-EM of ribosomal 80S complexes with termination factors reveals the translocated cricket paralysis virus IRES. *Mol. Cell* **2015**, *57*, 422–432. [CrossRef]
101. Kanamori, Y.; Nakashima, N. A tertiary structure model of the internal ribosome entry site (IRES) for methionine-independent initiation of translation. *RNA* **2001**, *7*, 266–274. [CrossRef]
102. Nishiyama, T.; Yamamoto, H.; Shibuya, N.; Hatakeyama, Y.; Hachimori, A.; Uchiumi, T.; Nakashima, N. Structural elements in the internal ribosome entry site of Plautia stali intestine virus responsible for binding with ribosomes. *Nucleic Acids Res.* **2003**, *31*, 2434–2442. [CrossRef] [PubMed]
103. Fernandez, I.S.; Bai, X.C.; Murshudov, G.; Scheres, S.H.; Ramakrishnan, V. Initiation of translation by cricket paralysis virus IRES requires its translocation in the ribosome. *Cell* **2014**, *157*, 823–831. [CrossRef] [PubMed]
104. Ruehle, M.D.; Zhang, H.; Sheridan, R.M.; Mitra, S.; Chen, Y.; Gonzalez, R.L.; Cooperman, B.S.; Kieft, J.S. A dynamic RNA loop in an IRES affects multiple steps of elongation factor-mediated translation initiation. *eLife* **2015**, *4*, e08146. [CrossRef] [PubMed]
105. Butcher, S.E.; Jan, E. tRNA-mimicry in IRES-mediated translation and recoding. *RNA Biol.* **2016**, *13*, 1068–1074. [CrossRef] [PubMed]
106. Fernandez, J.; Yaman, I.; Sarnow, P.; Snider, M.D.; Hatzoglou, M. Regulation of internal ribosomal entry site-mediated translation by phosphorylation of the translation initiation factor eIF2alpha. *J. Biol. Chem.* **2002**, *277*, 19198–19205. [CrossRef] [PubMed]
107. Ali, I.K.; McKendrick, L.; Morley, S.J.; Jackson, R.J. Activity of the hepatitis A virus IRES requires association between the cap-binding translation initiation factor (eIF4E) and eIF4G. *J. Virol.* **2001**, *75*, 7854–7863. [CrossRef] [PubMed]
108. Avanzino, B.C.; Fuchs, G.; Fraser, C.S. Cellular cap-binding protein, eIF4E, promotes picornavirus genome restructuring and translation. *Proc. Natl. Acad. Sci. USA* **2017**, *114*, 9611–9616. [CrossRef]
109. Sweeney, T.R.; Dhote, V.; Yu, Y.; Hellen, C.U. A distinct class of internal ribosomal entry site in members of the *Kobuvirus* and proposed *Salivirus* and *Paraturdivirus* genera of the Picornaviridae. *J. Virol.* **2012**, *86*, 1468–1486. [CrossRef]
110. Arhab, Y.; Bulakhov, A.G.; Pestova, T.V.; Hellen, C.U.T. Dissemination of internal ribosomal entry sites (IRES) between viruses by horizontal gene Transfer. *Viruses* **2020**, *12*, 612. [CrossRef]
111. Plank, T.D.; Whitehurst, J.T.; Kieft, J.S. Cell type specificity and structural determinants of IRES activity from the 5′ leaders of different HIV-1 transcripts. *Nucleic Acids Res.* **2013**, *41*, 6698–6714. [CrossRef] [PubMed]
112. Ohlmann, T.; Mengardi, C.; Lopez-Lastra, M. Translation initiation of the HIV-1 mRNA. *Translation (Austin)* **2014**, *2*, e960242.
113. Singh, G.; Seufzer, B.; Song, Z.; Zucko, D.; Heng, X.; Boris-Lawrie, K. HIV-1 hypermethylated guanosine cap licenses specialized translation unaffected by mTOR. *Proc. Natl. Acad. Sci. USA* **2022**, *119*, e2105153118. [CrossRef] [PubMed]
114. Lu, J.; Zhang, J.; Wang, X.; Jiang, H.; Liu, C.; Hu, Y. In vitro and in vivo identification of structural and sequence elements in the 5′ untranslated region of *Ectropis obliqua* picorna-like virus required for internal initiation. *J. Gen. Virol.* **2006**, *87*, 3667–3677. [CrossRef]
115. Neupane, R.; Pisareva, V.P.; Rodriguez, C.F.; Pisarev, A.V.; Fernandez, I.S. A complex IRES at the 5′-UTR of a viral mRNA assembles a functional 48S complex via an uAUG intermediate. *eLife* **2020**, *9*, e54575. [CrossRef] [PubMed]

116. Geng, G.; Wang, D.; Liu, Z.; Wang, Y.; Zhu, M.; Cao, X.; Yu, C.; Yuan, X. Translation of plant RNA viruses. *Viruses* **2021**, *13*, 2499. [CrossRef]
117. Lee, A.S.; Burdeinick-Kerr, R.; Whelan, S.P. A ribosome-specialized translation initiation pathway is required for cap-dependent translation of vesicular stomatitis virus mRNAs. *Proc. Natl. Acad. Sci. USA* **2013**, *110*, 324–329. [CrossRef]
118. Lee, A.S.; Kranzusch, P.J.; Doudna, J.A.; Cate, J.H. eIF3d is an mRNA cap-binding protein that is required for specialized translation initiation. *Nature* **2016**, *536*, 96–99. [CrossRef]
119. Liang, M.; Hody, C.; Yammine, V.; Soin, R.; Sun, Y.; Lin, X.; Tian, X.; Meurs, R.; Perdrau, C.; Delacourt, N.; et al. eIF4EHP promotes Ldh mRNA translation in and fruit fly adaptation to hypoxia. *EMBO Rep.* **2023**, *24*, e56460. [CrossRef]
120. Mueller, P.P.; Hinnebusch, A.G. Multiple upstream AUG codons mediate translational control of GCN4. *Cell* **1986**, *45*, 201–207. [CrossRef]
121. Mueller, P.P.; Jackson, B.M.; Miller, P.F.; Hinnebusch, A.G. The first and fourth upstream open reading frames in GCN4 mRNA have similar initiation efficiencies but respond differently in translational control to change in length and sequence. *Mol. Cell. Biol.* **1988**, *8*, 5439–5447. [PubMed]
122. Miller, P.F.; Hinnebusch, A.G. Sequences that surround the stop codons of upstream open reading frames in GCN4 mRNA determine their distinct functions in translational control. *Genes Dev.* **1989**, *3*, 1217–1225. [CrossRef] [PubMed]
123. Mueller, P.P.; Harashima, S.; Hinnebusch, A.G. A segment of GCN4 mRNA containing the upstream AUG codons confers translational control upon a heterologous yeast transcript. *Proc. Natl. Acad. Sci. USA* **1987**, *84*, 2863–2867. [CrossRef] [PubMed]
124. Dever, T.E.; Ivanov, I.P.; Hinnebusch, A.G. Translational regulation by uORFs and start codon selection stringency. *Genes Dev.* **2023**, *37*, 474–489. [CrossRef]
125. Ingolia, N.T.; Lareau, L.F.; Weissman, J.S. Ribosome profiling of mouse embryonic stem cells reveals the complexity and dynamics of mammalian proteomes. *Cell* **2011**, *147*, 789–802. [CrossRef] [PubMed]
126. Lee, S.; Liu, B.; Lee, S.; Huang, S.X.; Shen, B.; Qian, S.B. Global mapping of translation initiation sites in mammalian cells at single-nucleotide resolution. *Proc. Natl. Acad. Sci. USA* **2012**, *109*, E2424–E2432. [CrossRef] [PubMed]
127. Morris, D.R.; Geballe, A.P. Upstream open reading frames as regulators of mRNA translation. *Mol. Cell. Biol.* **2000**, *20*, 8635–8642. [CrossRef] [PubMed]
128. Uchiyama-Kadokura, N.; Murakami, K.; Takemoto, M.; Koyanagi, N.; Murota, K.; Naito, S.; Onouchi, H. Polyamine-responsive ribosomal arrest at the stop codon of an upstream open reading frame of the AdoMetDC1 gene triggers nonsense-mediated mRNA decay in Arabidopsis thaliana. *Plant Cell Physiol.* **2014**, *55*, 1556–1567. [CrossRef]
129. Dever, T.E.; Ivanov, I.P.; Sachs, M.S. Conserved upstream open reading frame nascent peptides that control translation. *Annu. Rev. Genet.* **2020**, *54*, 237–264. [CrossRef]
130. Calvo, S.E.; Pagliarini, D.J.; Mootha, V.K. Upstream open reading frames cause widespread reduction of protein expression and are polymorphic among humans. *Proc. Natl. Acad. Sci. USA* **2009**, *106*, 7507–7512. [CrossRef]
131. McGillivray, P.; Ault, R.; Pawashe, M.; Kitchen, R.; Balasubramanian, S.; Gerstein, M. A comprehensive catalog of predicted functional upstream open reading frames in humans. *Nucleic Acids Res.* **2018**, *46*, 3326–3338. [CrossRef] [PubMed]
132. Manske, F.; Ogoniak, L.; Jurgens, L.; Grundmann, N.; Makalowski, W.; Wethmar, K. The new uORFdb: Integrating literature, sequence, and variation data in a central hub for uORF research. *Nucleic Acids Res.* **2023**, *51*, D328–D336. [CrossRef]
133. Barbosa, C.; Peixeiro, I.; Romao, L. Gene expression regulation by upstream open reading frames and human disease. *PLoS Genet.* **2013**, *9*, e1003529. [CrossRef] [PubMed]
134. Silva, J.; Fernandes, R.; Romao, L. Translational regulation by upstream open reading frames and human diseases. *Adv. Exp. Med. Biol.* **2019**, *1157*, 99–116. [PubMed]
135. Zhang, H.; Wang, Y.; Lu, J. Function and evolution of upstream ORFs in eukaryotes. *Trends Biochem. Sci.* **2019**, *44*, 782–794. [CrossRef] [PubMed]
136. Ruiz-Orera, J.; Alba, M.M. Translation of small open reading frames: Roles in regulation and evolutionary innovation. *Trends Genet.* **2019**, *35*, 186–198. [CrossRef] [PubMed]
137. Vattem, K.M.; Wek, R.C. Reinitiation involving upstream ORFs regulates ATF4 mRNA translation in mammalian cells. *Proc. Natl. Acad. Sci. USA* **2004**, *101*, 11269–11274. [CrossRef] [PubMed]
138. Andreev, D.E.; Arnold, M.; Kiniry, S.J.; Loughran, G.; Michel, A.M.; Rachinskii, D.; Baranov, P.V. TASEP modelling provides a parsimonious explanation for the ability of a single uORF to derepress translation during the integrated stress response. *eLife* **2018**, *7*, e32563. [CrossRef]
139. Vasu, K.; Ramachandiran, I.; Chechi, A.; Khan, K.; Khan, D.; Kaufman, R.; Fox, P.L. Translational control of murine adiponectin expression by an upstream open reading frame element. *RNA Biol.* **2023**, *20*, 737–749. [CrossRef]
140. Johnstone, T.G.; Bazzini, A.A.; Giraldez, A.J. Upstream ORFs are prevalent translational repressors in vertebrates. *EMBO J.* **2016**, *35*, 706–723. [CrossRef]
141. Torrance, V.; Lydall, D. Overlapping open reading frames strongly reduce human and yeast STN1 gene expression and affect telomere function. *PLoS Genet.* **2018**, *14*, e1007523. [CrossRef] [PubMed]
142. Wright, B.W.; Yi, Z.; Weissman, J.S.; Chen, J. The dark proteome: Translation from noncanonical open reading frames. *Trends Cell Biol.* **2022**, *32*, 243–258. [CrossRef]
143. Vasu, K.; Khan, D.; Ramachandiran, I.; Blankenberg, D.; Fox, P.L. Analysis of nested alternate open reading frames and their encoded proteins. *NAR Genom. Bioinform.* **2022**, *4*, lqac076. [CrossRef] [PubMed]

144. Vanderperre, B.; Lucier, J.F.; Roucou, X. HAltORF: A database of predicted out-of-frame alternative open reading frames in human. *Database* **2012**, *2012*, bas025. [CrossRef] [PubMed]
145. Gunisova, S.; Hronova, V.; Mohammad, M.P.; Hinnebusch, A.G.; Valasek, L.S. Please do not recycle! Translation reinitiation in microbes and higher eukaryotes. *FEMS Microbiol. Rev.* **2018**, *42*, 165–192. [CrossRef] [PubMed]
146. Sherlock, M.E.; Baquero Galvis, L.; Vicens, Q.; Kieft, J.S.; Jagannathan, S. Principles, mechanisms, and biological implications of translation termination-reinitiation. *RNA* **2023**, *29*, 865–884. [CrossRef] [PubMed]
147. Gu, Y.; Mao, Y.; Jia, L.; Dong, L.; Qian, S.B. Bi-directional ribosome scanning controls the stringency of start codon selection. *Nat. Commun.* **2021**, *12*, 6604. [CrossRef] [PubMed]
148. Matsuda, D.; Dreher, T.W. Close spacing of AUG initiation codons confers dicistronic character on a eukaryotic mRNA. *RNA* **2006**, *12*, 1338–1349. [CrossRef]
149. Chirico, N.; Vianelli, A.; Belshaw, R. Why genes overlap in viruses. *Proc. Biol. Sci.* **2010**, *277*, 3809–3817. [CrossRef]
150. Ho, J.S.Y.; Zhu, Z.; Marazzi, I. Unconventional viral gene expression mechanisms as therapeutic targets. *Nature* **2021**, *593*, 362–371. [CrossRef]
151. Finkel, Y.; Stern-Ginossar, N.; Schwartz, M. Viral short ORFs and their possible functions. *Proteomics* **2018**, *18*, e1700255. [CrossRef] [PubMed]
152. Hwang, W.L.; Su, T.S. Translational regulation of hepatitis B virus polymerase gene by termination-reinitiation of an upstream minicistron in a length-dependent manner. *J. Gen. Virol.* **1998**, *79 Pt 9*, 2181–2189. [CrossRef] [PubMed]
153. Chen, A.; Kao, Y.F.; Brown, C.M. Translation of the first upstream ORF in the hepatitis B virus pregenomic RNA modulates translation at the core and polymerase initiation codons. *Nucleic Acids Res.* **2005**, *33*, 1169–1181. [CrossRef] [PubMed]
154. Zong, L.; Qin, Y.; Jia, H.; Ye, L.; Wang, Y.; Zhang, J.; Wands, J.R.; Tong, S.; Li, J. Differential regulation of hepatitis B virus core protein expression and genome replication by a small upstream open reading frame and naturally occurring mutations in the precore region. *Virology* **2017**, *505*, 155–161. [CrossRef]
155. Degnin, C.R.; Schleiss, M.R.; Cao, J.; Geballe, A.P. Translational inhibition mediated by a short upstream open reading frame in the human cytomegalovirus gpUL4 (gp48) transcript. *J. Virol.* **1993**, *67*, 5514–5521. [CrossRef] [PubMed]
156. Kronstad, L.M.; Brulois, K.F.; Jung, J.U.; Glaunsinger, B.A. Dual short upstream open reading frames control translation of a herpesviral polycistronic mRNA. *PLoS Pathog.* **2013**, *9*, e1003156. [CrossRef] [PubMed]
157. Kronstad, L.M.; Brulois, K.F.; Jung, J.U.; Glaunsinger, B.A. Reinitiation after translation of two upstream open reading frames (ORF) governs expression of the ORF35-37 Kaposi's sarcoma-associated herpesvirus polycistronic mRNA. *J. Virol.* **2014**, *88*, 6512–6518. [CrossRef]
158. Shabman, R.S.; Hoenen, T.; Groseth, A.; Jabado, O.; Binning, J.M.; Amarasinghe, G.K.; Feldmann, H.; Basler, C.F. An upstream open reading frame modulates ebola virus polymerase translation and virus replication. *PLoS Pathog.* **2013**, *9*, e1003147. [CrossRef]
159. Luukkonen, B.G.; Tan, W.; Schwartz, S. Efficiency of reinitiation of translation on human immunodeficiency virus type 1 mRNAs is determined by the length of the upstream open reading frame and by intercistronic distance. *J. Virol.* **1995**, *69*, 4086–4094. [CrossRef]
160. Krummheuer, J.; Johnson, A.T.; Hauber, I.; Kammler, S.; Anderson, J.L.; Hauber, J.; Purcell, D.F.; Schaal, H. A minimal uORF within the HIV-1 vpu leader allows efficient translation initiation at the downstream env AUG. *Virology* **2007**, *363*, 261–271. [CrossRef]
161. Wu, H.Y.; Guan, B.J.; Su, Y.P.; Fan, Y.H.; Brian, D.A. Reselection of a genomic upstream open reading frame in mouse hepatitis coronavirus 5′-untranslated-region mutants. *J. Virol.* **2014**, *88*, 846–858. [CrossRef] [PubMed]
162. Hofmann, M.A.; Senanayake, S.D.; Brian, D.A. A translation-attenuating intraleader open reading frame is selected on coronavirus mRNAs during persistent infection. *Proc. Natl. Acad. Sci. USA* **1993**, *90*, 11733–11737. [CrossRef] [PubMed]
163. van der Velden, G.J.; Klaver, B.; Das, A.T.; Berkhout, B. Upstream AUG codons in the simian immunodeficiency virus SIVmac239 genome regulate Rev and Env protein translation. *J. Virol.* **2012**, *86*, 12362–12371. [CrossRef] [PubMed]
164. van der Velden, G.J.; Vink, M.A.; Klaver, B.; Das, A.T.; Berkhout, B. An AUG codon upstream of rev and env open reading frames ensures optimal translation of the simian immunodeficiency virus Env protein. *Virology* **2013**, *436*, 191–200. [CrossRef] [PubMed]
165. Schepetilnikov, M.; Schott, G.; Katsarou, K.; Thiebeauld, O.; Keller, M.; Ryabova, L.A. Molecular dissection of the prototype foamy virus (PFV) RNA 5′-UTR identifies essential elements of a ribosomal shunt. *Nucleic Acids Res.* **2009**, *37*, 5838–5847. [CrossRef] [PubMed]
166. Pooggin, M.M.; Rajeswaran, R.; Schepetilnikov, M.V.; Ryabova, L.A. Short ORF-dependent ribosome shunting operates in an RNA picorna-like virus and a DNA pararetrovirus that cause rice tungro disease. *PLoS Pathog.* **2012**, *8*, e1002568. [CrossRef] [PubMed]
167. Luttermann, C.; Meyers, G. A bipartite sequence motif induces translation reinitiation in feline calicivirus RNA. *J. Biol. Chem.* **2007**, *282*, 7056–7065. [CrossRef] [PubMed]
168. Meyers, G. Characterization of the sequence element directing translation reinitiation in RNA of the calicivirus rabbit hemorrhagic disease virus. *J. Virol.* **2007**, *81*, 9623–9632. [CrossRef]
169. Zinoviev, A.; Hellen, C.U.T.; Pestova, T.V. Multiple mechanisms of reinitiation on bicistronic calicivirus mRNAs. *Mol. Cell* **2015**, *57*, 1059–1073. [CrossRef]
170. Putlyaeva, L.V.; Schwartz, A.M.; Korneev, K.V.; Covic, M.; Uroshlev, L.A.; Makeev, V.Y.; Dmitriev, S.E.; Kuprash, D.V. Upstream open reading frames regulate translation of the long isoform of SLAMF1 mRNA that encodes costimulatory receptor CD150. *Biochemistry (Moscow)* **2014**, *79*, 1405–1411. [CrossRef]

171. Sampath, P.; Mazumder, B.; Seshadri, V.; Fox, P.L. Transcript-selective translational silencing by gamma interferon is directed by a novel structural element in the ceruloplasmin mRNA 3′ untranslated region. *Mol. Cell. Biol.* **2003**, *23*, 1509–1519. [CrossRef] [PubMed]
172. Mukhopadhyay, R.; Ray, P.S.; Arif, A.; Brady, A.K.; Kinter, M.; Fox, P.L. DAPK-ZIPK-L13a axis constitutes a negative-feedback module regulating inflammatory gene expression. *Mol. Cell* **2008**, *32*, 371–382. [CrossRef] [PubMed]
173. Ray, P.S.; Fox, P.L. A post-transcriptional pathway represses monocyte VEGF-A expression and angiogenic activity. *EMBO J.* **2007**, *26*, 3360–3372. [CrossRef] [PubMed]
174. Vyas, K.; Chaudhuri, S.; Leaman, D.W.; Komar, A.A.; Musiyenko, A.; Barik, S.; Mazumder, B. Genome-wide polysome profiling reveals an inflammation-responsive post-transcriptional operon in IFN-gamma-activated monocytes. *Mol. Cell. Biol.* **2009**, *29*, 458–470. [CrossRef] [PubMed]
175. Sampath, P.; Mazumder, B.; Seshadri, V.; Gerber, C.A.; Chavatte, L.; Kinter, M.; Ting, S.M.; Dignam, J.D.; Kim, S.; Driscoll, D.M.; et al. Noncanonical function of glutamyl-prolyl-tRNA synthetase: Gene-specific silencing of translation. *Cell* **2004**, *119*, 195–208. [CrossRef] [PubMed]
176. Mazumder, B.; Sampath, P.; Seshadri, V.; Maitra, R.K.; DiCorleto, P.; Fox, P.L. Regulated release of L13a from the 60S ribosomal subunit as a mechanism of transcript-specific translational control. *Cell* **2003**, *115*, 187–198. [CrossRef] [PubMed]
177. Wells, S.E.; Hillner, P.E.; Vale, R.D.; Sachs, A.B. Circularization of mRNA by eukaryotic translation initiation factors. *Mol. Cell* **1998**, *2*, 135–140. [CrossRef]
178. Vicens, Q.; Kieft, J.S.; Rissland, O.S. Revisiting the closed-loop model and the nature of mRNA 5′-3′ communication. *Mol. Cell* **2018**, *72*, 805–812. [CrossRef]
179. Filbin, M.E.; Kieft, J.S. Linking Alpha to Omega: Diverse and dynamic RNA-based mechanisms to regulate gene expression by 5′-to-3′ communication. *F1000Research* **2016**, *5*. [CrossRef]
180. Kapasi, P.; Chaudhuri, S.; Vyas, K.; Baus, D.; Komar, A.A.; Fox, P.L.; Merrick, W.C.; Mazumder, B. L13a blocks 48S assembly: Role of a general initiation factor in mRNA-specific translational control. *Mol. Cell* **2007**, *25*, 113–126. [CrossRef]
181. Keene, J.D. RNA regulons: Coordination of post-transcriptional events. *Nat. Rev. Genet.* **2007**, *8*, 533–543. [CrossRef] [PubMed]
182. Mukhopadhyay, R.; Jia, J.; Arif, A.; Ray, P.S.; Fox, P.L. The GAIT system: A gatekeeper of inflammatory gene expression. *Trends Biochem. Sci.* **2009**, *34*, 324–331. [CrossRef] [PubMed]
183. Marquez-Jurado, S.; Nogales, A.; Zuniga, S.; Enjuanes, L.; Almazan, F. Identification of a gamma interferon-activated inhibitor of translation-like RNA motif at the 3′ end of the transmissible gastroenteritis coronavirus genome modulating innate immune response. *mBio* **2015**, *6*, e00105. [CrossRef] [PubMed]
184. Mazumder, B.; Poddar, D.; Basu, A.; Kour, R.; Verbovetskaya, V.; Barik, S. Extraribosomal L13a is a specific innate immune factor for antiviral defense. *J. Virol.* **2014**, *88*, 9100–9110. [CrossRef] [PubMed]
185. Basu, A.; Penumutchu, S.; Nguyen, K.; Mbonye, U.; Tolbert, B.S.; Karn, J.; Komar, A.A.; Mazumder, B. A structurally conserved RNA element within SARS-CoV-2 ORF1a RNA and S mRNA regulates translation in response to viral S protein-induced signaling in human lung cells. *J. Virol.* **2022**, *96*, e0167821. [CrossRef] [PubMed]
186. Khan, D.; Terenzi, F.; Liu, G.; Ghosh, P.K.; Ye, F.; Nguyen, K.; China, A.; Ramachandiran, I.; Chakraborty, S.; Stefan, J.; et al. A viral pan-end RNA element and host complex define a SARS-CoV-2 regulon. *Nat. Commun.* **2023**, *14*, 3385. [CrossRef] [PubMed]
187. Arif, A.; Yao, P.; Terenzi, F.; Jia, J.; Ray, P.S.; Fox, P.L. The GAIT translational control system. *Wiley Interdiscip. Rev. RNA* **2017**, *9*, e1441. [CrossRef] [PubMed]
188. Mazumder, B.; Seshadri, V.; Imataka, H.; Sonenberg, N.; Fox, P.L. Translational silencing of ceruloplasmin requires the essential elements of mRNA circularization: Poly(A) tail, Poly(A)-binding protein, and eukaryotic translation initiation factor 4G. *Mol. Cell. Biol.* **2001**, *21*, 6440–6449. [CrossRef]
189. Mazumder, B.; Seshadri, V.; Fox, P.L. Translational control by the 3′-UTR: The ends specify the means. *Trends Biochem. Sci.* **2003**, *28*, 91–98. [CrossRef]
190. Nicholson, B.L.; White, K.A. Functional long-range RNA-RNA interactions in positive-strand RNA viruses. *Nat. Rev. Microbiol.* **2014**, *12*, 493–504. [CrossRef]
191. Nicholson, B.L.; White, K.A. 3′ Cap-independent translation enhancers of positive-strand RNA plant viruses. *Curr. Opin. Virol.* **2011**, *1*, 373–380. [CrossRef] [PubMed]
192. Simon, A.E.; Miller, W.A. 3′ cap-independent translation enhancers of plant viruses. *Annu. Rev. Microbiol.* **2013**, *67*, 21–42. [CrossRef] [PubMed]
193. Miras, M.; Miller, W.A.; Truniger, V.; Aranda, M.A. Non-canonical translation in plant RNA viruses. *Front. Plant Sci.* **2017**, *8*, 494. [CrossRef] [PubMed]
194. Truniger, V.; Miras, M.; Aranda, M.A. Structural and functional diversity of plant virus 3′-cap-independent translation enhancers (3′-CITEs). *Front. Plant Sci.* **2017**, *8*, 2047. [CrossRef] [PubMed]
195. Bernardes, W.S.; Menossi, M. Plant 3′ regulatory regions from mRNA-encoding genes and their uses to modulate expression. *Front. Plant Sci.* **2020**, *11*, 1252. [CrossRef] [PubMed]
196. Banerjee, A.K.; Blanco, M.R.; Bruce, E.A.; Honson, D.D.; Chen, L.M.; Chow, A.; Bhat, P.; Ollikainen, N.; Quinodoz, S.A.; Loney, C.; et al. SARS-CoV-2 disrupts splicing, translation, and protein trafficking to suppress host defenses. *Cell* **2020**, *183*, 1325–1339. [CrossRef] [PubMed]

197. Khan, D.; Fox, P.L. Aminoacyl-tRNA synthetase interactions in SARS-CoV-2 infection. *Biochem. Soc. Trans.* **2023**, *51*, 2127–2141. [CrossRef]
198. Slobodin, B.; Sehrawat, U.; Lev, A.; Hayat, D.; Zuckerman, B.; Fraticelli, D.; Ogran, A.; Ben-Shmuel, A.; Bar-David, E.; Levy, H.; et al. Cap-independent translation and a precisely located RNA sequence enable SARS-CoV-2 to control host translation and escape anti-viral response. *Nucleic Acids Res.* **2022**, *50*, 8080–8092. [CrossRef]
199. Guo, L.; Allen, E.M.; Miller, W.A. Base-pairing between untranslated regions facilitates translation of uncapped, nonpolyadenylated viral RNA. *Mol. Cell* **2001**, *7*, 1103–1109. [CrossRef]
200. Rakotondrafara, A.M.; Polacek, C.; Harris, E.; Miller, W.A. Oscillating kissing stem-loop interactions mediate 5′ scanning-dependent translation by a viral 3′-cap-independent translation element. *RNA* **2006**, *12*, 1893–1906. [CrossRef]
201. Nicholson, B.L.; Wu, B.; Chevtchenko, I.; White, K.A. Tombusvirus recruitment of host translational machinery via the 3′ UTR. *RNA* **2010**, *16*, 1402–1419. [CrossRef] [PubMed]
202. Sharma, S.D.; Kraft, J.J.; Miller, W.A.; Goss, D.J. Recruitment of the 40S ribosome subunit to the 3′-untranslated region (UTR) of a viral mRNA, via the eIF4 complex, facilitates cap-independent translation. *J. Biol. Chem.* **2015**, *290*, 11268–11281. [CrossRef] [PubMed]
203. Kraft, J.J.; Treder, K.; Peterson, M.S.; Miller, W.A. Cation-dependent folding of 3′ cap-independent translation elements facilitates interaction of a 17-nucleotide conserved sequence with eIF4G. *Nucleic Acids Res.* **2013**, *41*, 3398–3413. [CrossRef] [PubMed]
204. Zhao, P.; Liu, Q.; Miller, W.A.; Goss, D.J. Eukaryotic translation initiation factor 4G (eIF4G) coordinates interactions with eIF4A, eIF4B, and eIF4E in binding and translation of the barley yellow dwarf virus 3′ cap-independent translation element (BTE). *J. Biol. Chem.* **2017**, *292*, 5921–5931. [CrossRef] [PubMed]
205. Batten, J.S.; Desvoyes, B.; Yamamura, Y.; Scholthof, K.B. A translational enhancer element on the 3′-proximal end of the Panicum mosaic virus genome. *FEBS Lett.* **2006**, *580*, 2591–2597. [CrossRef] [PubMed]
206. Wang, Z.; Treder, K.; Miller, W.A. Structure of a viral cap-independent translation element that functions via high affinity binding to the eIF4E subunit of eIF4F. *J. Biol. Chem.* **2009**, *284*, 14189–14202. [CrossRef] [PubMed]
207. Lewicka, A.; Roman, C.; Jones, S.; Disare, M.; Rice, P.A.; Piccirilli, J.A. Crystal structure of a cap-independent translation enhancer RNA. *Nucleic Acids Res.* **2023**, *51*, 8891–8907. [CrossRef]
208. Ojha, M.; Vogt, J.; Das, N.K.; Redmond, E.; Singh, K.; Banna, H.A.; Sadat, T.; Koirala, D. Structure of saguaro cactus virus 3′ translational enhancer mimics 5′ cap for eIF4E binding. *Proc. Natl. Acad. Sci. USA* **2024**, *121*, e2313677121. [CrossRef]
209. Stupina, V.A.; Meskauskas, A.; McCormack, J.C.; Yingling, Y.G.; Shapiro, B.A.; Dinman, J.D.; Simon, A.E. The 3′ proximal translational enhancer of Turnip crinkle virus binds to 60S ribosomal subunits. *RNA* **2008**, *14*, 2379–2393. [CrossRef]
210. McCormack, J.C.; Yuan, X.; Yingling, Y.G.; Kasprzak, W.; Zamora, R.E.; Shapiro, B.A.; Simon, A.E. Structural domains within the 3′ untranslated region of Turnip crinkle virus. *J. Virol.* **2008**, *82*, 8706–8720. [CrossRef]
211. Miras, M.; Sempere, R.N.; Kraft, J.J.; Miller, W.A.; Aranda, M.A.; Truniger, V. Interfamilial recombination between viruses led to acquisition of a novel translation-enhancing RNA element that allows resistance breaking. *New Phytol.* **2014**, *202*, 233–246. [CrossRef] [PubMed]
212. Du, Z.; Alekhina, O.M.; Vassilenko, K.S.; Simon, A.E. Concerted action of two 3′ cap-independent translation enhancers increases the competitive strength of translated viral genomes. *Nucleic Acids Res.* **2017**, *45*, 9558–9572. [CrossRef] [PubMed]
213. Yuan, X.; Shi, K.; Meskauskas, A.; Simon, A.E. The 3′ end of Turnip crinkle virus contains a highly interactive structure including a translational enhancer that is disrupted by binding to the RNA-dependent RNA polymerase. *RNA* **2009**, *15*, 1849–1864. [CrossRef] [PubMed]
214. Le, M.T.; Kasprzak, W.K.; Kim, T.; Gao, F.; Young, M.Y.; Yuan, X.; Shapiro, B.A.; Seog, J.; Simon, A.E. Folding behavior of a T-shaped, ribosome-binding translation enhancer implicated in a wide-spread conformational switch. *eLife* **2017**, *6*, e22883. [CrossRef]
215. Dreher, T.W. Viral tRNAs and tRNA-like structures. *Wiley Interdiscip. Rev. RNA* **2010**, *1*, 402–414. [CrossRef] [PubMed]
216. Yot, P.; Pinck, M.; Haenni, A.L.; Duranton, H.M.; Chapeville, F. Valine-specific tRNA-like structure in turnip yellow mosaic virus RNA. *Proc. Natl. Acad. Sci. USA* **1970**, *67*, 1345–1352. [CrossRef] [PubMed]
217. Sherlock, M.E.; Hartwick, E.W.; MacFadden, A.; Kieft, J.S. Structural diversity and phylogenetic distribution of valyl tRNA-like structures in viruses. *RNA* **2021**, *27*, 27–39. [CrossRef]
218. Dreher, T.W.; Uhlenbeck, O.C.; Browning, K.S. Quantitative assessment of EF-1alpha.GTP binding to aminoacyl-tRNAs, aminoacyl-viral RNA, and tRNA shows close correspondence to the RNA binding properties of EF-Tu. *J. Biol. Chem.* **1999**, *274*, 666–672. [CrossRef]
219. Colussi, T.M.; Costantino, D.A.; Hammond, J.A.; Ruehle, G.M.; Nix, J.C.; Kieft, J.S. The structural basis of transfer RNA mimicry and conformational plasticity by a viral RNA. *Nature* **2014**, *511*, 366–369. [CrossRef]
220. Hammond, J.A.; Rambo, R.P.; Kieft, J.S. Multi-domain packing in the aminoacylatable 3′ end of a plant viral RNA. *J. Mol. Biol.* **2010**, *399*, 450–463. [CrossRef]
221. Barends, S.; Bink, H.H.; van den Worm, S.H.; Pleij, C.W.; Kraal, B. Entrapping ribosomes for viral translation: tRNA mimicry as a molecular Trojan horse. *Cell* **2003**, *112*, 123–129. [CrossRef]
222. Matsuda, D.; Dreher, T.W. Cap- and initiator tRNA-dependent initiation of TYMV polyprotein synthesis by ribosomes: Evaluation of the Trojan horse model for TYMV RNA translation. *RNA* **2007**, *13*, 129–137. [CrossRef]

223. Bonilla, S.L.; Sherlock, M.E.; MacFadden, A.; Kieft, J.S. A viral RNA hijacks host machinery using dynamic conformational changes of a tRNA-like structure. *Science* **2021**, *374*, 955–960. [CrossRef]
224. Sherlock, M.E.; Langeberg, C.J.; Kieft, J.S. Diversity and modularity of tyrosine-accepting tRNA-like structures. *RNA* **2024**, *30*, 213–222. [CrossRef]

viruses

MDPI

Article

Factor-Dependent Internal Ribosome Entry Site and -1 Programmed Frameshifting Signal in the Bemisia-Associated Dicistrovirus 2

Yihang Chen [1], Subash Chapagain [1], Jodi Chien [1], Higor Sette Pereira [2], Trushar R. Patel [2], Alice K. Inoue-Nagata [3] and Eric Jan [1,*]

1 Department of Biochemistry and Molecular Biology, Life Sciences Institute, University of British Columbia, Vancouver, BC V6T 1Z3, Canada; yihangchen1013@gmail.com (Y.C.); chsubash@student.ubc.ca (S.C.); chienjodi@gmail.com (J.C.)
2 Alberta RNA Research and Training Institute, Department of Chemistry and Biochemistry, University of Lethbridge, Lethbridge, AB T1K 3M4, Canada; higor.pereira@uleth.ca (H.S.P.); trushar.patel@uleth.ca (T.R.P.)
3 Embrapa Vegetables, Brasília 70770-901, Brazil; alice.nagata@embrapa.br
* Correspondence: ej@mail.ubc.ca

Abstract: The dicistrovirus intergenic (IGR) IRES uses the most streamlined translation initiation mechanism: the IRES recruits ribosomes directly without using protein factors and initiates translation from a non-AUG codon. Several subtypes of dicistroviruses IRES have been identified; typically, the IRESs adopt two -to three overlapping pseudoknots with key stem-loop and unpaired regions that interact with specific domains of the ribosomal 40S and 60S subunits to direct translation. We previously predicted an atypical IGR IRES structure and a potential -1 programmed frameshift (-1 FS) signal within the genome of the whitefly Bemisia-associated dicistrovirus 2 (BaDV-2). Here, using bicistronic reporters, we demonstrate that the predicted BaDV-2 -1 FS signal can drive -1 frameshifting in vitro via a slippery sequence and a downstream stem-loop structure that would direct the translation of the viral RNA-dependent RNA polymerase. Moreover, the predicted BaDV-2 IGR can support IRES translation in vitro but does so through a mechanism that is not typical of known factorless dicistrovirus IGR IRES mechanisms. Using deletion and mutational analyses, the BaDV-2 IGR IRES is mapped within a 140-nucleotide element and initiates translation from an AUG codon. Moreover, the IRES does not bind directly to purified ribosomes and is sensitive to eIF2 and eIF4A inhibitors NSC1198983 and hippuristanol, respectively, indicating an IRES-mediated factor-dependent mechanism. Biophysical characterization suggests the BaDV-2 IGR IRES contains several stem-loops; however, mutational analysis suggests a model whereby the IRES is unstructured or adopts distinct conformations for translation initiation. In summary, we have provided evidence of the first -1 FS frameshifting signal and a novel factor-dependent IRES mechanism in this dicistrovirus family, thus highlighting the diversity of viral RNA-structure strategies to direct viral protein synthesis.

Keywords: virus; frameshifting; RNA; IRES

1. Introduction

RNA viruses carry genomes with limited coding capacity. As such, a key essential step in the viral replication cycle is commandeering the host translational machinery to direct viral protein synthesis. In eukaryotes, protein synthesis is initiated via the cap-dependent scanning model, whereby ~10 core translation factors mediate recruitment of the small 40S ribosomal subunit to the 5′cap and scanning [1]. RNA viruses have adapted unique strategies to recruit and manipulate the host ribosome, some of which utilize an internal ribosome entry site (IRES), which is a cis-acting RNA element, typically structured, that can mediate ribosome recruitment in a 5′-end independent manner and using a subset of translation initiation factors and/or other cellular proteins. Currently, viral IRESs can be grouped into six distinct classes based on the conservation of RNA structure and factor requirements; however, with the expanding viromes identified via metagenomic approaches [2,3], it is anticipated that many more classes have yet to be discovered.

The dicistrovirus intergenic IRES studied to date are the most streamlined translation mechanism to date [4]. Dicistrovirus contains a ~8–10 kb viral RNA genome encoding two main open reading frames, ORF1 and ORF2. ORF1 encodes the viral non-structural proteins such as replicase, protease, and RNA helicase, whereas ORF2 encodes the structural proteins. The ORFs are translated by distinct IRESs, the 5′UTR and the IGR IRES. The 5′UTR IRES utilizes a Type III-like IRES mechanism requiring several translation factors [5]. The model IGR IRES, exemplified by that in cricket paralysis virus (CrPV), adopts a triple-pseudoknot RNA structure that can directly recruit the 40S ribosomal subunit, and then the 60S, to form the 80S ribosome and initiate translation from a non-AUG codon. Some IGR IRESs can recruit preformed 80S to direct translation, including a fraction of CrPV IGR IRES [6–8]. Cryo-EM studies of IRES:ribosome complexes captured in the translocation cycles revealed that distinct domains of the IRES interact with specific parts of the ribosome in order to manipulate the ribosome into a primed elongation mode of translation [9–13]. Within the IGR IRESs, there are distinct subtypes with specific RNA structures; they adopt a similar RNA structure as the CrPV IGR IRES to direct translation [13,14]. However, it is evident that there may be other more atypical IGR IRES mechanisms within dicistroviridae. The IGR from Halastavi árva virus (HalV), isolated from the intestinal contents of freshwater carp (*Cyprinus carpio*) [15], was identified to contain novel IRES features [7]; the HalV IGR IRES lacks key domains including PKIII, SLIV, and SLV. Moreover, the HalV IGR IRES does not recruit the 40S subunit but instead binds to preformed 80S ribosomes to drive translation. The discovery of HalV IGR IRES and other similar IRESs leads to the possibility of alternative IGR IRES mechanisms within the family *Dicistroviridae*.

Virome metagenomics studies have identified novel dicistrovirus-like genomes, which provide an opportunity to address the repertoire of IRES mechanisms that drive protein synthesis in this viral family [2,3]. A recent report by Hedil et al. identified a novel dicistrovirus genome with unique features. Bemisia-associated dicistrovirus 2 (BaDV-2) was isolated from the sweet potato whitefly *Bemisia tabaci* [16]. The 8012 nt genomic sequence of BaDV-2 contains (Accession MN231041) three ORFs (ORF1a, ORF1b, and ORF2) with a 5′UTR, an intergenic region (IGR), a 3′UTR, and a poly(A) tail. ORF1a contains an RNA helicase (nt 1780-2121) and a 3C cysteine protease motif (nt 3541-3654). ORF1b is in -1 frame and contains an RdRp motif (nt 4047-5261). ORF2 contains two picornavirus capsid protein-like Rhv motifs (nt 5893 to 6177 and 6730–7002, respectively), a calicivirus coat-protein motif (nt 7021 to 7209), and a CrPV-capsid protein-like motif (nt 7336 to 7977). A putative -1 frameshifting signal (-1 FS) was proposed to drive the translation of the full-length ORF1 polyprotein, consisting of a slippery sequence (GUCUUUU, nt 3780-6) and a putative pseudoknot, whereas a putative IGR IRES was proposed to drive the translation of ORF2.

In this study, we demonstrated that the BaDV-2 IGR supports IRES activity using a bicistronic reporter in vitro in insect *Spodoptera frugiperda* (Sf) 21 and rabbit reticulocyte lysate (RRL) translation extracts. Deletion and mutagenesis analyses mapped the BaDV-2 IRES activity within a 140-nucleotide element containing a predicted stem-loop. Moreover, the IRES is sensitive to eIF2 and eIF4A inhibitors NSC1198983 and hippuristanol, respectively, indicating that this IRES is factor-dependent. Finally, we showed that the BADV-2 putative -1 FS uses a stem-loop structure and slippery sequence to support 4% and 8% -1 frameshifting in Sf-21 and RRL, respectively. In summary, we have provided evidence of a novel IRES and -1 FS mechanism in the dicistrovirus family, thus highlighting the diversity of viral strategies to direct viral protein synthesis.

2. Materials and Methods

2.1. Cell Culture

Drosophila Schneider 2 cell line (S2) cells (Invitrogen) were maintained in Shields and Sang M3 insect medium (Sigma-Aldrich, St. Louis, MI, USA) supplemented with 10% fetal bovine serum at 25 °C.

2.2. Plasmids

BaDV-2 IGR IRES fragment (nucleotides 5325 to nucleotides 5694, Accession No. MN231041) (Twist Biosciences) was cloned into bicistronic and monocistronic luciferase reporter plasmids using Gibson assembly. Deletion mutations and specific nucleotide mutations were generated by PCR-based methods. BaDV-2 IGR IRES with a downstream P2A peptide sequence (Twist Biosciences) was cloned into the CrPV infectious clone [17], replacing the CrPV IGR IRES.

2.3. In Vitro Transcription

Monocistronic, bicistronic, and infectious clones were linearized by NarI, BamHI, and Eco53KI (NEB) restriction enzymes, respectively. RNA was transcribed using T7 RNA polymerase and subsequently purified with an RNeasy kit (Qiagen). 5′capping and polyadenylation were performed post-transcriptionally (CellScript). The RNA integrity and purity was confirmed by denaturing agarose gel analysis, and concentration was measured with a spectrophotometer (Nanodrop).

2.4. Transfection

S2 cells (3×10^6 cells) were transfected with bicistronic or infectious clone RNA (3 μg) using Lipofectamine 2000 reagent (Thermo Fisher Scientific, Waltham, MA, USA). For monitoring reporter RNA translation under CrPV infection, S2 cells were transfected with bicistronic RNA for 1 h, followed by infection with CrPV (MOI = 20) for 4.5 h. For infectious clone transfection experiments, S2 cells were transfected and then harvested at the indicated times and lysed in 1× passive lysis buffer (Promega, Tokyo, Japan). Lysates were cleared and protein concentration was measured by Bradford assay (Bio-Rad, Hercules, CA, USA). Western blotting was performed using an anti-VP2 antibody.

2.5. In Vitro Translation Reactions

In vitro transcribed bicistronic (1 μg) and infectious clone (2 μg) RNAs were incubated in Sf-21 cell extract (Promega) or rabbit reticulocyte lysate (RRL) (Promega) for 2 h or 45 min, respectively, in the presence of [^{35}S]-methionine/cysteine. Inhibitors (hippuristanol or NSC119889) at appropriate concentrations were added to the reaction 5 min prior to the addition of RNA. Reactions were either loaded and resolved on 15% SDS-PAGE gels and analyzed by phosphorimager analysis (Typhoon, Amersham, Chicago, IL, USA) or analyzed for enzymatic luciferase activity (Promega) using a luminometer (Turner Designs TD-20/20, San Jose, CA, USA).

2.6. Ribosome Filter Binding Assay

The 40S and 60S ribosomal subunits were purified as previously described [18]. RNA was dephosphorylated by FastAP thermosensitive alkaline phosphatase (Thermo Fisher Scientific) then labeled with T4 polynucleotide kinase (NEB) and [γ^{32}P]-ATP. [γ^{32}P]-RNA (final concentration: 0.5 nM) was heated at 65 °C for 3 min prior to the addition of 1× buffer E (final concentration: 20 mM Tris pH 7.5, 100 mM KCl, 2.5 mM MgOAc, 0.25 mM Spermidine and 2 mM DTT) and gently cooled to room temperature for 20 min, allowing for proper folding. The RNA was then incubated with purified 40S ribosomal subunits from 0.1 nM to 100 nM for 5 min, followed by 60S ribosomal subunits from 0.15 nM to 150 nM for 15 min, with 50 ng/mL in vitro transcribed noncompetitor RNA prepared from pcDNA3 (linearized with EcoRV) to prevent non-specific binding [19]. Reactions were loaded to the Bio-Dot filtration apparatus (Bio-Rad) with nitrocellulose and nylon membranes (Zeta-Probe, Bio-Rad). The membranes were then washed three times with 1× buffer E, then dried and imaged by phosphoimager analysis (Amershan Typhoon, ImageQuant™ 800). The fraction bound and the dissociation constant (K_D) were calculated as previously described [18].

2.7. RNA Purification and Size Exclusion Chromatography with Multi-Angle Light Scattering (SEC-MALS)

In vitro transcribed BaDV-2 IGR IRES RNA was purified by size exclusion chromatography (SEC) with an ÄKTA pure fast protein liquid chromatography (FPLC) (Global Life Science Solutions, New York, NY, USA) as previously described [20]. RNA-containing fractions were then pooled and concentrated by ethanol precipitation with resuspension in HEPES folding buffer (final concentration: 50 mM HEPES, 150 mM NaCl, 15 mM MgCl2, 3% Glycerol, and at pH 7.4) for SEC-MALS analysis and small-angle X-ray scattering (SAXS) analysis at the concentration of 2 mg/mL. SEC-MALS analysis was performed as previously described with the Optilab Refractive Index System (Wyatt Technology) [20].

2.8. Small-Angle X-ray Scattering (SAXS) and Atomic Structure Calculation

SAXS was performed as previously described by D'Souza et al. [20]. Briefly, SAXS data for BaDV-2 "minimal" IRES RNA were collected at 2 mg/mL using HEPES folding buffer. Samples were run at Diamond Light Source Ltd. synchrotron (Didcot, UK) on the B21 SAXS beamline, with a high-performance liquid chromatography (HPLC) system attached upstream to ensure the monodispersity [21]. A specialized flow cell was employed in conjunction with an inline Agilent 1200 HPLC system (Agilent Technologies, Stockport, UK). Then, 50 μL of RNA samples were injected onto a Shodex KW403-4F (Showa Denko America Inc., New York, NY, USA) size exclusion column with a flow rate maintained at 0.160 mL/minute. The eluted samples were exposed to X-rays with 3 s exposure time and 600 frames. Analysis of scattering data was performed using the ATSAS suite as previously described [22,23]. Twenty RNA models were generated using DAMMIN, averaged and filtered to a single representative using DAMAVER.

We utilized MC-Fold [24] to anticipate multiple low-energy secondary structures for BaDV-2 IRES RNA. This was performed to produce secondary structure predictions for downstream use. In each case, we chose the structures with the lowest energy as input files for MC-Sym, an application that facilitates the reconstruction of three-dimensional structures using known structures based on fragments [24]. Using MC-Sym, we computed 100 all-atom structures for each RNA and minimized them using MC-Sym's implemented protocols. These minimized structures were then compared to experimental SAXS data with CRYSOL based on minimized structures, to determine Rg and the goodness-of-fit parameter ($\chi 2$) [25]. To subject the minimized structures to CRYSOL, we used the protocols implemented in MC-Sym. Finally, we ranked the MC-Sym-derived structures with low-resolution structures based on their $\chi 2$ values and aligned them using the program SUPCOMB [26].

2.9. SHAPE Probing

For SHAPE-MaP analysis of the "minimal" IRES, in vitro transcribed RNA for BaDV-2 IRES was used as described [27]. Briefly, 500 ng of RNA was heated to 95° for 3 min, followed by the addition of Buffer E (final concentration of 20 mM Tris, pH 7.5, 0.1 M KCl, pH 7.0, 2.5 mM MgOAc, 0.25 mM spermidine and 2 mM dithiothreitol (DTT)) and incubated at 30 °C for 20 min. Folded RNA was modified by adding N-methylisatoic anhydride (NMIA) dissolved in DMSO (final concentration 5 mM) and incubated for 45 min at 30 °C. Control reactions containing only DMSO (no NMIA), as well as additional denaturing control (DC) reactions were performed in parallel. The RNA was recovered by ethanol precipitation.

RT Primer extension of modified and control RNAs was performed with primer 5′-CCTTCACTTTTAACATGGTTGGCCTG-3′ using Superscript III Reverse Transcriptase

(SSIII, Thermo Fisher Scientific) with Mn^{2+}, followed by second-strand cDNA synthesis (NEB #E6111). The ds cDNA was then used for subsequent library preparation and sequencing. Libraries were sequenced on an Illumina MiSeq platform following the manufacturer's standard cluster generation and sequencing protocols (UBC Sequencing and Bioinformatics Consortium).

SHAPE-Mapper 2.0 was used to analyze the sequencing data as described [28]. RNAstructure [29] was used to predict and model the secondary structure using the SHAPE-MaP data.

3. Results

3.1. BaDV-2 Contains a Programmed -1 Frameshift (-1 FS) Signal

A pseudoknot structure with an upstream putative slippery sequence "GUCUUUU", which are characteristics of a -1 programmed frameshift signal (FS), was proposed within the BaDV-2 genome [16,30] (Figure 1A). To determine whether this element can support frameshifting, we cloned the putative -1 FS element (nucleotides 3742 to 3884, Accession MN231041) into a bicistronic reporter between the Renilla (RLuc, 0 frame) and Firefly luciferase (FLuc, -1 frame) open reading frames (Figure 1B). We cloned sequences upstream and downstream of the predicted -1 FS signal to ensure the entire element is present. We also generated an in-frame control reporter whereby the RLuc and FLuc ORFs are in the same frame and a mutation (UUU3783-85 to CCC) (SS MUT) that disrupts the putative slippery sequence from "GUCUUUU" to "GUCCCCU" (Figure 1B).

Bicistronic reporter RNAs were incubated in Sf-21 insect cell lysate or rabbit reticulocyte lysate (RRL) and monitored luciferase activities and expression by incorporation of [^{35}S]-methionine/cysteine (Figure 1C,D). FLuc activity was detected in Sf21 and RRL reactions with the bicistronic reporter containing the wild-type BADV-2 -1 FS. Moreover, a fusion RLuc-FLuc protein product (~100 kDa) was detected, supporting that -1 frameshifting occurred (Figure 1C,D, lane 2). An RLuc protein was also detected, indicating that a fraction of translating ribosomes did not undergo frameshifting (Figure 1C,D, lane 2). As expected, the in-frame control reporter RNA resulted in robust expression of the 100 kDa RLuc-FLuc fusion protein and loss of RLuc expression (Figure 1C,D, lane 3). Based on the relative luciferase expression, the BADV-2 -1 FS signal activity occurred at 4 and 8% efficiency in Sf-21 and RRL, respectively. Mutant bicistronic RNAs containing either a premature stop codon at the end of RLuc (Figure 1C,D, lane 1) or the slippery sequence mutant (SS Mut, UUU3783-5CCC) (Figure 1C,D, lane 4) abolished FLuc expression and activity, thus supporting the presence of a -1 FS signal. These results demonstrated that the BaDV-2 genome contains a bona fide -1 FS signal.

3.2. Characterization of the BADV-2 -1 FS Signal

The proposed BaDV-2 -1 FS signal contains a putative pseudoknot structure seven nucleotides downstream of the slippery sequence [31]. To delineate the minimal BADV-2 -1 FS signal, systematic 3-end deletions were created (Figure 2A). Progressive 3′ deletions significantly inhibited -1 FS activity, including 3′DEL 3 and 3′DEL 2 (Figure 2). These results suggested that the elements downstream of the putative -1 FS element may promote frameshifting.

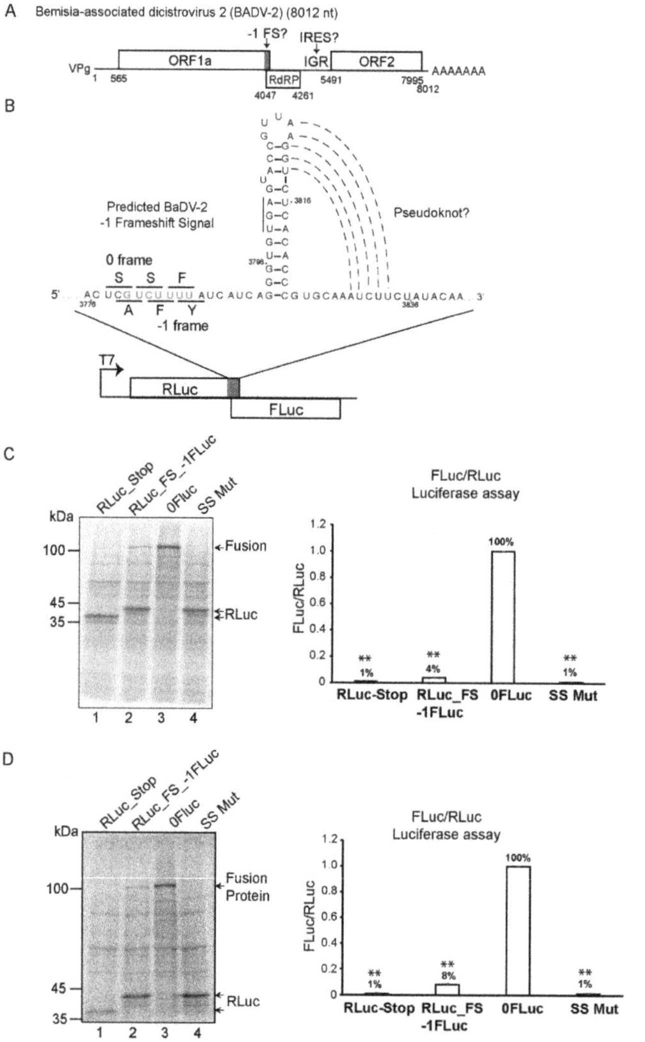

Figure 1. In vitro translation assays of BaDV-2 -1 FS in Sf-21 and RRL. (**A**) Simplified schematic view of the BaDV-2 genome. (**B**) The predicted secondary structure of BaDV-2 -1 frameshifting signal and schematic view of the bicistronic reporter with RLuc and FLuc. The slippery sequence is shown in red letters and the amino acids decoded in the -1 and 0 frames are shown in black letters. Bicistronic reporter RNAs containing wild-type or mutant BaDV-2 -1 FSS and in-frame controls were incubated in (**C**) Sf-21 or (**D**) RRL extracts. "RLuc Stop" contains a premature stop codon at the end of RLuc; "RLuc_FS_-1FLuc" contains the wild-type -1 FS with FLuc in -1 frame; "SS Mut" contains a mutant slippery sequence (UUU3783-5CCC); "0FLuc" is the in-frame control in which the RLuc and FLuc are both in the 0-reading frame. (left) Reactions were resolved on SDS-PAGE gel and monitored by phosphorimager analysis. Shown is a representative SDS-PAGE gel from at least three independent experiments. (right) Luciferase activities were measured and calculated as a ratio of FLuc/RLuc and normalized to the in-frame control (0FLuc). Groups were compared to 0FLuc. A one-way ANOVA statistical test was used to determine the *p* values (** $p < 0.01$). Shown are averages from at least three independent experiments ± standard deviation.

To determine whether the predicted stem-loop and pseudoknot are important for BaDV-2 -1 FS, we generated mutations that would disrupt base-pairing and tested-1 FS activity in RRL (Figure 2C). Mutations that disrupt the bottom base-paired domain of the stem-loop (M1, M2) abolished -1 FS activity to the same extent as the SS MUT; however, compensatory mutations (M1+M2) that restored base-pairing did not rescue activity completely, suggesting that the nucleotide identities may be important (Figure 2D). Mutations that disrupt the upper base-paired region (M4) inhibited -1 FS activity; however, the corresponding disrupted base-paired region (M5) had only a minor effect (~84% of the wild type) (Figure 2D). Compensatory mutations (M4 + M5) rescued -1 FS activity to wild-type or higher levels, suggesting that nucleotide identity of ACC (nt 3804 to 3806) may be important (Figure 2D). Of note, mutant (M5), which would disrupt the predicted pseudoknot base-pairing, had only a minor effect on -1 FS activity, thus indicating that the putative pseudoknot may not be critical for frameshifting (Figure 2D). Mutations that disrupt the apical loop (M3) also did not affect -1 FS activity (Figure 2D). In summary, we have demonstrated that the BaDV-2 genome contains a bona fide -1 FS signal and identified several key nucleotides and a stem-loop that are required for activity.

3.3. BaDV-2 IGR Supports Internal Ribosome Entry

We investigated whether the BaDV-2 IGR supports IRES activity. Our initial analysis predicted an atypical RNA structure that had features that were similar to known dicistrovirus IRES structures [16]. The predicted start site based on the proposed IRES structure starts at a GCG codon (nucleotide 5668 to 5670). We cloned the IGR of BaDV-2 (nucleotide 5325 to 5694) into the intergenic region of the RLuc-FLuc bicistronic reporter construct (Figure 3A). IRES activity was calculated as a ratio of FLuc/RLuc, where RLuc serves as a normalization control across experiments. We incubated in vitro transcribed RNAs into Sf-21 extracts and monitored scanning-dependent RLuc translation and IRES-mediated FLuc translation. As expected, CrPV IGR IRES directed FLuc expression and activity, whereas a mutant CrPV IGR IRES that disrupts PKI base-pairing (CC6214-5 to GG mutant) abolished IRES activity (Figure 3A). The BaDV-2 IGR also resulted in FLuc expression, ~1.5-fold translation compared to the wild-type CrPV IGR IRES (Figure 3A). To further confirm that the BaDV-2 IGR can support IRES activity and that the IRES is not linked to translation of RLuc, we added a strong hairpin loop within the 5′UTR upstream of RLuc to abolish scanning-dependent translation (Figure 3B). We also inserted the strong hairpin within the 5′UTR upstream of a BaDV-2 IRES-FLuc monocistronic reporter (Figure 3C). In reactions containing the bicistronic RNAs with the 5′UTR hairpin loop, only FLuc, but not RLuc, was translated, indicating the hairpin blocks' scanning-dependent translation of RLuc (Figure 3B,C). Further, addition of an upstream hairpin within the BaDV-2 IRES monocistronic RNA still resulted in FLuc activity. These results demonstrated conclusively that the BaDV-2 IGR contains an IRES.

3.4. "Mimimal" BADV-2 IGR That Directs IRES Activity

The initial predicted BaDV-2 IRES structure adopts several stem-loop and pseudoknot structures that resembled the dicistrovirus IGR IRES such as the CrPV IGR IRES [16]. A series of mutations was generated that disrupt predicted base-pairing, and BaDV-2 IRES activity was examined in vitro (Figure S1A). In summary, the majority of mutants did not affect BaDV-2 IRES activity (Figure S1B), suggesting an alternative RNA structure for BaDV-2 IRES translation initiation.

We next generated a series of 5′ and 3′ deletion mutations within the IGR region to pinpoint the minimal sequence for BaDV-2 IRES translation. Initially, nucleotides 5325 to 5694 of BaDV-2 was cloned and showed IRES activity, which we denote as wild type (full-length) (Figure 4). Compared to the wild-type BaDV-2 IRES, 5′ deletions from nucleotide 5325–5421 (5′DEL 2) but not from nucleotide (nt) 5325–5366 (5′Del 1) inhibited IRES activity (Figure 4A), thus setting the boundaries from the 5′end of the BaDV-2 IRES. Using a similar approach from the 3′end, the majority of 3′ deletions from 5506–5694 (3′DEL 1-5) resulted in

IRES activity as similar to or higher as compared to the wild-type IRES activity (Figure 4B). The only exception was 3′DEL 3, which showed ~30% of wild-type IRES activity (Figure 4B). To further pinpoint the minimal IRES, we generated combinations of 5′ and 3′ deletions. Combining 5′DEL 1 and 3′DEL 4 or 3′DEL 5 still showed IRES activity, whereas combining 5′DEL 2 and 3′DEL 4 inhibited IRES activity to ~40% of wild-type IRES activity (Figure 4B). Thus, the deletion analysis showed that the "minimal" BaDV-2 IRES is contained within a 140-nucleotide region from nt 5366–5506.

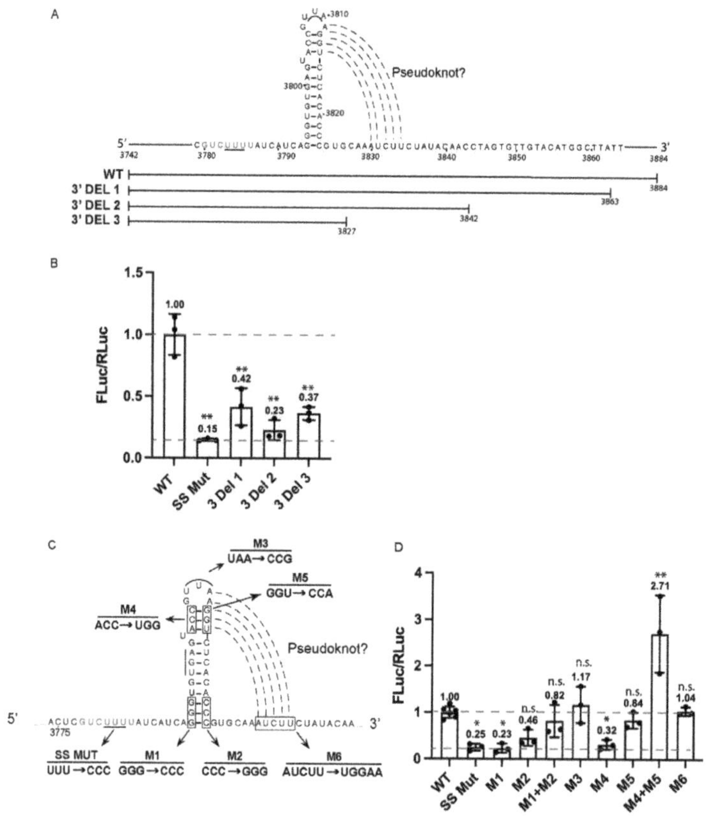

Figure 2. Structural analysis of BaDV-2 -1 FS element. (**A**) Schematic of deletion mutants surrounding the putative BaDV-2 -1 FS. The slippery sequence is shown in red letters. (**B**) In vitro transcribed RNAs were incubated in RRL and luciferase activities were measured and normalized to wild-type BaDV-2 -1 FS (BaDV-2 FS WT). Groups were compared to the wild-type BaDV-2 -1 FS. (**C**) Schematic view of BaDV-2 -1 FS mutations. (**D**) In vitro transcribed RNAs were incubated in RRL, and luciferase activities were measured and normalized to wild-type BaDV-2 -1 FSS (BaDV-2 WT). The top red dotted line represents Fluc/RLuc ratio of the wild-type -1 FS activity normalized to 1. The lower red dot line represents the background characterized by a mutation in the slippery site (SS MUT, UUU3783-5CCC) that abolishes -1 FS activity. Groups were compared to BaDV-2 WT. A one-way ANOVA statistical test was used to determine the p values (* $p < 0.05$, ** $p < 0.01$). "n.s." denotes the difference is not significant between the experimental groups and WT ($p > 0.05$). Shown are averages from at least three independent experiments ± standard deviation.

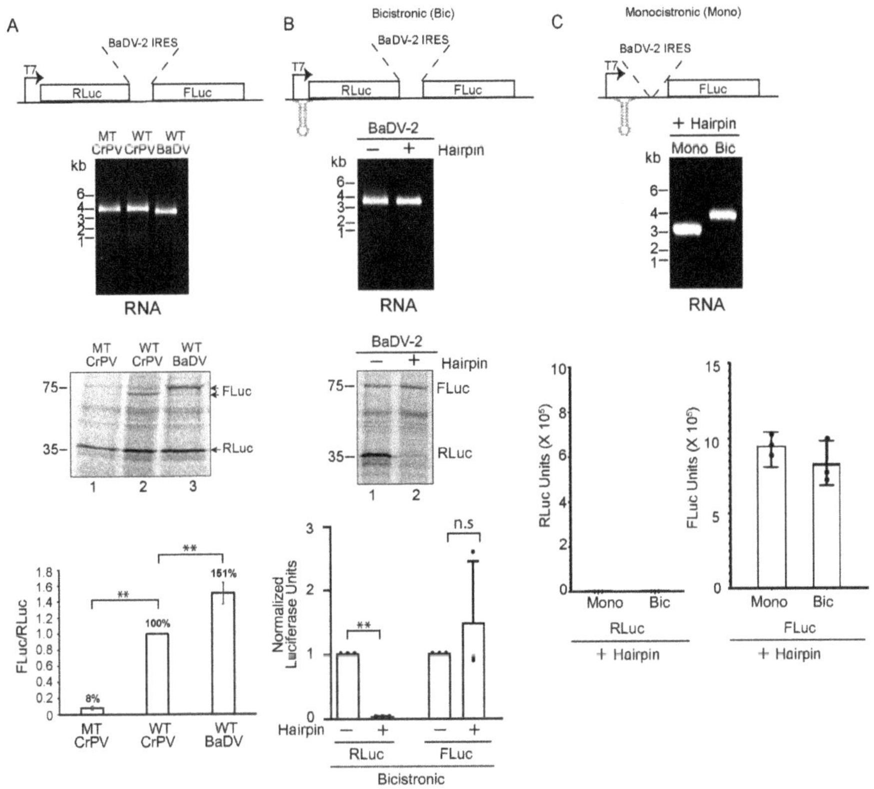

Figure 3. BaDV-2 IGR supports internal ribosome entry. (**A**) Bicistronic reporter RNAs containing either the CrPV IGR or the BaDV-2 IGR, (**B**) bicistronic RNAs with a 5′UTR strong hairpin, or (**C**) monocistronic reporter RNAs with a 5′UTR hairpin and the BaDV-2 IGR (top schematics) were incubated in Sf-21 extracts. In vitro transcribed RNAs were analyzed by agarose gel analysis (middle). Translation of FLuc and RLuc was measured by either radioisotope [^{35}S]-methionine/cysteine incorporation followed by SDS-PAGE and phosphorimager analysis or luciferase assays (graphs shown bottom). A paired t-test was used to determine the p values. ** $p < 0.01$. "n.s." denotes the difference is not significant between the control groups and the experimental groups ($p > 0.05$). Shown are representative SDS-PAGE gels and the averages from at least three independent experiments ± standard deviation.

3.5. Structural Analysis of the BaDV-2 "Minimal" IRES

Small-angle X-Ray scattering (SAXS) is a method to resolve the size and shape of monodispersed macromolecules [32]. A prerequisite for SAXS is a pure, monodispersed sample that displays similar hydrodynamic properties [33]. We first purified BaDV-2 "minimal" IRES RNA made from T7 promoter by FPLC Superdex 200 (S200). The elution profile of the RNAs is as shown in Figure 5A for pure monomeric BaDV-2 IRES RNA collected at approximately 12 to 13 mL. Purified RNA was analyzed on SEC-MALS to determine the absolute molecular weight. SEC-MALS confirmed BaDV-2 IGR IRES had a molecular weight of 47.53 (±0.79%) kDa (Figure 5B), consistent with the predicted molecular weight (49.22 kDa) and indicating it is a monomer in solution. Next, we assessed BaDV-2 IRES structural envelope using SAXS. The absence of aggregation in the solution is indicated by the linearity observed in the Guinier plots (Figure 5C). The low angle region can be used to calculate Rg at inverted space, which is equal to 49.63 Å. The Guinier plot provided reciprocal information, which was then converted into a real-space electron paired distribution using indirect Fourier transformations, resulting in the derivation of the P(r) plot (Figure 5D). The P(r) plot indicates that BaDV-2 "minimal" IRES assumes an elongated structure, with a Dmax of approximately 160 Å and a real-space Rg calculated at 49.87 ± 0.22 Å, as observed in the Gaussian distribution. Here, we showed a representative model in Figure 5C, in which the light blue is the SAXS low-resolution structure, and the superimposed model is the lowest energy atomistic structure determined; the corresponding secondary structure model in represented in Figure 5E.

To examine the secondary structure predicted by the SAXS IRES model, we re-examined the IRES translational activities by generating new mutations at key positions that would be predicted to alter base-pairing (Figure 5F, bottom).

Mutation of UAGU to GUUA (M4), which would disrupt an apical stem-loop reduced IRES translation by ~40%, whereas all other mutations that disrupted other predicted stem-loops did not disrupt IRES translation.

To further examine the structure of the BADV-2 IRES, we performed SHAPE structural probing (SHAPE-MaP) (Figure S3). SHAPE reactivities of each nucleotide was mapped onto the BADV-2 IRES. In general, the majority of nucleotides showed high reactivities throughout the IRES. The UAGU (nt 5447–5450), which when mutated to GUUA (M4) (Figure 5, bottom) leads to a decrease in translation by 40%, showed high SHAPE reactivity (Figure S3). In summary, the mutational analyses and structural probing analyses strongly hinted that the BaDV-2 IRES may adopt an unstructured RNA element that directs translation.

3.6. *BaDV-2 IGR Does Not Support IRES Activity in S2 Cells*

Next, we examined whether the BaDV-2 IGR IRES functions in mock- and dicistrovirus-infected S2 cells. We tested in both mock and infected cells as it has been shown that IRES translation is stimulated under virus infection [34]. We transfected in vitro transcribed bicistronic RNAs containing wild-type or mutant CrPV IRES or full-length wild-type, "minimal" or the 3′deletion (3′ DEL 5) mutant BaDV-2 IRES) into S2 cells, then infected with CrPV (MOI = 20) and measured FLuc activity at 6 h after transfection. In CrPV-infected cells, only the reporter RNAs containing the CrPV IRES displayed FLuc activity, whereas reporter RNAs containing the mutant CrPV IRES or any of the BaDV-2 IRES did not, suggesting that the BaDV-2 IRES translation is not supported in S2 cells. (Figure 6A).

We reasoned that the BaDV-2 IRES may only be functional under a viral genome context. We previously showed that the heterologous dicistrovirus IRESs can support IRES translation and infection using a chimeric CrPV infectious clone [18,35]. Briefly, we replaced the CrPV IGR IRES with the BaDV-2 IGR IRES (full length, nt 5325 to 5694) in the CrPV infectious clone (Figure 6B). We also generated a clone with an inserted P2A "stop-go" peptide site between BaDV-2 IGR IRES and ORF2 to ensure proper expression of CrPV ORF2 [36]. We first determined whether the BaDV-2 IRES could support translation of the infectious clone in Sf-21 extracts. Our results showed that the CrPV but not BaDV-2 IGR IRES supported translation of ORF2 proteins, as evidenced of VP2 expression (Figure 6C). To determine whether the CrPV-BaDV-2 chimeric clone is infectious, we transfected in vitro transcribed infectious clone RNAs into S2 cells and then followed VP2 expression by immunoblotting at 144 h after transfection. We previously showed that detection of VP2 after transfection of CrPV infectious clone RNA reflects productive CrPV infection in S2 cells [17]. VP2 was detected in S2 cells transfected with wild-type CrPV clone but not a mutant CrPV clone containing a stop codon in ORF1 or with the CrPV-BaDV-2 IGR IRES chimeric clone (both with and without P2A peptide site) (Figure 6D). Moreover, cytopathic effects were observed in S2 cells transfected with the wild-type CrPV clone but not with the other clones. In summary, the BaDV-2 IGR IRES is not active in *Drosophila* cells or under CrPV infection.

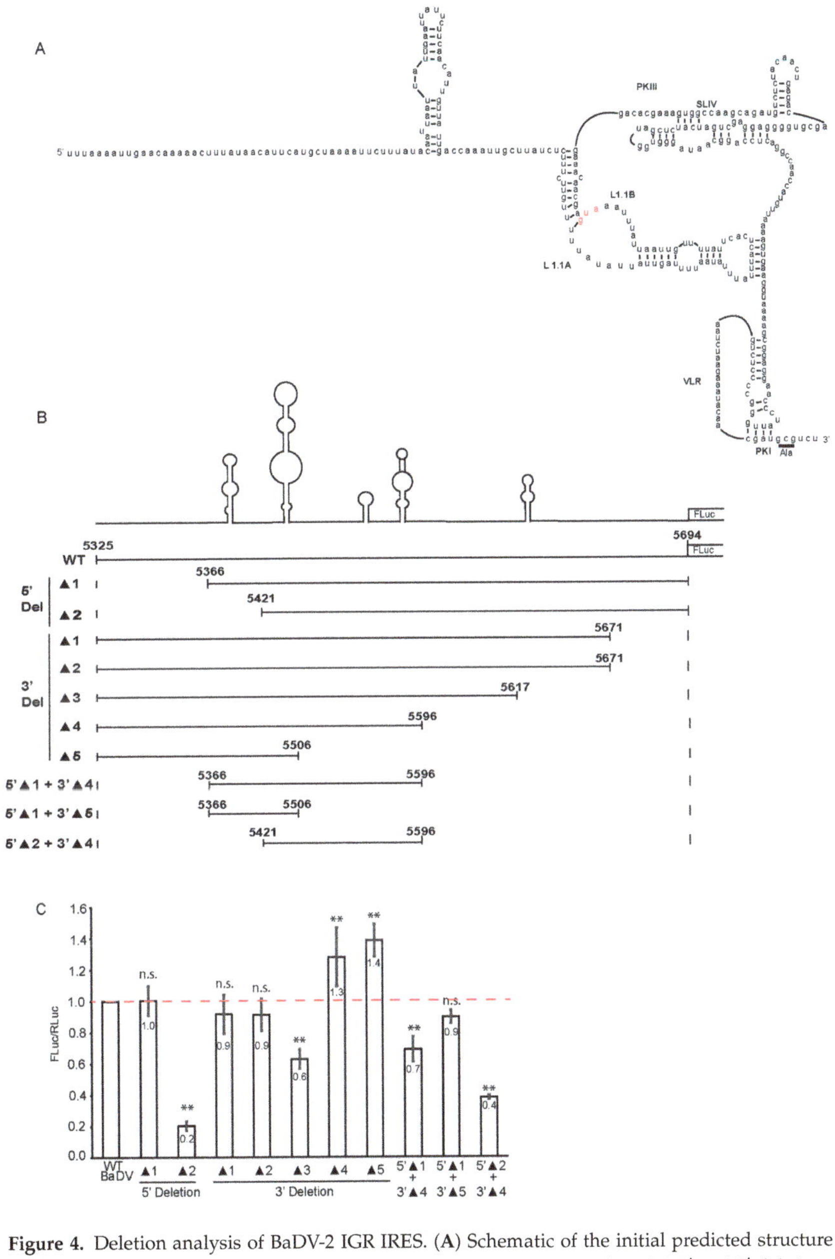

Figure 4. Deletion analysis of BaDV-2 IGR IRES. (**A**) Schematic of the initial predicted structure of the BaDV-2 IGR IRES. The red AUG denotes the predicted start codon. (**B**) 5′ and 3′ deletions were generated within the BaDV-2 IGR. "▲" = DEL = deletion. (**C**) Bicistronic RNAs containing the corresponding deletion mutants were incubated in Sf-21 extracts and RLuc and FLuc activities measured and normalized to the wild-type BaDV-2 IRES. The red line denotes the FLuc:RLuc ratio of the wild type IRES for comparison. A one-way ANOVA statistical test was used to determine the p value and thus the significance levels. Groups were compared to WT BaDV. ** $p < 0.01$. "n.s." denotes the difference is not significant between the experimental groups and WT ($p > 0.05$). Shown are the averages from at least three independent experiments ± standard deviation.

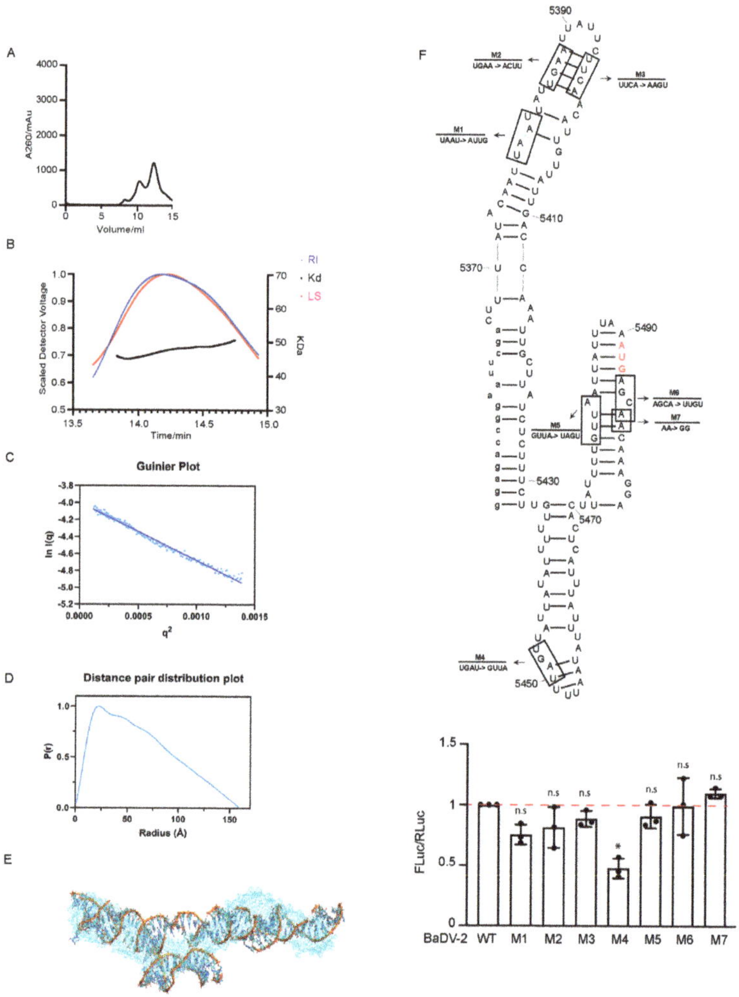

Figure 5. Structural analysis of BaDV-2 IGR IRES. (**A**) FPLC profiles associated with BaDV-2 IGR "minimal" IRES IVT RNA purification. (**B**) SEC-MALS traces of the peak from the BaDV-2 IRES RNA run and the absolute molecular weight across them. (**C**) The absence of aggregation and the presence of a liner Guinier region in the sample indicates its purity, which enables the calculation of Rg from the low-angle region data and signifies the homogeneity of the sample. (**D**) Normalized pair distance distribution plots for BaDV-2 IRES RNA permit the determination of Rg derived from the SAXS dataset and including each molecule's Dmax. P(r) versus radial distance plot indicates an elongated RNA structure in solution. (**E**) Predicted BaDV-2 IRES RNA structure model from SAXS data (light blue) overlapped with atomistic structure, represented in ribbons. (**F**) The secondary structural model of BaDV-2 IRES corresponding to the predicted model as shown in (**E**). The red letters denote the predicted AUG start codon. Mutation analysis of BaDV-2 "minimal" IRES with bicistronic reporter Sf-21 translation assays were performed and the results (M1-7) were measured by luciferase assays and normalized to wild-type (WT) BaDV-2 IGR IRES "minimal" IRES (bottom). Groups were compared to WT BaDV-2 IRES (red dotted line). An ANOVA statistical test was used to determine the p value and thus the significance levels. * $p < 0.05$. "n.s." denotes the difference is not significant between the experimental groups and WT ($p > 0.05$). Shown are the averages from at least three independent experiments ± standard deviation.

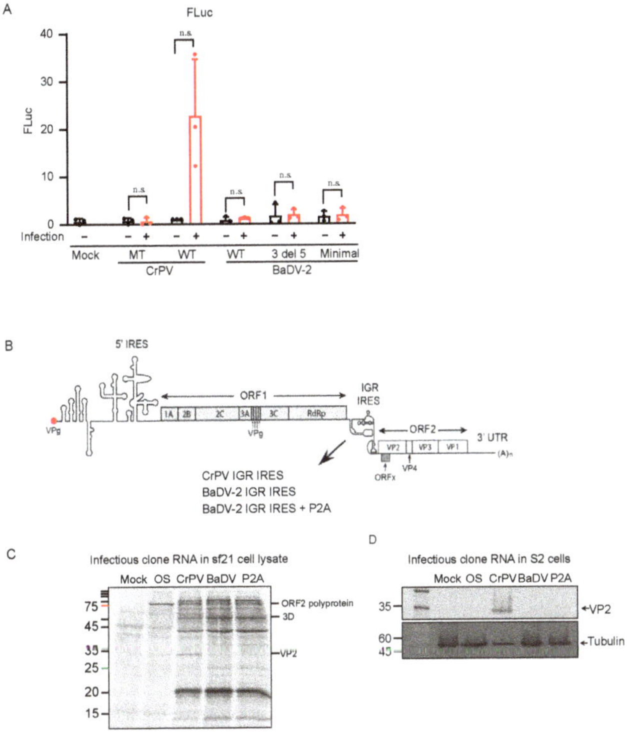

Figure 6. BaDV-2 IGR IRES activity in infectious clone and in cells. (**A**) Bicistronic reporters were transfected into S2 cells followed by mock infection or CrPV infection (MOI = 20). Cells were collected 6 h post transfection. Bicistronic RNAs tested contain the wild-type CrPV, mutant CrPV (CC6214-5 to GG to disrupt PKI base-pairing), wild-type BaDV-2 full-length IGR IRES (WT BaDV-2), BaDV-2 3′ DEL 5, and BaDV "minimal" IRES (double deletion at both 3′ and 5′ end. Luciferase activities were measured and normalized to wild-type CrPV with mock infection. Red and black dots represent data points from three independent experiments. A paired t-test was used to determine the p value and thus the significance levels. "n.s." denotes the difference is not significant between the control groups and the experimental groups ($p > 0.05$). Shown are the averages from at least three independent experiments ± standard deviation. (**B**) Schematic of infectious clone CrPV (CrPV, CrPV_IGR_IRES) and chimeric clones with CrPV IGR IRES replaced with BaDV-2 IGR IRES (BaDV-2 IGR IRES) and with P2A-site added after BaDV-2 IGR IRES (BaDV-2 IGR IRES+ P2A). (**C**) Infectious clone RNAs were incubated in Sf-21 extracts containing [35S]-methionine/cysteine and then monitored by SDS-PAGE analysis. Mock = mock transfection; OS = ORF1Stop; CrPV = pCrPV; BaDV = BaDV-2 IGR IRES; P2A = BaDV-2 IGR IRES + P2A. Shown is a representative gel from at least three independent experiments. (**D**) In vitro transcribed infectious clone RNAs were transfected into S2 cells for 144 h. VP2 expression was detected by immunoblotting. Shown are a representative SDS PAGE gel and the averages from at least three independent experiments ± standard deviation.

3.7. BaDV-2 IGR IRES Does Not Bind to Purified Ribosomes

A unique feature of dicistrovirus IGR IRESs is its ability to recruit ribosomes directly without the aid of initiation factors [37]. The CrPV IGR IRES can assemble ribosomes by first recruiting 40S subunits followed by 60S subunit joining; however, ~8% of the time, the CrPV IRES can also bind to pre-formed 80S ribosomes [8,38]. We assessed ribosomal binding to the IRES using an established filter binding assay [35]. Incubating increasing concentration of purified salt-washed ribosomes with wild-type CrPV IRES resulted in

CrPV IRES: ribosome complexes with an apparent K_D of 1.4 ± 0.3 nM (Figure 7), which is similar to previous reports [35]. By contrast, a mutant CrPV IRES containing disruptions that disrupt all three pseudoknot structures showed little to no 80S binding (Figure 7). The BaDV-2 IRES did not bind to purified ribosomes (Figure 7, K_D = 940.3 ± 4377 nM), thus indicating that this IRES is likely not using a factorless mechanism for translation.

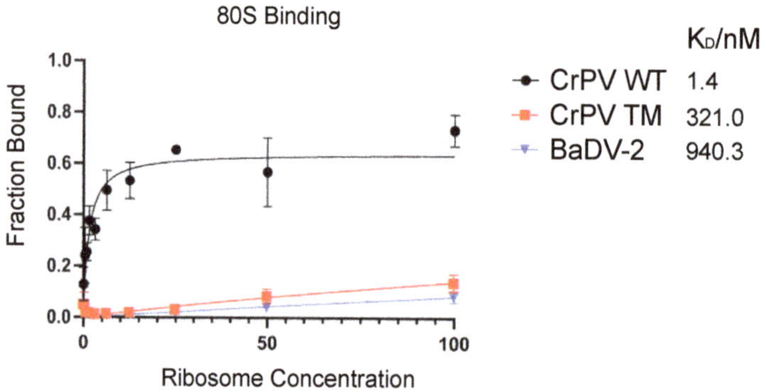

Figure 7. Affinity of 80S-BaDV-2 IGR IRES complexes. Filter binding assays. [32P]-BaDV-2 IGR IRES, [32P]-(CrPV) IGR IRES or mutant (ΔPKI/II/III) CrPV IGR IRES were incubated with increasing amounts of purified salt-washed 80S. The fractions bound were quantified by filter binding assay followed by phosphorimager analysis. CrPV WT has a K_D value of 1.4 nM, CrPV mutant (MT, mutations that disrupt all three pseudoknot base-pairings) has a K_D value of 321.0 nM and WT BaDV-2 has a K_D value of 940.3 nM. Shown are the averages from at least three independent experiments ± standard deviation.

3.8. BaDV-2 IGR IRES Initiates Translation from an AUG Start Codon

Our results so far point to an atypical IRES mechanism that drives BaDV-2 IGR IRES translation. We next focused on determining the start site of the BaDV-2 IGR IRES. In-frame stop codons (SC) were inserted across the full-length BaDV-2 IGR (Figure 8A). Stop codon insertions at nt 5503, 5608, and 5668 (SC1, 2, 3) all decreased IRES activity to ~5–25% of wild-type IRES activity (Figure 8, thus suggesting that the start site is upstream of nt 5503. Moreover, these experiments are in line with the deletion analysis that the "minimal" IRES is contained between nt 5366 and 5506. A stop codon insertion at nt 5444 (SC4) did not affect IRES activity, thus indicating that the start codon is between nts 5444 and 5503. We hypothesized that the AUG (nt 5491–5493) codon may be the initiation codon of the IRES. We mutated the AUG (nt 5491–5493) codon to UAG (M1) or GCG (M2) or deleted the AUG (DEL). Deletion of the AUG codon or mutating to UAG (M1) dramatically inhibited IRES translation by ~75–80% compared to wild-type IRES activity (Figure 8), albeit some residual ~20–25% IRES activity was observed. We reasoned that the residual IRES activities with the AUG codon mutations may be due to ribosomes recruited to the BaDV-2 IRES and then scanned downstream to initiate translation from the AUG codon of FLuc. To address this, we deleted the AUG codon of FLuc in combination with the deletion of the mutant AUG (nt 5491–5493) codons. Deleting the AUG codon of FLuc within the wild-type BaDV-2 IGR IRES inhibited IRES activity by ~10–15% compared to the wild-type IRES with the AUG codon of FLuc. By contrast, combining the deleted AUG codon of FLuc with deletion of AUG (nt 5491–5493) or M1 or M2 mutants abolished IRES-mediated FLuc activity completely. These results strongly demonstrated that the BaDV-2 IGR IRES initiates from the AUG (nt 5491–5493) codon and that a fraction of the 40S ribosomal subunits recruited to the IRES can scan to initiate translation from a downstream AUG codon.

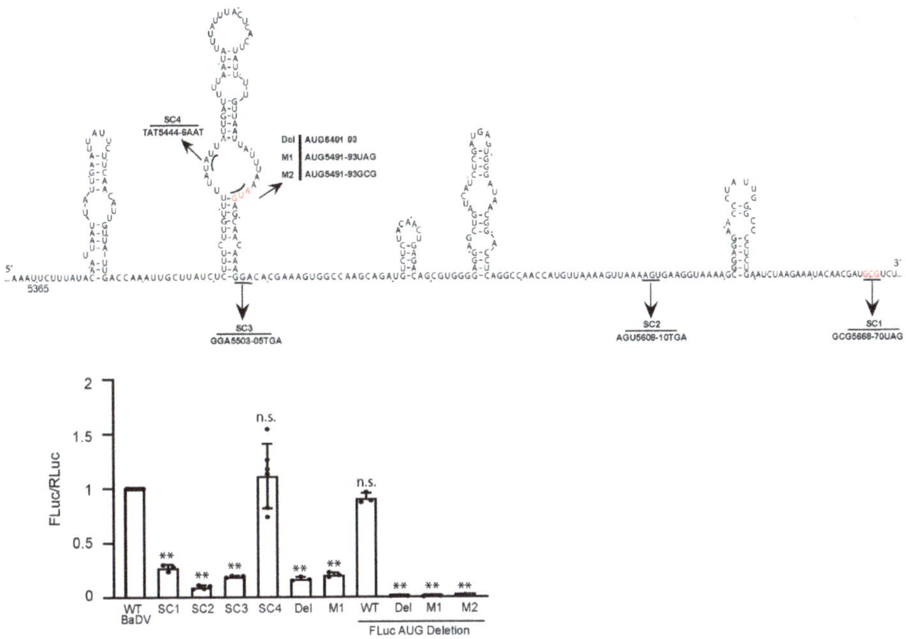

Figure 8. Stop codon mutations to confirm the BaDV-2 IGR IRES initiation site. (**A**) Schematic of stop codon mutations along the BaDV-2 IGR IRES (upper). The red letters denote the predicted AUG start codon. In vitro translation assays were performed in Sf-21 cell lysate with the indicated mutations using bicistronic reporter RNAs. RLuc and FLuc activities were measured and normalized to the wild-type BaDV-2 bicistronic RNA (bottom). An ANOVA statistical test was used to determine the p value and thus the significance levels. Groups were compared to WT BaDV. ** represents $p < 0.01$. "n.s." denotes the difference is not significant between the experimental groups and WT ($p > 0.05$). Shown are the averages from at least three independent experiments ± standard deviation.

3.9. *BaDV-2 IGR IRES Translation Requires eIF4A and eIF2*

To determine whether the BaDV-2 IRES uses translation factors, we tested IRES translation in extract incubated with inhibitors that target specific translation initiation factors (Figure 9A). Specifically, hippuristanol inhibits eIF4A helicase activity and NSC119889 blocks eIF2 from binding to initiator Met-tRNA$_i$ [39,40]. We pre-incubated these inhibitors in RRL or Sf-21 extracts followed by addition of IRES-containing reporter RNAs. Incubating with hippuristanol (10 μM and 15 μM) or NSC119889 (12.5 μM) significantly decreased scanning-dependent RLuc translation, as expected (Figure 9B,C). By contrast, CrPV IRES-mediated FLuc translation was relatively resistant to hippuristanol or NSC119889, in line with the CrPV IRES driving factorless translation (Figure 9B,C) [41]. Interestingly, BaDV-2 IRES translation was sensitive to both drug treatments inhibiting translation as much as scanning dependent, thus demonstrating that eIF4A and eIF2 ternary complexes are required for BaDV-2 IRES translation.

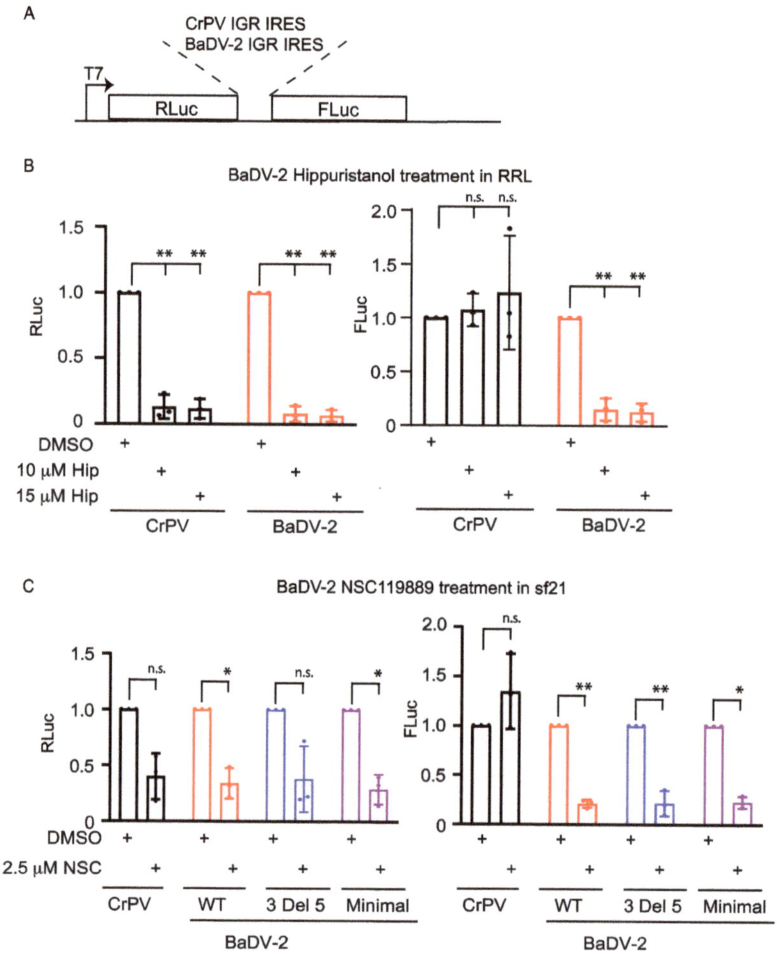

Figure 9. Effects of translation inhibitors on BaDV-2 IRES translation. (**A**) Schematic illustration of bicistronic reporter RNAs with CrPV or BaDV-2 IGR IRESs. In vitro transcribed reporter RNAs were incubated (**B**) in RRL with hippuristanol or (**C**) in Sf21 extracts with NSC119889 at the indicated concentrations. RLuc and FLuc activities were measured and normalized to DMSO-treated lysates with wild-type CrPV or BaDV-2 IGR IRES bicistronic RNAs. A one-way ANOVA statistical test was used to determine the p values. * $p < 0.05$, ** $p < 0.01$. "n.s." denotes the difference is not significant between the control groups and the experimental groups ($p > 0.05$). Shown are the averages from at least three independent experiments ± standard deviation.

4. Discussion

To date, the dicistrovirus family contains the most streamlined mechanisms of eukaryotic translation utilizing an RNA structure to directly recruit ribosomes and initiate translation. Recently, variations in dicistrovirus IRESs have come to light, showing that the dicistrovirus IRESs can have distinct ribosome assembly pathways (ex. IRES binds to preformed 80S) and can initiate from ribosomal P or A sites [6,7]. Here, in this study, we focused on a mechanism of a whitefly dicistrovirus virus, BaDV-2. Like other dicistroviruses, the BaDV-2 intergenic region can support IRES translation in vitro; however, several evidence demonstrate that this mechanism is distinct from other known dicistrovirus IRES mechanisms. One, the BaDV-2 IRES does not bind to ribosomes directly and likely requires a subset of translation initiation factors and other factors. Our results showed that eIF2 and eIF4A are required for BaDV-2 IRES translation. Two, the BaDV-2 IRES is within a 140-nucleotide element and starts translation from an AUG codon, in line with the requirement of eIF2 for IRES translation. Three, the BaDV-2 IRES likely does not utilize an RNA structure or may adopt different conformations to direct ribosome recruitment. Extensive mutational analysis and SAXS analysis and the requirement of the RNA helicase eIF4A support this model. Our results demonstrate that the BaDV-2 contains a novel IRES mechanism that is distinct from other dicistrovirus IRES mechanisms to date and may represent a novel class of IRES.

Our studies highlight the diverse IRES mechanisms that may be used within the family *Dicistroviridae*. It is likely that the BaDV-2 IRES evolved mechanisms that require factors in order to adapt to and drive viral protein synthesis in BaDV-2 infection in whiteflies. Thus, there may be specific cellular infection conditions that promote BaDV-2 IRES translation. Indeed, we showed that the BaDV-2 IRES is not functional in *Drosophila* cells and under CrPV infection (Figure 6). However, it is interesting to note that the BADV-2 IRES translation is supported in vitro in Sf21 extracts and not in *Drosophila* cells, thus pointing to specific host factors required for BADV-2 IRES translation. To date, until this study, dicistrovirus IRES mechanisms in general all utilize a factorless ribosome assembly mechanism; however, it is clear that the diverse translation mechanisms are utilized under specific dicistrovirus infections. The BaDV-2 IRES represents one of the most divergent IRES mechanisms within the family *Dicistroviridae*, which likely means that there are many alternative viral translation strategies that are utilized by dicistroviruses. It will be interesting to follow the evolutionary trajectory of dicistroviruses and whether they utilize a factorless IRES mechanism, a BaDV-2 IRES-like mechanism, or a yet undiscovered translation mechanism.

The BaDV-2 IRES utilizes at least eIF2 and eIF4A for IRES translation. Moreover, the exact structure of BaDV-2 that supports IRES function needs further investigation. Although SAXS analysis predicted a structure containing several stem-loops, our mutational analysis and SHAPE structural probing did not support this model. The data point to an unstructured or potential remodeling of the RNA structure that supports BaDV-2 IRES function, which is supported by its requirement for eIF4A. This evidence suggests a mechanism that is a bit reminiscent of some cellular IRESs that are thought to use unstructured elements that may be ribosome landing elements [42,43]. Further investigation into the specific BaDV-2 IRES elements and/or RNA conformations is warranted.

Our results have for the first time provided biochemical evidence of a bona fide -1 FS signal within a dicistrovirus genome. The BaDV-2 -1 FS signal contains standard elements of a frameshift element including a stem-loop and an upstream slippery sequence [28]. We ruled out the requirement of a predicted pseudoknot for BaDV-2 -1 FS activity in vitro. The -1 FS signal is between the 3C protease and RNA-dependent RNA polymerase (RdRP) [16]. Thus, it is likely that the -1 FS signal regulates the exact stoichiometry of the RdRP to other viral non-structural proteins for precise replication activity during the viral life cycle. It will be of interest to identify other -1 FS signals within discistroviruses and whether this is more widespread within this viral family. It is noteworthy that BaDV-1 has a typical dicistronic genome, while BaDV-2 is primarily tricistronic.

The whitefly complex *Bemisia tabaci* is a critical agricultural pest that occurs in most tropical and subtropical countries and affects important crops, like cotton, cassava, beans, soybeans, and vegetables [44–47]. The identification of viruses such as BaDV-1 and BaDV-2 that are associated with whiteflies may provide a direction for biopesticide approaches to control these pests [16,48], for example, those baculoviruses used for controlling caterpillars [49–51]. Understanding the molecular mechanisms, such as IRES-mediated viral protein synthesis, will provide valuable insights into the viral infection strategies of these pest-associated viruses that may be exploited for therapeutic approaches in agriculture.

Supplementary Materials: The following supporting information can be downloaded at: https://www.mdpi.com/article/10.3390/v16050695/s1, Figure S1: Predicted secondary model of BaDV-2

IGR IRES; Figure S2: BaDV-2 IGR IRES activity in mock-infected and CrPV-infected cells; Figure S3: SHAPE reactivity of BaDV-2 IRES.

Author Contributions: Conceptualization, Y.C. and E.J.; methodology, Y.C., S.C., J.C. and H.S.P.; validation, Y.C., S.C., J.C. and H.S.P.; formal analysis, Y.C., S.C., H.S.P., T.R.P. and E.J.; investigation, Y.C., S.C., J.C. and H.S.P.; resources, Y.C., S.C., J.C., H.S.P., T.R.P. and E.J.; writing—original draft preparation, Y.C. and E.J.; writing—review and editing, Y.C., S.C., H.S.P., T.R.P., A.K.I.-N. and E.J.; visualization, Y.C., S.C., H.S.P., T.R.P. and E.J.; supervision, T.R.P. and E.J.; funding acquisition, T.R.P., A.K.I.-N. and E.J. All authors have read and agreed to the published version of the manuscript.

Funding: This study was supported by a CIHR Project Grant (PJT-178342) and an NSERC Discovery grant (RGPIN-2023-03658) to E.J. H.S.P. is supported partly by NSERC CREATE Postdoctoral Fellowship and Canada Research Chair support to T.R.P. T.R.P. acknowledges the Canada Research Chair program, NSERC Research Tools and Infrastructure, and the Canada Foundation for Innovation grants. SAXS data were collected at the B21 beamline at the DIAMOND synchrotron. A.K.I.-N. is supported by Brazilian Agricultural Research Corporation: 20.22.03.023.00.00 and Fundacao de Apoio a Pesquisa do Distrito Federal: 00193-00002409/2023-17.

Institutional Review Board Statement: Not applicable.

Data Availability Statement: The original contributions presented in the study are included in the article/supplementary material, further inquiries can be directed to the corresponding author/s.

Acknowledgments: We thank the Jan lab for helpful discussions. Hippuristanol was a generous gifted by Jerry Pelletier.

Conflicts of Interest: The authors declare no conflict of interest.

References

1. Merrick, W.C.; Pavitt, G.D. Protein Synthesis Initiation in Eukaryotic Cells. *Cold Spring Harb. Perspect. Biol.* **2018**, *10*, a033092. [CrossRef] [PubMed]
2. Edgar, R.C.; Taylor, B.; Lin, V.; Altman, T.; Barbera, P.; Meleshko, D.; Lohr, D.; Novakovsky, G.; Buchfink, B.; Al-Shayeb, B.; et al. Petabase-scale sequence alignment catalyses viral discovery. *Nature* **2022**, *602*, 142–147. [CrossRef] [PubMed]
3. Shi, M.; Lin, X.-D.; Tian, J.-H.; Chen, L.-J.; Chen, X.; Li, C.-X.; Qin, X.-C.; Li, J.; Cao, J.-P.; Eden, J.-S.; et al. Redefining the invertebrate RNA virosphere. *Nature* **2016**, *540*, 539–543. [CrossRef]
4. Warsaba, R.; Sadasivan, J.; Jan, E. Dicistrovirus-Host Molecular Interactions. *Curr. Issues Mol. Biol.* **2020**, *34*, 83–112. [CrossRef] [PubMed]
5. Gross, L.; Vicens, Q.; Einhorn, E.; Noireterre, A.; Schaeffer, L.; Kuhn, L.; Imler, J.-L.; Eriani, G.; Meignin, C.; Martin, F. The IRES5′UTR of the dicistrovirus cricket paralysis virus is a type III IRES containing an essential pseudoknot structure. *Nucleic Acids Res.* **2017**, *45*, 8993–9004. [CrossRef] [PubMed]
6. Miścicka, A.; Lu, K.; Abaeva, I.S.; Pestova, T.V.; Hellen, C.U. Initiation of translation on nedicistrovirus and related intergenic region IRESs by their factor-independent binding to the P site of 80S ribosomes. *RNA* **2023**, *29*, 1051–1068. [CrossRef] [PubMed]
7. Abaeva, I.S.; Vicens, Q.; Bochler, A.; Soufari, H.; Simonetti, A.; Pestova, T.V.; Hashem, Y.; Hellen, C.U. The Halastavi árva Virus Intergenic Region IRES Promotes Translation by the Simplest Possible Initiation Mechanism. *Cell Rep.* **2020**, *33*, 108476. [CrossRef]
8. Petrov, A.; Grosely, R.; Chen, J.; O'leary, S.E.; Puglisi, J.D. Multiple Parallel Pathways of Translation Initiation on the CrPV IRES. *Mol. Cell* **2016**, *62*, 92–103. [CrossRef]
9. Pisareva, V.P.; Pisarev, A.V.; Fernández, I.S. Dual tRNA mimicry in the Cricket Paralysis Virus IRES uncovers an unexpected similarity with the Hepatitis C Virus IRES. *eLife* **2018**, *7*, e34062. [CrossRef]
10. Murray, J.; Savva, C.G.; Shin, B.-S.; E Dever, T.; Ramakrishnan, V.; Fernández, I.S.; Biology, M.L.O.M.; States, U. Structural characterization of ribosome recruitment and translocation by type IV IRES. *eLife* **2016**, *5*, e13567. [CrossRef]
11. Fernández, I.S.; Bai, X.-C.; Murshudov, G.; Scheres, S.H.; Ramakrishnan, V. Initiation of Translation by Cricket Paralysis Virus IRES Requires Its Translocation in the Ribosome. *Cell* **2014**, *157*, 823–831. [CrossRef] [PubMed]
12. Muhs, M.; Hilal, T.; Mielke, T.; Skabkin, M.A.; Sanbonmatsu, K.Y.; Pestova, T.V.; Spahn, C.M. Cryo-EM of Ribosomal 80S Complexes with Termination Factors Reveals the Translocated Cricket Paralysis Virus IRES. *Mol. Cell* **2015**, *57*, 422–432. [CrossRef] [PubMed]
13. Abeyrathne, P.D.; Koh, C.S.; Grant, T.; Grigorieff, N.; Korostelev, A.A. Ensemble cryo-EM uncovers inchworm-like translocation of a viral IRES through the ribosome. *eLife* **2016**, *5*, e14874. [CrossRef] [PubMed]
14. Acosta-Reyes, F.; Neupane, R.; Frank, J.; Fernández, I.S. The Israeli acute paralysis virus IRES captures host ribosomes by mimicking a ribosomal state with hybrid tRNAs. *EMBO J.* **2019**, *38*, e102226. [CrossRef] [PubMed]
15. Boros, Á.; Pankovics, P.; Simmonds, P.; Reuter, G. Novel Positive-Sense, Single-Stranded RNA (+ssRNA) Virus with Di-Cistronic Genome from Intestinal Content of Freshwater Carp (*Cyprinus carpio*). *PLoS ONE* **2011**, *6*, e29145. [CrossRef] [PubMed]

16. Hedil, M.; Nakasu, E.Y.; Nagata, T.; Wen, J.; Jan, E.; Michereff-Filho, M.; Inoue-Nagata, A.K. New features on the genomic organization of a novel dicistrovirus identified from the sweet potato whitefly Bemisia tabaci. *Virus Res.* **2020**, *288*, 198112. [CrossRef]
17. Kerr, C.H.; Wang, Q.S.; Keatings, K.; Khong, A.; Allan, D.; Yip, C.K.; Foster, L.J.; Jan, E. The 5′ Untranslated Region of a Novel Infectious Molecular Clone of the Dicistrovirus Cricket Paralysis Virus Modulates Infection. *J. Virol.* **2015**, *89*, 5919–5934. [CrossRef]
18. Wang, X.; Vlok, M.; Flibotte, S.; Jan, E. Resurrection of a Viral Internal Ribosome Entry Site from a 700 Year Old Ancient Northwest Territories Cripavirus. *Viruses* **2021**, *13*, 493. [CrossRef]
19. Jan, E.; Sarnow, P. Factorless Ribosome Assembly on the Internal Ribosome Entry Site of Cricket Paralysis Virus. *J. Mol. Biol.* **2002**, *324*, 889–902. [CrossRef]
20. D'souza, M.H.; Mrozowich, T.; Badmalia, M.D.; Geeraert, M.; Frederickson, A.; Henrickson, A.; Demeler, B.; Wolfinger, M.T.; Patel, T.R. Biophysical characterisation of human LincRNA-p21 sense and antisense Alu inverted repeats. *Nucleic Acids Res.* **2022**, *50*, 5881–5898. [CrossRef]
21. Cowieson, N.P.; Edwards-Gayle, C.J.C.; Inoue, K.; Khunti, N.S.; Doutch, J.; Williams, E.; Daniels, S.; Preece, G.; Krumpa, N.A.; Sutter, J.P.; et al. Beamline B21: High-throughput small-angle X-ray scattering at Diamond Light Source. *J. Synchrotron Radiat.* **2020**, *27*, 1438–1446. [CrossRef] [PubMed]
22. Krahn, N.; Zhang, J.; Melnikov, S.V.; Tharp, J.M.; Villa, A.; Patel, A.; Howard, R.J.; Gabir, H.; Patel, T.R.; Stetefeld, J.; et al. tRNA shape is an identity element for an archaeal pyrrolysyl-tRNA synthetase from the human gut. *Nucleic Acids Res.* **2023**, *52*, 513–524. [CrossRef] [PubMed]
23. Pereira, H.S.; Gemmill, D.L.; Siddiqui, M.Q.; Vasudeva, G.; Patel, T.R. Mapping and characterization of G-quadruplexes in monkeypox genomes. *J. Med. Virol.* **2023**, *95*, e28783. [CrossRef] [PubMed]
24. Parisien, M.; Major, F. The MC-Fold and MC-Sym pipeline infers RNA structure from sequence data. *Nature* **2008**, *452*, 51–55. [CrossRef] [PubMed]
25. Svergun, D.; Barberato, C.; Koch, M.H.J. CRYSOL—A Program to Evaluate X-ray Solution Scattering of Biological Macromolecules from Atomic Coordinates. *J. Appl. Crystallogr.* **1995**, *28*, 768–773. [CrossRef]
26. Kozin, M.B.; Svergun, D.I. Automated matching of high- and low-resolution structural models. *J. Appl. Crystallogr.* **2001**, *34*, 33–41. [CrossRef]
27. A Wilkinson, K.; Merino, E.J.; Weeks, K.M. Selective 2′-hydroxyl acylation analyzed by primer extension (SHAPE): Quantitative RNA structure analysis at single nucleotide resolution. *Nat. Protoc.* **2006**, *1*, 1610–1616. [CrossRef] [PubMed]
28. Busan, S.; Weeks, K.M. Accurate detection of chemical modifications in RNA by mutational profiling (MaP) with ShapeMapper 2. *RNA* **2017**, *24*, 143–148. [CrossRef]
29. Reuter, J.S.; Mathews, D.H. RNAstructure: Software for RNA secondary structure prediction and analysis. *BMC Bioinform.* **2010**, *11*, 129. [CrossRef]
30. Brierley, I.; Pennell, S. Structure and Function of the Stimulatory RNAs Involved in Programmed Eukaryotic-1 Ribosomal Frameshifting. *Cold Spring Harb. Symp. Quant. Biol.* **2001**, *66*, 233–248. [CrossRef]
31. Atkins, J.F.; Loughran, G.; Bhatt, P.R.; Firth, A.E.; Baranov, P.V. Ribosomal frameshifting and transcriptional slippage: From genetic steganography and cryptography to adventitious use. *Nucleic Acids Res.* **2016**, *44*, 7007–7078. [CrossRef] [PubMed]
32. Stetefeld, J.; McKenna, S.A.; Patel, T.R. Dynamic light scattering: A practical guide and applications in biomedical sciences. *Biophys. Rev.* **2016**, *8*, 409–427. [CrossRef] [PubMed]
33. Dzananovic, E.; Patel, T.R.; Chojnowski, G.; Boniecki, M.J.; Deo, S.; McEleney, K.; Harding, S.E.; Bujnicki, J.M.; McKenna, S.A. Solution conformation of adenovirus virus associated RNA-I and its interaction with PKR. *J. Struct. Biol.* **2014**, *185*, 48–57. [CrossRef] [PubMed]
34. Wang, Q.S.; Jan, E. Switch from Cap- to Factorless IRES-Dependent 0 and +1 Frame Translation during Cellular Stress and Dicistrovirus Infection. *PLoS ONE* **2014**, *9*, e103601. [CrossRef] [PubMed]
35. Au, H.H.T.; Elspass, V.M.; Jan, E. Functional Insights into the Adjacent Stem-Loop in Honey Bee Dicistroviruses That Promotes Internal Ribosome Entry Site-Mediated Translation and Viral Infection. *J. Virol.* **2018**, *92*, 10–128. [CrossRef] [PubMed]
36. Wang, Q.S.; Au, H.H.; Jan, E. Methods for studying IRES-mediated translation of positive-strand RNA viruses. *Methods* **2013**, *59*, 167–179. [CrossRef]
37. Wilson, J.E.; Pestova, T.V.; Hellen, C.U.; Sarnow, P. Initiation of Protein Synthesis from the A Site of the Ribosome. *Cell* **2000**, *102*, 511–520. [CrossRef] [PubMed]
38. Jan, E.; Kinzy, T.G.; Sarnow, P. Divergent tRNA-like element supports initiation, elongation, and termination of protein biosynthesis. *Proc. Natl. Acad. Sci. USA* **2003**, *100*, 15410–15415. [CrossRef]
39. Novac, O.; Guenier, A.-S.; Pelletier, J. Inhibitors of protein synthesis identified by a high throughput multiplexed translation screen. *Nucleic Acids Res.* **2004**, *32*, 902–915. [CrossRef]
40. Cencic, R.; Pelletier, J. Hippuristanol—A potent steroid inhibitor of eukaryotic initiation factor 4A. *Translation* **2016**, *4*, e1137381. [CrossRef]
41. Jang, C.J.; Jan, E. Modular domains of the *Dicistroviridae* intergenic internal ribosome entry site. *RNA* **2010**, *16*, 1182–1195. [CrossRef] [PubMed]

42. Komar, A.A.; Hatzoglou, M. Internal Ribosome Entry Sites in Cellular mRNAs: Mystery of Their Existence. *J. Biol. Chem.* **2005**, *280*, 23425–23428. [CrossRef] [PubMed]
43. Mitchell, S.A.; Spriggs, K.A.; Bushell, M.; Evans, J.R.; Stoneley, M.; Le Quesne, J.P.; Spriggs, R.V.; Willis, A.E. Identification of a motif that mediates polypyrimidine tract-binding protein-dependent internal ribosome entry. *Genes Dev.* **2005**, *19*, 1556–1571. [CrossRef] [PubMed]
44. Byrne, D.N.; Bellows, T.S.J. Whitefly Biology. *Annu. Rev. Entomol.* **1991**, *36*, 431–457. [CrossRef]
45. Legg, J.P.; Sseruwagi, P.; Boniface, S.; Okao-Okuja, G.; Shirima, R.; Bigirimana, S.; Gashaka, G.; Herrmann, H.-W.; Jeremiah, S.; Obiero, H.; et al. Spatio-temporal patterns of genetic change amongst populations of cassava Bemisia tabaci whiteflies driving virus pandemics in East and Central Africa. *Virus Res.* **2014**, *186*, 61–75. [CrossRef]
46. Robertson, J.L.; Stansly, P.A.; Naranjo, S.E. Bemisia: Bionomics and Management of a Global Pest. *J. Econ. Èntomol.* **2014**, *107*, 482. [CrossRef]
47. Schutze, I.X.; E Naranjo, S.; Yamamoto, P.T. Impact of *Bemisia tabaci* MEAM1 (Hemiptera: Aleyrodidae) on Soybean Yield and Quality Under Field Conditions. *J. Econ. Èntomol.* **2022**, *115*, 757–766. [CrossRef] [PubMed]
48. Nakasu, E.Y.; Hedil, M.; Nagata, T.; Michereff-Filho, M.; Lucena, V.S.; Inoue-Nagata, A.K. Complete genome sequence and phylogenetic analysis of a novel dicistrovirus associated with the whitefly Bemisia tabaci. *Virus Res.* **2018**, *260*, 49–52. [CrossRef] [PubMed]
49. Moscardi, F. Assessment of the application of baculoviruses for control of lepidoptera. *Annu. Rev. Èntomol.* **1999**, *44*, 257–289. [CrossRef]
50. Ferreira, B.; Melo, F.; Silva, A.; Sanches, M.; Moscardi, F.; Ribeiro, B.; Souza, M. Biological and molecular characterization of two Anticarsia gemmatalis multiple nucleopolyhedrovirus clones exhibiting contrasting virulence. *J. Invertebr. Pathol.* **2019**, *164*, 23–31. [CrossRef]
51. Arthurs, S.; Dara, S.K. Microbial biopesticides for invertebrate pests and their markets in the United States. *J. Invertebr. Pathol.* **2019**, *165*, 13–21. [CrossRef] [PubMed]

viruses

MDPI

Review

The Repurposing of Cellular Proteins during Enterovirus A71 Infection

Sudeshi M. Abedeera [1], Jesse Davila-Calderon [2], Christina Haddad [2], Barrington Henry [1], Josephine King [2], Srinivasa Penumutchu [1] and Blanton S. Tolbert [1,3,*]

[1] Department of Biochemistry and Biophysics, Perelman School of Medicine, University of Pennsylvania, Philadelphia, PA 19104, USA; sudeshi.abedeera@pennmedicine.upenn.edu (S.M.A.); barrington.henry@pennmedicine.upenn.edu (B.H.); srinivasa.penumutchu@pennmedicine.upenn.edu (S.P.)

[2] Department of Chemistry, Case Western Reserve University, Cleveland, OH 44106, USA; jxd519@case.edu (J.D.-C.); christinahaddad621@gmail.com (C.H.); jak362@case.edu (J.K.)

[3] Howard Hughes Medical Institute, Chevy Chase, MD 20815, USA

* Correspondence: blanton.tolbert@pennmedicine.upenn.edu

Abstract: Viruses pose a great threat to people's lives. Enterovirus A71 (EV-A71) infects children and infants all over the world with no FDA-approved treatment to date. Understanding the basic mechanisms of viral processes aids in selecting more efficient drug targets and designing more effective antivirals to thwart this virus. The 5′-untranslated region (5′-UTR) of the viral RNA genome is composed of a cloverleaf structure and an internal ribosome entry site (IRES). Cellular proteins that bind to the cloverleaf structure regulate viral RNA synthesis, while those that bind to the IRES also known as IRES trans-acting factors (ITAFs) regulate viral translation. In this review, we survey the cellular proteins currently known to bind the 5′-UTR and influence viral gene expression with emphasis on comparing proteins' functions and localizations pre- and post-(EV-A71) infection. A comprehensive understanding of how the host cell's machinery is hijacked and reprogrammed by the virus to facilitate its replication is crucial for developing effective antivirals.

Keywords: enterovirus A71 (EV-A71); ITAF; IRES; 5′-UTR

1. Introduction

Viral infections can cause life-threatening illness. Enterovirus A71 (EV-A71), the causative agent of hand, foot, and mouth disease (HFMD), has resulted in high morbidity and in some cases mortality rates in infants and children under five years old [1]. Mild symptoms of HFMD include rashes on the hands, feet, and mouth, fever, and painful blister-like sores. In addition, infection with EV-A71 has been known to cause more severe complications such as brainstem encephalitis, cardiopulmonary failure, acute flaccid paralysis, and death in severe cases [1,2].

According to the World Health Organization (WHO), EV-A71 was first isolated in 1969 in California, USA; however, some studies showed that the virus could have emerged in 1963 when a worldwide epidemic occurred in the late twentieth century [3]. Outbreaks of EV-A71 have occurred primarily in East Asian countries as well as in Europe and America [4]. In China, there were approximately 7.2 million infections and 2 thousand deaths between 2008 and 2012 [5]. Waves of infections occurred in the Netherlands, France, and Germany from 2007–2013 [5]. Most recent incidences of EV-A71 infection include several cases in Spain in 2016 and around 13,000 cases in Gansu Province, China in 2018 [6,7]. Also in 2018, the Children's Hospital of Colorado reported a 2.75 percent increase in confirmed cases as compared to the average from the past 5 years (2013–2017) [8]. In the same year, Vietnam reported 53,000 clinical cases of HFMD and 6 deaths [9]. Approximately 2500 Taiwanese individuals were infected with EV-A71 in 2020 and needed emergency or outpatient visits to the hospital [2]. Notably, the National Institute of Allergy and Infectious

Diseases (NIAD) listed EV-A71 as an emerging infectious disease [10]. Due to the severity of EV-A71 infection and its life-threatening complications to children, researchers and scientists have been working on ways to combat this virus. To date, there is no antiviral treatment for EV-A71 infection; however, progress has been made in the development of vaccines. There are three inactivated, whole EV-A71 vaccines available in China (efficacy was at least 90% as confirmed in phase III clinical trials). Other types of vaccines are still in the development process [11,12].

EV-A71 belongs to the Picornaviridae family, genus Enterovirus, and species Enterovirus A [13]. It is a non-enveloped RNA virus with a positive sense, single-stranded RNA [14]. The viral genome of EV-A71 is around 7400 nucleotides long and composed of a 5′-untranslated region (5′-UTR), an open reading frame (ORF) that codes for viral polyprotein, and a 3′-UTR [14]. The 5′-UTR of the genome folds into a stable secondary structure that has six stem-loops (SLI-VI), as determined by structure prediction software. The SLI is also known as the cloverleaf structure while SLII-VI forms a type I internal ribosome entry site (IRES) (Figures 1b and 2) [4,15,16]. The 5′-UTR region is a control hub for EV-A71 genome replication and translation.

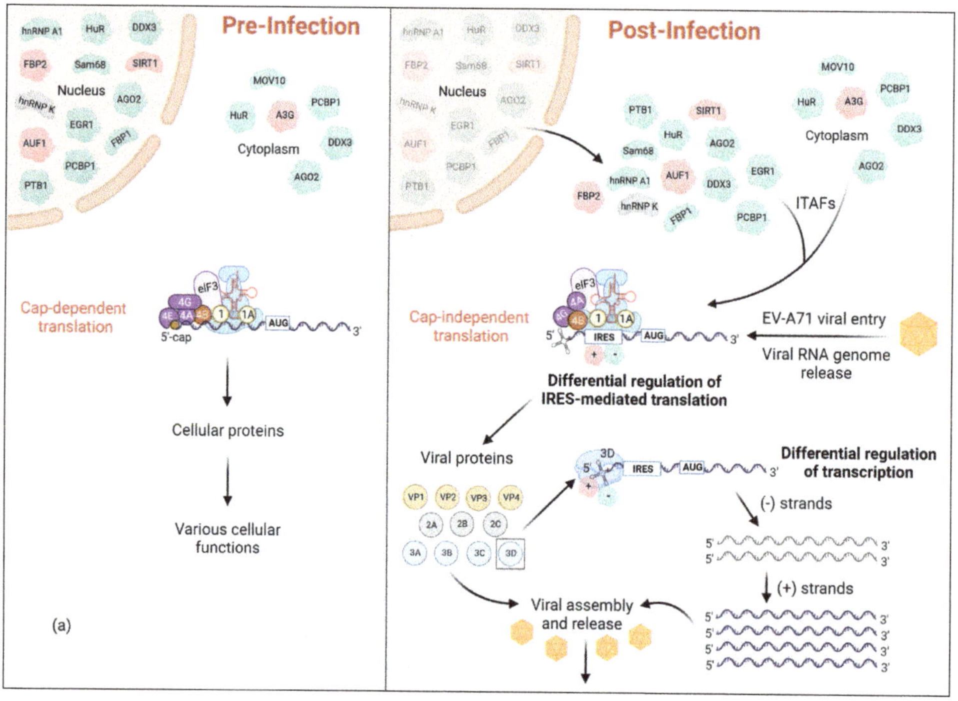

Figure 1. *Cont.*

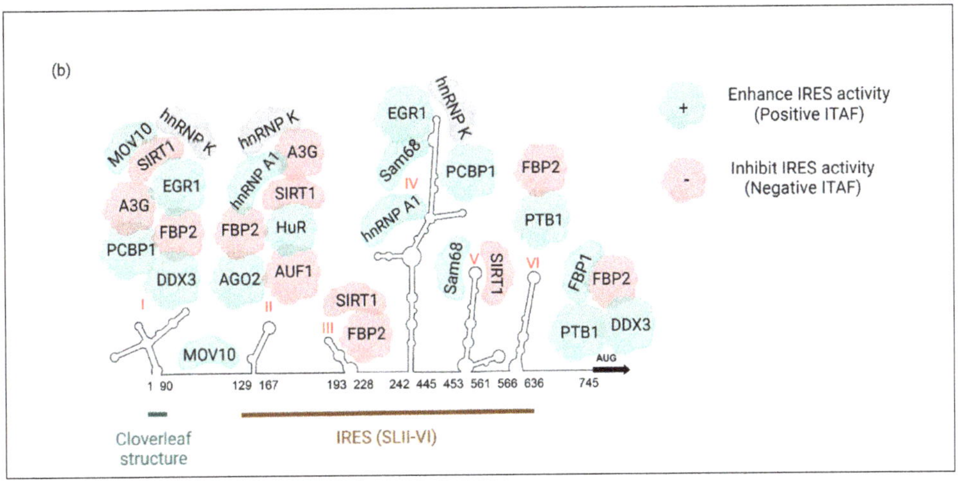

Figure 1. Reprogramming of the cellular environment upon EV-A71 infection. Following viral infection, several host RNA binding proteins, collectively referred to as ITAFs, change their subcellular localization. (**a**) These cellular ITAFs are primarily localized in the nucleus or cytoplasm while some of them can be shuttled between nucleus and cytoplasm based on their cellular functions. Regardless, upon EV-A71 infection, all ITAFs will be relocalized into the cytoplasm to regulate IRES-mediated translation. (**b**) The 5′-UTR of EV-A71 genome consists of the cloverleaf structure (SLI) and the IRES (SLII-VI). Cellular ITAFs bind to specific region(s) within IRES to either positively or negatively regulate IRES-mediated translation. Positive and negative ITAFs are indicated in green and red, respectively. The ITAFs, of which the effect on translation is not discovered yet, are indicated in grey. Some ITAFs can also bind the SLI region to regulate viral replication.

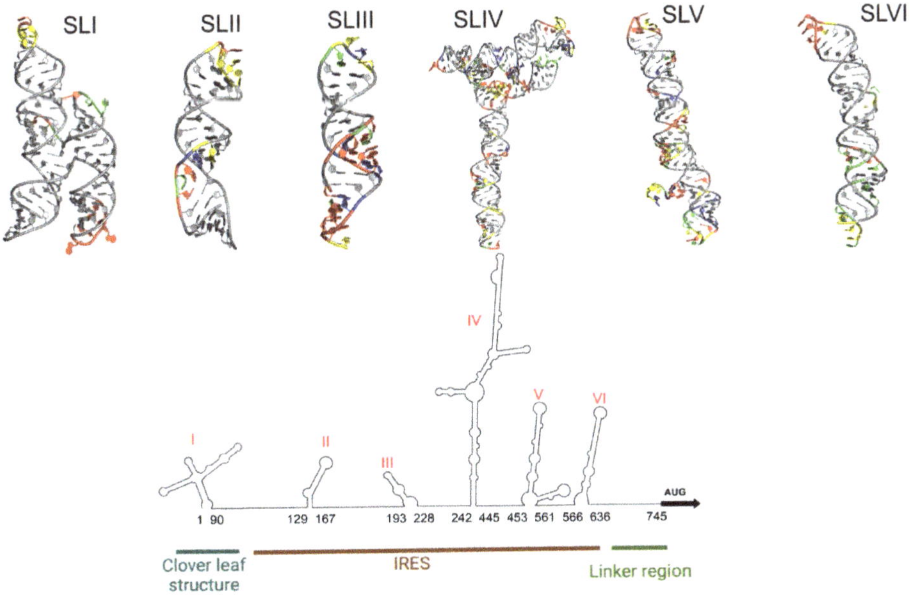

Figure 2. The EV-A71 5′-UTR contains phylogenetically conserved stem-loops (SLs) that adopt well-defined 3D structures. Depicted is a predicted secondary structure of the 5′-UTR of EV-A71,

and 3D models of each individual stem-loop. The four nucleotides: adenosine (red), uridine (green), guanosine (blue), cytidine (yellow) are color coded in the regions that are predicted to be single stranded and most likely to interact with listed ITAFs. The base-paired nucleotides are shown in grey. 3D structures of SL I and II are published PDB structures while SLs III–VI were produced using the FAFFAR module of the ROSETTA software suit. The lowest energy structure for each stem-loop has been selected from a pool of 10,000 structures. The structures of SLI (8DP3) and SLII (5V17) were obtained from the PDB. Note, the structure of SLI of EV-A71 has not been published yet. Shown here is the SLI of the coxsackievirus B3 (CVB3) that was selected due to the high secondary structural homology between SLI regions of EV-A71 and CVB3.

Upon infection, EV-A71 causes rapid inhibition of the cap-dependent translation machinery of the host cell and utilizes host translation machinery for cap-independent translation of viral proteins mediated by its IRES element within the 5′-UTR (Figure 1a) [17]. Once the viral particle binds to the host cell receptor and the viral genome is released into the cytoplasm, translation results in a single polyprotein encoded by the ORF. This resultant polyprotein is cleaved by cellular and/or viral proteases, mainly 3C protease ($3C^{pro}$) to generate a total of 11 viral proteins VP4, VP2, VP3, VP1, 2A, 2B, 2C, 3A, 3B, 3C, and 3D. As shown in the case of several enteroviruses that are related to EV-A71 such as polioviruses (PVs), coxsackieviruses (CVs), and rhinoviruses (RVs), $3C^{pro}$ along with $2A^{pro}$ also cleaves host translation initiation factors (eIFs): eIF4G and eIF5B, and PABP, resulting in the shutdown of host cell's cap-dependent translation [17–20]. Inducing stress granule formation by $2A^{pro}$ and deactivation of eIF2α in response to the endoplasmic reticulum (ER) stress induced by EV-A71 infection further contributes to the shutdown of host translation [17]. It has been reported that PVs use a fragment of eIF5B resulting from $3C^{pro}$ cleavage as a substitute for eIF2α in order to continue viral RNA translation [21]. In addition, $3C^{pro}$ cleaves many other cellular factors to evade cellular responses against viral infections and also to induce cell apoptosis [22–25]. However, the information on the cleavage of diverse host factors specifically upon infection by EV-A71 awaits discovery. For IRES-mediated translation to take place, 40S ribosomal subunit (43S ribosomal preinitiation complex) must be recruited on to the IRES elements mediated by eIFs [26], while auxiliary RNA-binding proteins (RBPs), which are known as IRES trans-acting factors (ITAFs), bind and differentially stabilize IRES elements to positively or negatively regulate the translation. Initiation of IRES-mediated translation occurs using only a subset of eIFs (as certain eIFs are degraded in order to inhibit host cap-dependent translation. The binding of ITAFs at specific regions of IRES can act as RNA chaperones and modulate IRES structure to make translation initiation feasible with the available subset of eIFs [27,28].

Picornaviruses have 5 types of IRESs (Type 1–5) and these IRESs are categorized based on the mechanism of recruiting 40S subunit to the start codon, RNA secondary structure, and ITAF requirements. Comparisons between the different IRES secondary structures and host factors required to recruit the ribosome have been reviewed elsewhere [27,29–31]. EV-A71, along with PV, human rhinovirus type 2 (HRV-2), bovine enterovirus (BEV), and coxsackievirus type B3 (CVB3), contains a Type 1 IRES structure. Type I IRES-containing viruses utilize a "land and scan" mechanism for start codon recognition by the ribosome, where ribosome is recruited at a landing site located at or immediately upstream of IRES SLVI followed by the translation initiation at the landing site or at a downstream site ribosome recognizes as it scans through the linker region between IRES and principal ORF [26,32–34]. The sequence and secondary structure of Type 1 IRES of a given virus is unique while it can have shared structural similarities with other Type 1 IRESs from different viruses. Hence, the host ITAF requirements for EV-A71 IRES differ and are unique from those of other related viruses containing Type 1 IRES elements. Exceptions can exist as in the case of PCBP2 which is implicated in the translation initiation of PV, EV-A71, and BEV [26].

Most ITAFs that are redistributed to the cytoplasm in response to viral infection and cellular stress are either nuclear or cycle between the nucleus and cytoplasm at varying degrees (Figure 1). This cytoplasmic redistribution of ITAFs is achieved mainly via $2A^{pro}$-

and 3C^{pro}-mediated or caspase-mediated pathways that target the nuclear pore complex to perturb nucleocytoplasmic trafficking [35]. Nucleocytoplasmic transport of proteins and RNA occurs through the nuclear pore complexes (NPC). These complexes are made of stationary and soluble components. Stationary components are different nucleoporins (Nups) where phenylalanine-glycine (FG)-Nups that contain intrinsically disordered FG-repeat domains are actively involved in the nucleocytoplasmic transport via interactions with soluble components of NPC, mainly nuclear transport receptors (karyopherins; including importins and exportins). Proteins that contain a nuclear localization signal (NLS) or nuclear export signal (NES) will be recognized by importins or exportins, respectively, to be imported into or exported out of the nucleus, respectively [35].

One of the main strategies of enteroviruses for the perturbation of nucleocytoplasmic trafficking is via the cleavage of specific components of NPC mainly FG-Nups mediated by 2A^{pro} while 3C^{pro} contributes to a lesser extent [35]. EV-A71 is also found to cleave karyopherin subunit α1 (KPNA1), a soluble component of NPC, in a 2A^{pro}- and 3C^{pro}-independent but caspase-dependent manner [36]. While the nuclear proteins are re-localized in the cytoplasm via perturbation of NPC, enterovirus proteases can also cleave NLS of proteins to prevent their import into the nucleus as observed in the case of 3C^{pro}-mediated cleavage of the NLS of La autoantigen upon PV infection [37]. It is noteworthy that the perturbation of protein and RNA trafficking also results in cytoplasmic retention of transcription factors and blocks the mRNA export into the cytoplasm, thereby preventing host antiviral responses (e.g., interferon signaling) and making ribosomes available for efficient translation of viral RNA [35].

The majority of the ITAFs that have been discovered so far are positive regulators of EV-A71 translation. Some of the ITAFs also regulate viral replication via binding to the cloverleaf structure (SLI) of the 5′-UTR. Given the fact that the cellular localization and functional roles of most of these ITAFs have been changed upon EV-A71 infection, it is clear that EV-A71 naturally reprograms the cellular environment to achieve optimal viral replication. Here, we review up-to-date information on ITAFs of EV-A71 that have been identified so far (Table 1) and how they are repurposed during EV-A71 replication, which will aid in the process of developing more effective antiviral therapies against EV-A71 infection.

Table 1. Cellular proteins (ITAFs) that are repurposed upon EV-A71 infection. The ITAFs, their cellular distributions pre- and post-infection, their regulatory site(s) within the IRES (if known), and their effect on IRES-mediated translation are indicated. The PDB IDs of the proteins (bound to RNA) shown in Figure 3 are indicated in bold.

ITAF	Other Names	Cellular Distribution		Regulatory Site	IRES Activity	PDB ID(s)
		Pre-Infection	Post-Infection			
Heterogeneous nuclear ribonucleoprotein A1	hnRNP A1	Nucleus	Cytoplasm	SLII and IV	Enhancement	**4YOE: 1.92 Å** 5MPL: NMR 5MPG: NMR
AU-rich element RNA-binding protein 1	AUF1, hnRNP D	Nucleus	Cytoplasm	SLII	Inhibition	1WTB: NMR 1 × 0F: NMR
Heterogeneous nuclear ribonucleoprotein K	hnRNP K	Nucleus	Cytoplasm	SLI, II, and IV	-	1J5K: NMR 1ZZI: 1.80Å 1ZZJ: 2.30 Å 7CRE: 3.00 Å
Poly(c)-binding protein 1	PCBP1, hnRNP E1	Nucleus, Cytoplasm	Cytoplasm	SLI and IV	Enhancement	1ZTG: 3 Å 3VKE: 1.77 Å
Polypyrimidine tract-binding protein 1	PTB, PTB1 PTBP1 hnRNP I	Nucleus	Cytoplasm	SLVI + linker (564–742 nt)	Enhancement	**2N3O: NMR** 2AD9: NMR 2ADB: NMR 2ADC: NMR

Table 1. *Cont.*

ITAF	Other Names	Cellular Distribution		Regulatory Site	IRES Activity	PDB ID(s)
		Pre-Infection	Post-Infection			
Far upstream binding protein 2	FBP2, FUBP2 KSRP, KHSRP	Nucleus	Cytoplasm	SLI-SLII (1-167) SLII-SLIII (91-228) SLVI + linker (566–745)	Inhibition	**4B8T: NMR**
C-terminus cleaved FBP2	$FBP2_{1-503}$	-	-	5′-UTR	Enhancement	
N-terminus cleaved FBP2	$FBP2_{190-711}$	-	-	5′-UTR	Inhibition	
Far upstream binding protein 1	FBP1 FUBP1	Nucleus	Cytoplasm	Linker (686–714 nt)	Enhancement	1J4W: NMR
Cleaved FBP1	$FBP1_{1-371}$	-	-	Linker (656–674 nt)	Enhancement	
Argonaute 2	Ago2	Nucleus, Cytoplasm (P-bodies)	-	SLII	Enhancement	**5KI6: 2.15 Å**
Human Antigen R	HuR, ELAVL1	Nucleus, shuttles to Cytoplasm	Cytoplasm	SLII	Enhancement	4ED5: 2.00 Å **6G2K: 2.00 Å** 6GC5: 1.90 Å 6GD2: 1.90 Å
Moloney leukemia virus 10 (C-terminus domain)	MOV10	Cytoplasm	Cytoplasm (P-bodies & aggregates perinuclear)	SLI and IRES (Excluding the linker region)	Enhancement	
Silent mating type information regulation 2 homolog 1	SIRT1	Nucleus	Cytoplasm	SLI, II, III and V	Inhibition	
68-kDa Src-associated protein in mitosis	Sam68, KHDRBS1	Nucleus	Cytoplasm	SLIV and V	Enhancement	
Far upstream element-binding protein 3	FUBP3	Nucleus	Cytoplasm	5′-UTR	Enhancement	
Growth arrest and DNA damage-inducible protein 34	GADD34, PPP1R15A	ER membrane Mitochondial membrane	-	5′-UTR	Enhancement	
DEAD-box protein 3	DDX3	Nucleus Cytoplasm	-	Full 5′-UTR	Enhancement	**6O5F: 2.5 Å**
Apolipoprotein B mRNA-editing enzyme, catalytic polypeptide-like 3G	APOBEC3G, A3G	Cytoplasm (mainly) Nucleus P-bodies	Cytoplasm Virions	SLI and II	Inhibition	5ZVA: 2.30 Å 5ZVB: 2.00 Å 6BUX: 1.86 Å 7UXD: 1.50 Å
Staufen homolog 1	Staufen1	Rough ER Cytoplasm	-	5′-UTR	Enhancement	**6HTU: 2.89 Å**
Heterogeneous nuclear ribonucleoprotein H	HNRNP H	Nucleus	Cytoplasm	5′-UTR	Enhancement	
Heterogeneous nuclear ribonucleoprotein F	HNRNP F	Nucleus	Cytoplasm	5′-UTR	Enhancement	**2KFY: NMR** 2KG0: NMR 2KG1: NMR
Early growth response-1	EGR1	Nucleus	Cytoplasm	SLI and IV	Enhancement	4R2A: 1.59 Å

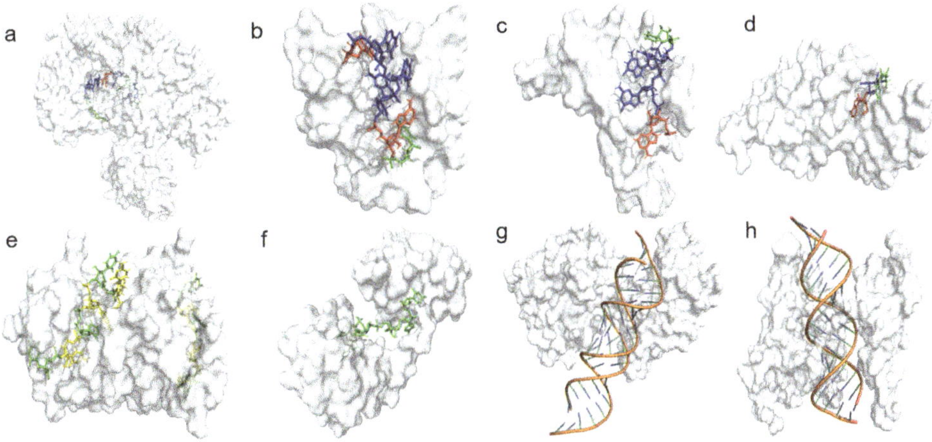

Figure 3. Cellular ITAFs bind with specificity to short and degenerate sequence motifs. Depicted are published PDB structures of cellular ITAFs bound to RNA. These structures have been experimentally determined by both X-ray and nuclear magnetic resonance (NMR) spectroscopy. RNA has been color coded: adenosine (red), uridine (green), guanosine (blue), cytidine (yellow). ITAFs that bind to double-stranded RNA have the RNA depicted as a cartoon structure. The identity and PDB ids of the depicted ITAFs are as follows: (**a**) Argonaute 2 (5KI6), (**b**) hnRNP F (2KFY), (**c**) Far -Upstream Binding Protein 2 (4B8T), (**d**) hnRNP A1 (4YOE), (**e**) Polypyrimidine tract-binding protein 1 (2N3O), (**f**) Human Antigen R (6G2K), (**g**) DEAD-Box Protein 3 (6O5F), (**h**) Staufen homolog 1 (6HTU). Note, the crystal structure of the hnRNP A1-RNA complex (4YOE) was solved with a 5′-AGU-3′ sequence bound to only its RRM1 domain, confirming that a single-strand AG dinucleotide is the minimal specificity motif.

2. Cellular IRES Trans-Acting Factors (ITAFs) That Regulate EV-A71 Translation

2.1. hnRNP A1

Heterogeneous nuclear ribonucleoproteins (hnRNPs) are a family of diverse cellular proteins that play important roles in most aspects of RNA metabolism [38]. The quintessential member of this family of proteins is the abundantly expressed hnRNP A1. HnRNP A1 is primarily involved in the regulation of alternative splicing [39]; however, it is also known to modulate transcription, translation, stability, and nucleo-cytoplasmic export of mRNAs [40–43]. In addition to its regulatory functions in mRNA metabolism, hnRNP A1 is involved in microRNA (miRNA) processing, telomere maintenance, and modulation of transcription factor activity [40,44,45]. HnRNP A1 localizes to the nucleus but has the ability to transport mature mRNAs to the cytoplasm under stress-induced conditions [40,46,47]. HnRNP A1 achieves its varied functions in part due to its ubiquity and preference to bind short degenerate sequences, consisting of a 5′-YAG-3′ motif [48]. The original crystal structure of an hnRNP A1-RNA complex showed that its UP1 domain interacts specifically with the AG dinucleotide sequence via a "nucleobase pocket" formed by the β-sheet surface of RRM1 and the inter-RRM linker (Figure 3d) [49]. Other structures have shown that hnRNP A1 specifically binds 5′-YAG-3′ containing RNA sequences using various binding modes, revealing that it achieves cognate RNA recognition using context-dependent mechanisms [48,50,51].

Upon EV-A71 infection, hnRNP A1 re-localizes to the cytoplasm where it binds specifically to the SLII and SLVI domains of the viral IRES [15,52–55]. Individual knockdowns of hnRNP A1 and its homolog hnRNP A2 showed no significant effect on IRES activity, viral RNA synthesis, and viral 3C protease ($3C^{pro}$). Interestingly, however, simultaneous knockdown of both hnRNP A1 and A2 resulted in a significant attenuation of IRES activity and reduction of viral titers by 4 log_{10} units, indicating that these two proteins are functionally interchangeable [52]. The SLII-hnRNP A1 interaction has been thoroughly characterized biophysically and functionally. HnRNP A1 binds to a phylogenetically conserved bulge loop to change the structure of SLII, which in turn stimulates IRES activity. Mutations or deletions to the bulge loop abrogate hnRNP A1 binding, significantly attenuate IRES translation and inhibit EV-A1 replication by ~5 $\log_{10}$ units [15,53,56]. Thus, the recruitment of hnRNP A1 to the EV-A71 5′-UTR is essential for efficient viral replication.

2.2. AUF1 (hnRNP D)

AU-rich element RNA-binding protein 1 (AUF1) also known as hnRNP D is found in 4 isoforms, p37, p40, p42, and p45, which are generated by alternative splicing and numbered based on their molecular weight [57]. HnRNP D is primarily responsible for mRNA decay via binding to AU-rich elements (AREs) in order to regulate mRNA turnover [57,58]. AREs are destabilizing elements found in most mRNA [59]. Additionally, AUF1 is involved in activating telomerase expression, repressing senescence, and maintaining normal aging [60]. AUF1 is also known to associate with several protein complexes [61–63].

Upon EV-A71 cell infection, AUF1 is transported from the nucleus to the cytoplasm and it associates with the EV-A71 5′-UTR [60]. AUF1 specifically binds the SLII of the IRES region [60,64]. The knockdown of AUF1 showed an increase in IRES activity, production of viral $3C^{pro}$, and viral titers with no effect on viral RNA synthesis or cap-dependent translation [43]. Hence, AUF1 is a negative regulator of IRES-dependent translation and viral replication.

Both AUF1 and hnRNP A1 bind to the same bulge region within the SLII domain. HnRNP A1 binding enhances IRES activity whereas AUF1 binding inhibits IRES activity in a putative mechanism that tunes the levels of IRES-dependent translation to meet the replication needs of EV-A71 [64]. A competitive binding assay between hnRNP A1 and AUF1 demonstrated that the levels of SLII-bound hnRNP A1 reduced upon increasing levels of AUF1 [65]. However, the combined knockdowns of hnRNP A1/A2 and AUF1 had no effect on IRES activity showing that IRES has an intrinsic function that can be tuned by these proteins as suggested by Lin et al. [65]. In addition to EV-A71, AUF1 has been reported to negatively regulate the replication of three related picornaviruses; PV, HRV, and CV in mammalian cells [65,66].

Small virus-derived RNAs (at least vsRNA1-4) are generated in EV-A71 infected cells due to the cleavage of EV-A71 5′-UTR by Dicer. Out of these vsRNAs, vsRNA1 is found to bind the SLII of the IRES to reduce IRES-mediated translation. It has been discovered that vsRNA1 promotes the association of AUF1 with SLII as well as the association of 2 other positive ITAFs, Ago2 and HuR, which will be discussed in detail later in this review. It is possible that vsRNA1 controls EV-A71 translation and replication via its ability to modulate selected ITAF-IRES interactions at SLII [17,67].

2.3. hnRNP K

HnRNP K is a versatile protein with functional roles in the nucleus as well as the cytoplasm [58]. It plays diverse functions in mRNA metabolism including regulation of transcription (via binding to CT-rich promoter regions), alternative splicing, mRNA silencing during cell differentiation (via interaction with GSK3β), mRNA stability (during cellular stress conditions), and translation [68–72]. HnRNP K interacts with RNA via recognizing CU-rich RNA patches while it can also interact with proteins using its Lysine (K)-rich domains [73].

Upon EV-A71 infection, hnRNP K that was localized in the nucleus will be enriched in the cytoplasm [54]. Lin et al. showed that hnRNP K binds to the 5′-UTR, specifically at the regions SL I-II and SLIV to regulate EV-A71 replication. The knockdown of hnRNP K in EV-A71 infected cells reduced viral replication and delayed the synthesis of positive and negative RNA strands [54]. These results indicate that hnRNP K positively regulates EV-A71 replication. The results by Lin et al. [54] do not show that hnRNP K enhances IRES-mediated translation even though it binds at the SLIV domain. However, the interaction of hnRNP K at SLIV may have a stabilization effect on EV-A71 that will promote viral RNA synthesis.

2.4. PCBP1 (hnRNP E1)

Poly(C)-binding protein 1 (PCBP1), also known as hnRNP E1, is a host protein located primarily in the nucleus and can be shuttled between the nucleus and cytoplasm. It has three K-homology (KH) domains that mediate RNA binding [74,75]. PCBP1 binds to CU-rich regions within the 3′-UTR of mRNA to increase its stability [76]. PCBP1 also has roles in modulating alternative splicing, mRNA silencing, transcription, and translation [77–80]. It is known that PCBP1 interacts with the 5′-UTR of PV RNA and facilitates viral RNA replication [75]. However, the mechanisms by which PCBP1 facilitates viral replication and translation of many viruses are not known.

Upon EV-A71 infection, PCBP1 primarily localizes in the cytoplasm and co-localizes with the EV-A71 RNA in the ER-derived membrane. PCBP1 specifically binds to the 5′-UTR at SLI and IV using its KH1 domain. Knockdown of PCBP1 decreased the level of VP1 viral protein while the overexpression of PCBP1 increased the VP1 production. Further, the knockdown of PCBP1 resulted in reduced EV-A71 viral titers while the overexpression of PCBP1 increased the titers. Hence, PCBP1 binds the EV-A71 5′-UTR and positively regulates viral protein expression and virus production [75].

2.5. PTB (hnRNP I)

The polypyrimidine tract-binding protein 1 (PTB or PTBP1), also known as hnRNP I, is an RNA-binding protein where it interacts with polypyrimidine stretches (CU repeats or CU-rich elements) on RNA mediated by four RNA recognition motifs (RRMs) (Figure 3e) [58,81]. Even though PTB is mostly known for its role in the regulation of alternative splicing of pre-mRNAs [82–84], it is also involved in various other cellular processes including mRNA stabilization, regulation of mRNA translation, and miRNA-mediated regulation of gene expression [85–87].

Upon cell infection, PTB is translocated from the nucleus to the cytoplasm to participate in viral processes. A study by Xi et al. discovered that PTB specifically binds to the SL VI (stem-loop VI and linker region, nt 564–742) of IRES using RRM1-2. Their results further demonstrate that PTB is a positive ITAF that enhances the EV-A71 IRES-mediated initiation of translation, viral protein expression, and virus production [88].

2.6. FBP2

The far upstream element binding protein 2 (FBP2 or FUBP2), which is also known as KH-type splicing regulatory protein (KHSRP or KSRP), is a host protein that plays diverse roles including regulation of transcription, pre-mRNA splicing, and mRNA editing [89–91]. FBP2 also regulates mRNA degradation via binding to AREs at 3′-ends as well as the maturation of miRNA precursors (Figure 3c) [92–94]. FBP2 can shuttle between the nucleus and the cytoplasm, and its localization can vary in a tissue- and context-dependent manner [93].

An extensive study by Lin et al. identified FBP2 as a cellular ITAF that binds to the 5′-UTR of EV-A71, in vitro and in vivo [55]. A pull-down assay suggests that FBP2 interacts with the regions SLI-II (nt 1–167), SLII-III (nt 91–228), and SLVI and spacer region (nt 566–745) (stem-loop VI and spacer region) in EV-A71 5′-UTR via a region that involves at least KH2 and KH4 domains. They further showed that FBP2 is relocalized from the nucleus to the cytoplasm during EV-A71 infection. Knockdown of FBP2 resulted in an increase in viral protein synthesis while FBP2 overexpression resulted in a decrease in viral protein synthesis. Further, FBP2 was shown to negatively regulate IRES-dependent translation to confirm the role of FBP2 as a negative regulator (negative ITAF) of IRES function. The competitive binding assay between PTB and FBP2 suggests that FBP2 acts as a negative ITAF via its ability to bind IRES competitively with positive ITAFs, such as PTB [55].

N- and C-Terminus Cleaved FBP2

Upon investigating FBP2, truncated products of the protein were detected in EV-A71-infected cells. The nonstructural proteins of EV-A71: 2A, 2B, 2C, 2BC, 3A, 3AB, 3C, and 3D are not responsible for the cleavage FBP2. However, the truncation of FBP2 is a result of viral replication that was found to occur through various virus-induced mechanisms involving caspase activation, proteasomes, and autophagy [95]. Five truncated products were detected, two of which, $FBP2_{1–503}$ and $FBP2_{190–711}$, bind to the 5′-UTR. Like full-length FBP2, $FBP2_{190–711}$ negatively regulates viral translation, while $FBP2_{1–503}$ is a positive regulator of IRES-driven translation and viral protein synthesis [95]. Hence, EV-A71 infection cleaves FBP2 at the C-terminus to reverse its function from a negative to a positive regulator of viral translation while retaining the ability of FBP2 to bind to the 5′-UTR [95].

2.7. FBP1

The far upstream element binding protein 1 (FBP1 or FUBP1), which is highly homologous to FBP2, can bind to RNA or ssDNA [96]. FBP1 activates the transcription of a proto-oncogene *c-myc* (cellular myelocytomatosis oncogene) that mediates cell growth by binding to the far upstream element (FUSE) upstream of the *c-myc* promoter [97]. It also inhibits the translation of nucleophosmin by binding to the 3′-UTR of its mRNA [98]. Additionally, FBP1 plays a role in post-transcriptional regulation of a growth-associated protein 43 (GAP43) in neural development. It promotes GAP43 mRNA degradation by binding to a pyrimidine-rich region at the 3′-UTR of the transcript [96]. FBP1 is known to interact with the UTRs Hepatitis C virus (HCV) and Japanese encephalitis virus (JEV) RNA genomes to positively and negatively regulate viral replication, respectively [99,100].

A study by Huang et al. identified FBP1 as a positive ITAF for EV-A71 replication [101]. Similar to FBP2, FBP1 naturally resides in the nucleus. However, upon EV-A71 infection, FBP1 is redistributed in the cytoplasm, where most of the steps in viral replication take place [101]. In vitro and in vivo studies show that FBP1 binds to the EV-A71 5′-UTR specifically at the linker region downstream of the IRES (nt 686–714). FBP1 contains four KH-type RNA binding domains (KH1-4) flanked by N- and C-terminal domains. However, KH3-4 is enough for FBP1 to bind the EV-A71 5′-UTR. FBP1 has demonstrated that it can positively regulate the IRES-dependent translation of EV71 via binding the 5′-UTR. FBP1 and FBP2 both bind to the linker region downstream of IRES but have opposing roles in IRES-mediated translation by acting as a positive and a negative regulator, respectively. An in vitro competition binding assay between FBP1 and FBP2 revealed that FBP1 outcompeted FBP2 in binding to the IRES linker region. This result suggests that FBP1 may act as a positive ITAF by preventing the negative ITAF, FBP2 binding to the IRES linker region [101].

C-Terminus Cleaved FBP1

EV-A71 induces the cleavage of FBP1 during the middle stages of infection. The viral protease 2A ($2A^{pro}$) is responsible for the truncation of FBP1 into its primary cleavage product, $FBP1_{1-371}$. However, unlike in the case of FBP2, the FBP1 cleavage is not due to proteasome, lysosome, or caspase activity [102]. Similar to intact FBP1, $FBP1_{1-371}$ binds to the linker region of the 5′-UTR but at a different nucleotide sequence (nt 656–674). The non-competitive and simultaneous binding of FBP1 and $FBP1_{1-371}$ additively enhance IRES-mediated translation as well as EV-A71 viral yield [102].

2.8. Ago2

The Argonaute 2 (Ago2) protein is implicated in transcriptional and post-transcriptional gene silencing [103]. Ago2 is a component of the RNA-induced silencing complex (RISC), where Ago2 binds to a guide RNA, such as a microRNA or a short interfering RNA (Figure 3a) [103,104]. Silencing occurs through binding of this guide RNA to a complementary strand on the target mRNA [103,104]. This will either cause an endonucleolytic cleavage of the mRNA by Ago2 or the inhibition of translation [103–105]. The silencing pathway chosen is based on the degree of complementarity between the guide RNA and its target mRNA [103]. In addition, Ago 2 was found to upregulate translation via binding to AU-rich elements at the 3′-UTR [106].

Ago2 is cellularly localized in the nucleus and the cytoplasm [105]. However, Ago2's localization upon EV-A71 infection has not been reported. Ago2 was identified as an ITAF that specifically binds at the SLII of EV-A71 5′-UTR. The knockdown of Ago2 resulted in reduced IRES activity, demonstrating that Ago2 is a positive ITAF. Further, Ago2 knockdown resulted in a reduction of $3C^{pro}$ expression (25–35%) and viral yields (10^3 fold) [67].

2.9. HuR

Human antigen R (HuR), also known as ELAV-like RNA-binding protein 1 (ELAVL1), binds various mRNAs via recognition of AU-rich elements to regulate their stability and translation (Figure 3f) [107,108]. HuR is predominantly located in the nucleus but shuttles between the nucleus and cytoplasm [109]. The N-terminal RRM1 and RRM2 of HuR mediates its interactions with poly-U or AU-rich elements while RRM3 mediates its interactions with the poly-A tail of target mRNAs [109,110]. HuR has also been shown to positively regulate the expression of Ago2 [111].

Upon EV-A71 infection, HuR previously localized primarily in the nucleus is translocated into the cytoplasm [112]. HuR was identified as an ITAF that specifically binds at the SLII of EV-A71 5′-UTR. The knockdown of HuR resulted in reduced IRES activity, demonstrating that HuR is a positive ITAF. Further, HuR knockdown resulted in a reduction of $3C^{pro}$ expression (25–35%) and viral yields (10^3 fold) [67].

Additive Effects of Ago2 and HuR

The simultaneous knockdown of Ago2 and HuR did not show an additive effect on the reduction of IRES activity, while the effect was additive on the reduction of $3C^{pro}$ expression (2.5 fold) and viral yields (10^6 fold). However, neither independent nor simultaneous knockdown of Ago2 and HuR had an effect on viral RNA synthesis. These results suggest that Ago2 and HuR influence EV-A71 replication by acting as a positive regulator of IRES-dependent translation [67].

As described here, there are several ITAFs that bind SLII of EV-A71 5′-UTR, including Ago2, HuR, AUF1, and hnRNP A1/A2 [52,65,67]. Competitive binding studies were carried out to study the dependency of binding of one protein to another. Based on the results from knockdown assays, AUF1, Ago2, or HuR did not depend on one another to bind to SLII. However, an increase in vsRNA1 caused an increase in the binding of AUF1, Ago2, and HuR to SLII, while the binding of hnRNP A1/A2 was unaffected. Given that AUF1 is a negative regulator and while HuR and Ago2 are positive regulators of IRES activity, it is speculated that vsRNA1 binding at SLII fine-tunes the binding of AUF1, HuR, and Ago2 at SLII to regulate viral IRES-mediated translation [67]. Thus, it can be deduced that SLII acts as a coordination hub to recruit different ITAFs which in turn modulate viral protein synthesis, and that vsRNA1 further sculpts the SLII-ITAF interactions to regulate viral gene expression.

2.10. MOV10

The Moloney Leukemia Virus 10 (MOV10) is a cellular RNA helicase protein found in the cytoplasm [113,114]. It colocalizes with Ago2 in the RISC complex to mediate microRNA-guided mRNA cleavage [115]. MOV10 mediates mRNA degradation of thousands of mRNAs by initially binding to the 3′-UTR and then translocating along the 3′-UTR to unfold the structure and disassemble proteins prior to mRNA degradation [114]. The C-terminus of MOV10 consists of seven highly conserved helicase motifs that account for its RNA binding activity, while its N-terminal CH-domain mediates protein–protein interactions [116].

A study by Wang et al. has demonstrated that MOV10 positively regulates EV-A71 replication [113]. The knock down of MOV10 drastically reduced the levels of viral protein as well as positive-strand RNA. MOV10 was found to interact with SLI, the cloverleaf-like structure, and the IRES of E-AV71 5′-UTR to facilitate viral RNA replication and IRES-dependent translation, respectively. Further, it was shown that MOV10 brings out the positive regulation of EV-A71 replication through its C-terminus (the site of RNA helicase activity), while the N-terminus possesses a potentially inhibitory effect on viral production by inhibiting viral translation via an unknown mechanism. Upon EV-A71 infection, MOV10 formed distinct perinuclear aggregates and co-localized with processing bodies (P-bodies). In addition to its interaction with Ago2, MOV10 was shown to interact with HuR, another positive ITAF that interacts with the EV-A71 5′-UTR [113].

2.11. SIRT1

Silent mating-type information regulation 2 homolog 1 (SIRT1) is a member of the sirtuin family. It is a NAD^+-dependent deacetylase and is involved in a broad range of physiological functions including control of gene expression, metabolism, and aging [117]. SIRT1 has been implicated in obesity-associated metabolic diseases, cancer, aging, cellular senescence, neurodegeneration, and inflammatory signaling in response to environmental stress [117–119]. SIRT1 has been known to positively regulate human immunodeficiency virus 1 (HIV-1) transcription and hepatitis B virus (HBV) replication through the deacetylation of Tat and transcription factor AP-1, respectively [120,121].

SIRT1 is localized in the nucleus; however, upon EV-A71 infection, the protein is translocated from the nucleus to the cytoplasm [122]. According to Han et al., SIRT1 interacts with the SLI (cloverleaf structure) and the SLs II, III, and V within IRES of EV-A71 5′-UTR to repress viral RNA replication and IRES-mediated translation, respectively. At the same time, SIRT1 binds viral $3D^{pol}$ protein and results in the repression of viral genome replication [122].

2.12. Sam68

The 68 kDa Src-associated protein in mitosis (Sam68), also known as the KH domain-containing, RNA-binding, signal transduction-associated 1 (KHDRBS1) protein, is implicated in many processes including cell cycle and signaling, cell growth, alternative splicing, pre-mRNA splicing, and trafficking [123–125]. Sam68 belongs to the signal transduction and activation of RNA (STAR) protein family as well as the hnRNP K homology (KH) domain family of RNA-binding proteins [126]. It contains one KH domain that interacts with RNA and several proline-rich sequences that facilitate protein–protein interactions with SH3- and WW domain-containing proteins [127]. The KH domain of Sam68 recognizes and binds RNA using U(U/A)AA direct repeat motifs [128,129].

Sam68 is an ITAF that resides in the nucleus and gets relocated to the cytoplasm upon EV-A71 infection. Similar redistribution of Sam from the nucleus to the cytoplasm has also been reported in cases of FMDV and PV infections [130,131]. Sam68 specifically interacts with the SLIV and V of 5′-UTR of EV-A71 RNA using its hnRNP K homology (KH) domain. Sam68 was found to interact with other ITAFs including Poly(C)-binding protein 2 (PCBP2) and Poly(A)-binding protein (PABP) to facilitate viral replication. Sam68 is a positive regulator of IRES activity, viral protein expression, and viral titer. However, Sam68 does not have any regulatory effect on viral genome replication [132].

2.13. FUBP3

The far upstream element-binding protein 3 (FUBP3) is a single-stranded NA-binding protein that recognizes only one strand of the far upstream element (FUSE) [90]. The structure of FUBP3 consists of four regularly spaced K homology (KH) domains that recognize similar sequences in single-stranded DNA or RNA targets [133]. FUBP3 has roles in the regulation of transcription, splicing, and translation [134]. FUBP3 is known to bind the 3′-UTR of the Japanese encephalitis virus (JEV) to regulate RNA replication and promote subsequent viral translation and viral particle production [133].

A study by Tsai et al. on the inhibition of EV-A71 replication and internal ribosome entry site (IRES) activity by Kaempferol (a flavonoid) has discovered FUBP3 as a new ITAF that associates with EV71 5′-UTR to enhance the IRES-dependent translation [135]. Another study by Huang et al. has reported the interaction of FUBP3 with the 5′-UTR of EV-A71 to regulate its replication in differentiated neuronal cells [136]. Upon EV-A71 infection, FBP3 in the nucleus will be relocated into the cytoplasm [136].

2.14. GADD34

The growth arrest and DNA damage-inducible protein 34 (GADD34), also known as PPP1R15A, is a protein that is upregulated in response to various cell stress-inducing stimuli. GADD34 interacts with serine/threonine protein phosphatase 1 (PP1) to dephosphorylate eIF2α, thereby restarting protein synthesis for cells to recover from an integrated stress response (ISR) [137–139]. GADD34 is known to play an important role in regulating the interferon response of virus-induced innate immunity [140,141]. While there are many reports regarding the role of GADD34 in the inhibition of viral replication [140–142], GADD34 has been shown to promote the replication of infectious bronchitis virus (IBV) [143]. GADD34 attenuated HIV-1 replication via inhibition of viral protein expression in a mechanism mediated by 5′-UTR/TAR RNA, probably by modulating TAR RNA structure [144].

A study by Li et al. demonstrated that EV-A71 activates GADD34 via viral precursor protein 3CD to promote IRES-mediated viral translation. GADD34 is a short-lived protein and is highly expressed only under the conditions of cellular stress [145]. Upon EV-A71 infection, 3CD upregulates GADD34 translation via the upstream open reading frame (uORF) within the 5′-UTR of GADD34. GADD34 is unable to bind directly to EV-A71 5′-UTR. Hence, 3CD binds at the SLI (cloverleaf structure) and recruits GADD34 to the 5′-UTR of EVA-A71. Once bound at the 5′-UTR, GADD34 promotes the EV-A71 IRES activity through its PEST (P: proline, E: glutamate, S: serine, and T: threonine) repeats (1, 2, and 3). However, the dephosphorylation of eIF2α by GADD34 was unrelated to these observed effects in the upregulation of EV-A71 replication [146].

2.15. DDX3

DEAD-box protein 3 is a DEAD-box RNA helicase that regulates translation and is encoded by the X- and Y-linked paralogs *DDX3X* and *DDX3Y*. DDX3X is ubiquitously expressed and essential for viability while DDX3Y is male specific and shows lower and more variable expression in somatic tissues compared to DDX3X. However, the roles of DDX3X and DDX3Y in translation are functionally redundant [147]. DDX3 has a relaxed substrate specificity and is implicated in many cellular processes such as gene expression including transcription, splicing, mRNA export, translation, cell cycle control, regulation apoptosis, and innate immune signaling (Figure 3g) [148]. It has been known that DDX3 is a prime target for viral manipulation during Hepatitis C virus (HCV), Hepatitis B virus (HBV), Human Immunodeficiency Virus (HIV), and poxvirus infections as viral proteins interact with DDX3 to utilize its function for the process of viral replication [148]. It has been shown that DDX3X is specifically required for the translation initiation of transcripts that possess highly stable secondary structures within their 5′-UTR that resist the unwinding activity of eIF4A. DDX3X binds the 5′-UTR via its interactions with eIF4G and PABP, and works in corporation with eIF4A to destabilize the secondary structure and facilitate ribosome entry [149].

DDX3X shuttles between the nucleus and the cytoplasm and possesses RNA-dependent ATPase/helicase activity [148]. Su et al. demonstrated that DDX3X binds the 5′-UTR of EV-A71 and enhances IRES-dependent translation, partly mediated by viral $2A^{pro}$ and $3C^{pro}$ protease activity. DDX3X can bind efficiently to IRES + spacer, SLI-III, SLIV-VI, and SLVI + spacer regions in the EV-A71 5′-UTR, irrespective of EV71 infection. Their results strongly suggest that the truncated eIF4G (cleaved by viral $2A^{pro}$) binds specifically to SLV and recruits DDX3X to SLVI or a region nearby to locally unwind the secondary structure of SLVI, thereby facilitating ribosome entry and scanning. They also showed that DDX3X also enhances the IRES activity of coxsackievirus A16, Echovirus 9, EMCV, and HCV implicating DDX3X is a general cellular factor for the translation of these highly structured viral IRESs [150].

2.16. APOBEC3G

APOBEC3G (apolipoprotein B mRNA-editing enzyme, catalytic polypeptide-like 3G) or A3G is a member of the APOBEC superfamily. A3G is a cytidine deaminase that contains a conserved His-X-Glu and Cys-X-X-Cys Zn^{2+} coordination motif. A3G is an interferon-inducible cellular protein and plays an important role in defending against viral infections. It has been demonstrated to inhibit the infection of several viruses such as human immunodeficiency virus-1 (HIV-1), T-cell leukemia virus type 1 (HTLV-1), hepatitis B virus (HBV), and Hepatitis C virus (HCV) [151,152]. A3G can inhibit viral replication in a cytidine deaminase activity-dependent (C-terminal domain) manner, as well as in a deaminase activity-independent manner, where it is the N-terminal domain's RNA binding ability that mediates incorporation of A3G into viral particles, thereby disrupting reverse transcription or genome encapsidation [153–155].

Li et al. demonstrated that A3G inhibits EV-A71 virus replication via competitive binding to the 5′-UTR of EV-A71 and inhibiting the 5′-UTR activity. A3G binding impaired the interaction between the 5′-UTR and the host protein poly(C)-binding protein 1 (PCBP1), an ITAF that enhances the viral RNA replication and IRES-mediated translation. A3G was shown to bind the SLI and II while PCBP1 is known to bind to SLI and IV of EV-A71 5′-UTR [75,156]. Hence, A3G binds at SLI and competitively inhibits the binding of PCBP1 at the same location, as suggested by the higher binding affinity of A3G compared to PCBP1 towards the 5′-UTR and the reduction of 5′UTR-PCBP1 interactions with increasing A3G expression levels. However, EV-A71 has developed mechanisms to overcome the suppression by A3G via its degradation through the autophagy–lysosome pathway mediated by viral protein $2C^{pro}$ [156].

2.17. Staufen1

Staufen1 is one of the two homologs of Staufen, a double-stranded RNA (dsRNA) and tubulin-binding protein [157]. Staufen1 contains four dsRNA-binding domains (RBDs) of which dsRBD2-4 are shown to bind dsRNA (Figure 3h) [158]. Staufen1 is known for its ability to regulate cellular mRNA translation, trafficking, and degradation via Staufen1-mediated RNP formation. Staufen1 enhances the translation efficiency via binding to the 5′-UTR while it can bind to the 3′-UTR to promote mRNA degradation [27,159]. A study by Chen et al. has demonstrated that Staufen1 facilitates both translation and replication of the EV-A71 genome. The RBD2-3 of Staufen1 was identified to interact with the 5′-UTR EV-A71 to enhance the IRES activity as well as the translation efficiency. Further, the binding of Staufen1 at the 5′-UTR increased the stability of viral RNA [160].

Staufen1 has been reported to play roles in the life cycles of other RNA viruses, including Hepatitis C virus (HCV), influenza A virus, and HIV-1. During the infection of HCV, Staufen1 demonstrated roles in the viral replication, translation, and trafficking of the HCV genome [161], while it facilitated the viral particle assembly of HIV-1 and Influenza A viruses [162,163].

2.18. hnRNP H and hnRNP F

The two hnRNP proteins, hnRNP F and hnRNP H, are closely related and fall under the hnRNP F/H subfamily. They are found to act as activators as well as repressors in regulating alternative splicing, depending on the context of the binding site [164]. HnRNP F/H possess three quasi-RNA recognition motifs (qRRMs) that preferably bind to poly(G)-rich sequences in the target exons and/or adjacent introns in order to regulate alternative splicing and 3′-end processing of numerous genes (Figure 3b) [165].

Both hnRNP H and F are located in the nucleus and are relocated into the cytoplasm upon EV-A71 infection. A study by Tsai et al. on the inhibition of EV-A71 by Kaempferol (a flavonoid) via impairing its replication and internal ribosome entry site (IRES) activity has discovered hnRNP F and hnNNP H as new ITAFs that associates with EV-A71 5′-UTR to enhance the IRES activity. Both hnRNP H and F were discovered initially as kaempferol-induced cellular factors associated with the 5′-UTR of the EV-A71 genome. However, they have carried out individual knockdown of endogenous HNRH1 and HNRPF proteins to show that it resulted in decreased EV-A71 IRES activity. Unfortunately, the specific binding sites of hnRNP H and F on the EV-A71 5′-UTR have not been mapped [135].

2.19. EGR1

Early growth response-1 (EGR1) is a C_2H_2-type zinc finger protein, and it is a transcription factor that activates many genes essential for growth, proliferation, or differentiation via binding to GC-rich recognition motifs. EGR1 expression is induced in response to various extracellular stimuli including growth factors, hormones, and neurotransmitters. EGR1 then couples extracellular signals to long-term cellular responses by altering the gene expression of its target genes [166]. EGR1 is found to activate microRNA-141 (miR-141) expression to suppress eukaryotic initiation factor 4E (eIF4E) production, resulting in the facilitation of EV-A71 replication via shutting off host protein synthesis [167].

A study by Song et al. has revealed that the first two zinc fingers of EGR1 bind directly to SLI and SLIV of the EV-A71 5′-UTR to regulate viral replication via enhancing both IRES-mediated translation and RNA replication. Further, this study demonstrated that EGR1 facilitates EV-A71 replication in a manner independent of miR-141 and eIF4E, via its interactions with the 5′-UTR [168].

2.20. TIA-1 and TIAR

T-cell intracellular antigen 1 (TIA-1) and TIA-1-related protein (TIAR) have been identified as nucleating components of mammalian stress granules (SGs) [169]. TIA proteins consist of three RNA-binding domains and a glutamine-rich carboxyl-terminal domain that enables aggregation to insoluble aggregates. TIA proteins have many important roles in the regulation of mRNA metabolism, especially during environmental stress conditions, including the regulation of splicing, mRNA stability, storage, and translation efficiency. The TIAR/TIA-1 proteins are known to selectively target specific mRNAs via binding to specific AREs located at the 3′ end of the target mRNAs [170]. TIA-1 and TIAR have been reported to be translocated from the nucleus to the cytoplasm to form SG-like granules as a result of PV and enterovirus (CVB3 or EV-A71) infections [171,172].

Wang et al. demonstrated that TIA1 and TIAR were recruited into SGs following EV-A71 infection and were localized to the sites of viral replication. Both TIA-1 and TIAR interact only with the SLI (nt 1–105) of EV-A71 5′-UTR and result in positive regulation of viral replication. The silencing of either TIA-1 or TIAR expression significantly reduced not only viral replication but also viral progeny production, via regulation of the level of viral RNA. Even though TIA-1 and TIAR bind to AREs, RNA pull-down assays proved that they did not bind to 5′-UTR or 3′-UTR of EV-A71 genome, except for SLI, making both TIA1 and TIAR distinct from many ITAFs that are shown to interact with multiple loops within 5′-UTR [173].

2.21. Additional Cellular Proteins

In addition to the proteins listed in Table 1, there are many additional cellular proteins that were discovered to associate with EV-A71 5′-UTR while their specific binding sites and roles in EV-A71 viral replication have not yet been determined. These proteins were identified by pull-down assays utilizing biotin-labeled 5′-UTR followed by MALDI-TOF MS characterization. Such additional cellular proteins that were discovered by individual studies carried out by Lin et al. and Xi et al. are as follows: TBP-associated factor TAFII 150 (TBP-AFII 150), N-ras upstream protein (unr), glycyl-tRNA synthetase (GARS), keratin, IGF-II mRNA-binding protein 1 (IMP-1), polypyrimidine tract-binding protein 2 (PTB-2), ErbB3-binding protein (EBP1), poly(rC)-binding protein 2 (PCBP2), pro alpha (I) collagen, Lrp protein, PTB-associated splicing factor (PSF), splicing factor, proline and glutamine rich (SFPQ), cationic trypsinogen, CGI-55 protein, elongation factor 1-gamma (EEF1G), 2-phosphopyruvate-hydratase alpha-enolase, collagen-binding protein 2, mutant beta-actin, and DNA-binding protein [54,55,88].

3. Discussion

Upon EV-A71 infection, the host cellular pathways will be repurposed to facilitate viral replication. Some of the main pathways that are perturbed include innate immunity, RNA metabolism, nuclear-cytoplasmic transport, stress granule formation, and autophagy. EV-A71 uses two main strategies to achieve the repurposing of cellular machinery. The first strategy is the redistribution of cellular proteins from their original locations into the cytoplasm where viral replication takes place. This will perturb their original cellular functions while being utilized solely in different steps of viral replication. The second strategy is the cleavage of key cellular proteins involved in above mentioned cellular pathways, mediated by viral proteases (mainly 3C^{pro} and 2A^{pro}) or other virus-induced pathways (caspases, proteosome, and autophagy) (Figure 4) [17,27].

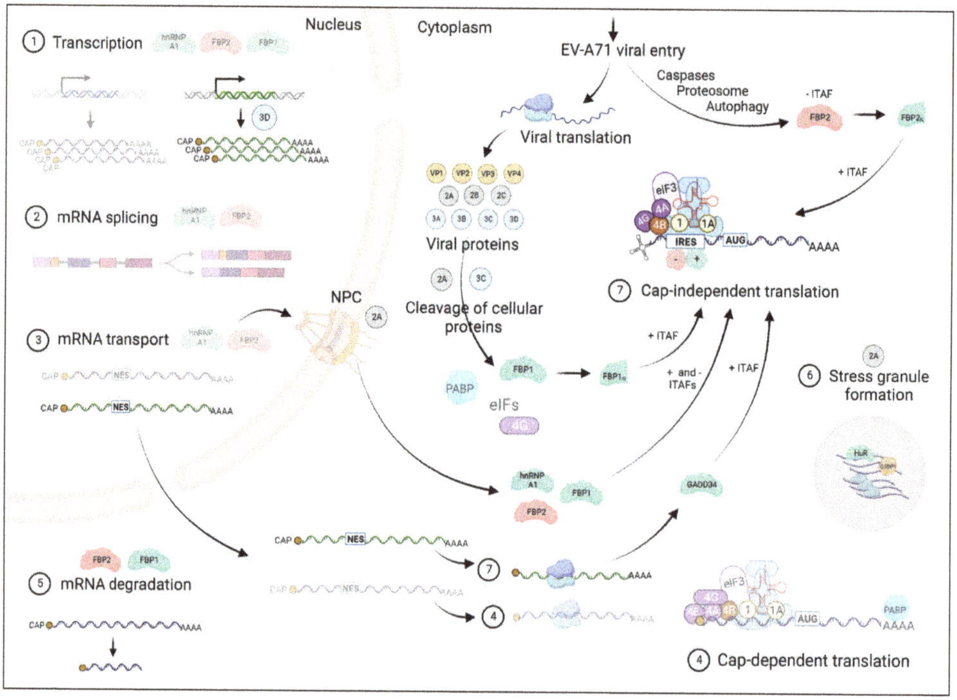

Figure 4. EV-A71 infection perturbs normal RNA metabolism, which leads to a shift in cellular homeostasis. Shown in steps 1–7 are some of the main cellular mechanisms that occur in the host cell. The mechanisms that are illustrated as faded are the ones that are affected by EV-A71 infection and are inhibited specifically due to the relocalization of cellular proteins to be used as ITAFs in EV-A71 IRES-mediated (cap-independent) translation. A few of the ITAFs, hnRNPA1, FBP1, FBP2, GADD34, and HuR, are used to represent the whole repertoire of ITAFs to help explain different viral mechanisms adapted by EV-A71 to facilitate its IRES-mediated translation. Positive ITAFs are colored in green while negative ITAFs are colored in red. These ITAFs are shown at the cellular functions that they regulate prior to EV-A71 infection. Note, a cleaved product of eIF4G is still involved in translation initiation, even though it is shown as faded due to cleavage by viral protease 3C^{pro}.

As the viral RNA genome is released into the cytoplasm upon EV-A71 infection, viral proteins are translated in an IRES-dependent manner while the viral genome is replicated via mechanisms mediated by RNA-dependent RNA polymerase (RdRp) $3D^{pol}$ [174]. The 5′-UTR of the viral RNA genome, which is composed of the cloverleaf structure (SLI) and the IRES (SLII-VI), functions as a control hub for viral RNA replication and translation via recruiting viral and cellular proteins (Figure 1). Cellular RNA-binding proteins that are recruited to the IRES and regulate viral translation are called ITAFs. The regulatory effect of these ITAFs on viral IRES-mediated translation can either be positive or negative [15,16]. Table 1 summarizes the ITAFs associated with the EV-A71 5′-UTR indicating their effect on viral translation and how they are redistributed in the cell upon EV-A71 infection.

It is noteworthy that hnRNP proteins are a key target of EV-A71, possibly due to the fact that hnRNP family proteins are central players in regulating various aspects of mRNA metabolism including alternative splicing, stability, transcription, and translation, while each of them is structurally and functionally diverse [58]. The hnRNP proteins are primarily localized in the nucleus and upon EV-A71 infection, they are relocated into the cytoplasm where viral replication takes place. Out of the seven hnRNP proteins listed in this review, hnRNP A1, hnRNP E1 (PCBP1), hnRNP I (PTB), hnRNP H, and hnRNP F act as positive ITAFs that bind EV-A71 5′-UTR to enhance IRES-mediated viral translation. The only hnRNP that acts as a negative ITAF is hnRNP D (AUF1). The hnRNP proteins that act as positive ITAFs also play simultaneous roles in other steps of EV-A71 replication such as viral genome replication.

In addition to the hnRNP proteins, FBP1, Ago2, HuR, MOV10, Sam68, FBP3, GADD34, DDX3, HSPA6, Staufen1, and EGR1 act as positive ITAFs, while only FBP2, SIRT1, and APOBEC3G act as negative ITAFs. However, EV-A71 has strategies to repurpose even the negative ITAFs into positive ITAFs. One such example is the cleavage of the C-terminus of FBP2 to convert the negative ITAF into a positive ITAF [102]. ITAFs can have different and unique cellular localizations that are related to their cellular function; however, they will be redistributed into the cytoplasm to bind the EV-A71 5′-UTR and enhance the IRES-mediated translation. Figure 1 visually represents the re-localization of cellular proteins before and after EV-A71 infection as well as their binding sites at the 5′ UTR of the EV-A71 RNA genome.

Apart from the relocation of cellular ITAFs into the cytoplasm, the levels of mRNA as well as the translation of positive ITAFs can also be enhanced upon EV-A71 infection, as observed in the case of GADD34 [146]. Further, EV-A71 viral proteins such as $2A^{pro}$ or $3C^{pro}$ play important roles in the process of repurposing cellular proteins to facilitate viral replication, mainly via their protease activity. Their activity is known to facilitate the redistribution of positive ITAFs from the nucleus to the cytoplasm via cleaving certain components of the nuclear pore complex [35]. Sometimes, their activity can truncate a cellular protein to give it a new feature that is beneficial for viral replication as observed in the case of eIF4G, where a truncated version of eIF4G (cleaved by viral $2A^{pro}$) binds specifically to SLV of IRES and recruits positive ITAF DDX3X facilitating ribosome entry and scanning [150,175]. In some cases, even positive ITAFs are cleaved by viral proteases as observed in the cleavage of FBP1 by $2A^{pro}$ to give rise to an additive enhancement of IRES activity [102] (Figure 4).

It is noteworthy that proteins such as poly(C)-binding protein 2 (PCBP2) and glycyl-tRNA synthetase (GARS) that are known to play important roles in IRES-mediated translation initiation by Type 1 IRES structures are not well studied on EV-A71. It has been shown that the translation initiation by PV, EV-A71, and BEV Type 1 IRESs necessitate PCBP2 at SLIV in addition to different eIFs [26,176–178]. GARS has been shown to bind SLV of the PV IRES adjacent to the binding site of the key initiation factor eIF4G, thereby enhancing the IRES activity [179]. In addition to PCBP2 and GARS, there are many other ITAFs such as La, SRp20, and upstream of N-Ras (unr) that have been implicated in the initiation of Type 1 IRES-mediated translation initiation during other enterovirus infections [180–182]. Most of these ITAFs have been discovered to associate with EV-A71 5′-UTR as indicated in

Section 2.21: Additional Cellular Proteins, using pull-down followed by MS analysis, while their specific binding sites and roles in EV-A71 viral replication are yet to be determined.

The mechanism of fine-tuning the IRES activity to regulate EV-A71 viral replication is a very complex process involving the interplay between the positive and negative cellular ITAFs as well as the viral proteins. Several recent studies have expanded the repertoire of positive and negative ITAFs suggesting new targets for EV-A71 inhibition. Most importantly, studies that identify the interplay between positive and negative ITAFs that bind at the same site, not only help in understanding the means of IRES-mediated translation regulation but also highlight these sites as better candidates for drug targeting. We believe that a comprehensive and generalized understanding of the mechanism of how EV-A71 hijacks and repurposes host cellular proteins for its gain can guide the development of novel and efficacious approaches for drug targeting.

4. Future Perspectives

Considering the increased interest in discovering novel small molecule inhibitors of RNA viruses like EV-A71, it is important to implement robust strategies to identify biologically relevant targets for therapeutic intervention. The 5′-UTR of EV-A71 represents a promising target given its central importance in regulating the cellular stages of viral replication cycles. Moreover, the EV-A71 5′-UTR is under high selective pressure to maintain its overall structure because it associates with a collection of ITAFs to differentially control viral gene expression. The next generation of EV-A71 antivirals to be discovered should therefore function by selectively perturbing specific 5′-UTR-ITAF interactions to drive viral replication towards predictable outcomes (Figures 1 and 2). Hence, small molecules that successfully perturb specific SLI-ITAF and IRES-ITAF interactions will function as novel antivirals against EV-A71 via inhibition of viral RNA replication and translation, respectively.

To date, SLII is the only RNA structure of the EV-A71 5′-UTR where specific binding sites of several of its ITAFs have been biophysically investigated to reveal the competitive interplay between hnRNP A1 and AUF1 for its bulge loop [53,65,183]. Further, this region has been successfully targeted by a small molecule antiviral DMA-135 (2-log and 5-log reduction of viral titers at 0.5 and 50 μM of DMA-135, respectively; IC_{50} of 7.54 ± 0.0024 μM) where its mechanism of action is to allosterically increase the binding affinity of AUF1 to SLII, thus shifting the SLII regulatory axis towards translation repression [64]. Notably, EV-A71 can evolve drug resistance to DMA-135 by changing nucleotides that form part of the AUF1 binding site, while not perturbing the binding activity of hnRNP A1 [56]. These collective observations suggest that other small molecules with the capacity to selectively perturb specific 5′-UTR-ITAF interactions await discovery.

To realize this outcome, we believe that it is prudent to better understand the structural-based mechanisms by which ITAFs assemble onto the EV-A71 5′-UTR, determine the extent to which subsets of ITAFs cooperate to enhance or suppress IRES-dependent translation and how they interact with eIFs, characterize the differential cell-type expression patterns of ITAFs upon viral infection, and perform comparative virological and structural studies of related enteroviruses that contain Type 1 IRES elements, such as EV-D68. Lastly, we must also remember that ITAFs and viral RNA elements are covalently modified so it should be a priority to understand the extent to which such modifications modulate EV-A71 gene expression at a molecular level [184]. We believe that the coordination of these studies will accelerate the discovery of the next generation of antiviral agents that target viral RNA complexes and produce a library of chemical biology reagents that can be deployed to selectively modulate conserved host–virus interactions with the potential to reveal new biological functions.

Author Contributions: The authors are listed in the alphabetical order of their last names. Conceptualization, B.S.T.; RNA 3D structure predictions and preparation of representative protein–RNA complexes, B.H. and S.P.; writing—original draft preparation, J.D.-C., C.H. and J.K.; writing—review and editing, S.M.A., B.H., S.P. and B.S.T. All authors have read and agreed to the published version of the manuscript.

Funding: This research was funded by NIAID grant 7R01 AI150830 and NIGMS grant 5R01 GM126833.

Data Availability Statement: Not applicable.

Conflicts of Interest: The authors declare no conflicts of interest.

References

1. Nayak, G.; Bhuyan, S.K.; Bhuyan, R.; Sahu, A.; Kar, D.; Kuanar, A. Global Emergence of Enterovirus 71: A Systematic Review. *Beni-Suef Univ. J. Basic Appl. Sci.* **2022**, *11*, 78. [CrossRef] [PubMed]
2. Hu, Y.-L.; Chen, C.-M.; Wang, E.-T.; Kuo, H.-W.; Shih, W.-L.; Fang, C.-T.; Liu, D.-P.; Chang, L.-Y. The Secular Trend of Enterovirus A71 after the Implementation of Preventive Measures in Taiwan. *BMC Public Health* **2022**, *22*, 1483. [CrossRef] [PubMed]
3. Enterovirus 71. Available online: https://www.who.int/teams/health-product-policy-and-standards/standards-and-specifications/vaccine-standardization/enterovirus-71 (accessed on 8 November 2023).
4. Huang, P.-N.; Shih, S.-R. Update on Enterovirus 71 Infection. *Curr. Opin. Virol.* **2014**, *5*, 98–104. [CrossRef] [PubMed]
5. Ibba, R.; Carta, A.; Madeddu, S.; Caria, P.; Serreli, G.; Piras, S.; Sestito, S.; Loddo, R.; Sanna, G. Inhibition of Enterovirus A71 by a Novel 2-Phenyl-Benzimidazole Derivative. *Viruses* **2021**, *13*, 58. [CrossRef] [PubMed]
6. Liu, H.; Zhang, Y.; Zhang, H.; Zheng, Y.; Gou, F.; Yang, X.; Cheng, Y.; McClymont, H.; Li, H.; Liu, X.; et al. Prototypes Virus of Hand, Foot and Mouth Disease Infections and Severe Cases in Gansu, China: A Spatial and Temporal Analysis. *BMC Infect. Dis.* **2022**, *22*, 408. [CrossRef] [PubMed]
7. Wörner, N.; Rodrigo-García, R.; Antón, A.; Castellarnau, E.; Delgado, I.; Vazquez, È.; González, S.; Mayol, L.; Méndez, M.; Solé, E.; et al. Enterovirus-A71 Rhombencephalitis Outbreak in Catalonia: Characteristics, Management and Outcome. *Pediatr. Infect. Dis. J.* **2021**, *40*, 628. [CrossRef]
8. Messacar, K.; Burakoff, A.; Nix, W.A.; Rogers, S.; Oberste, M.S.; Gerber, S.I.; Spence-Davizon, E.; Herlihy, R.; Dominguez, S.R. Notes from the Field: Enterovirus A71 Neurologic Disease in Children—Colorado, 2018. *MMWR Morb. Mortal. Wkly. Rep.* **2018**, *67*, 1017–1018. [CrossRef]
9. Nhan, L.N.T.; Hong, N.T.T.; Nhu, L.N.T.; Nguyet, L.A.; Ny, N.T.H.; Thanh, T.T.; Han, D.D.K.; Van, H.M.T.; Thwaites, C.L.; Hien, T.T.; et al. Severe Enterovirus A71 Associated Hand, Foot and Mouth Disease, Vietnam, 2018: Preliminary Report of an Impending Outbreak. *Eur. Surveill.* **2018**, *23*, 1800590. [CrossRef]
10. NIAID Emerging Infectious Diseases/Pathogens | NIH: National Institute of Allergy and Infectious Diseases. Available online: https://www.niaid.nih.gov/research/emerging-infectious-diseases-pathogens (accessed on 8 November 2023).
11. Li, M.-L.; Shih, S.-R.; Tolbert, B.S.; Brewer, G. Enterovirus A71 Vaccines. *Vaccines* **2021**, *9*, 199. [CrossRef]
12. Yi, E.-J.; Shin, Y.-J.; Kim, J.-H.; Kim, T.-G.; Chang, S.-Y. Enterovirus 71 Infection and Vaccines. *Clin. Exp. Vaccine Res.* **2017**, *6*, 1–14. [CrossRef]
13. Chan, Y.-F.; Sam, I.-C.; AbuBakar, S. Phylogenetic Designation of Enterovirus 71 Genotypes and Subgenotypes Using Complete Genome Sequences. *Infect. Genet. Evol.* **2010**, *10*, 404–412. [CrossRef] [PubMed]
14. Brown, B.A.; Pallansch, M.A. Complete Nucleotide Sequence of Enterovirus 71 Is Distinct from Poliovirus. *Virus Res.* **1995**, *39*, 195–205. [CrossRef] [PubMed]
15. Tolbert, M.; Morgan, C.E.; Pollum, M.; Crespo-Hernández, C.E.; Li, M.-L.; Brewer, G.; Tolbert, B.S. HnRNP A1 Alters the Structure of a Conserved Enterovirus IRES Domain to Stimulate Viral Translation. *J. Mol. Biol.* **2017**, *429*, 2841–2858. [CrossRef] [PubMed]
16. Thompson, S.R.; Sarnow, P. Enterovirus 71 Contains a Type I IRES Element That Functions When Eukaryotic Initiation Factor eIF4G Is Cleaved. *Virology* **2003**, *315*, 259–266. [CrossRef] [PubMed]
17. Lai, M.-C.; Chen, H.-H.; Xu, P.; Wang, R.Y.L. Translation Control of Enterovirus A71 Gene Expression. *J. Biomed. Sci.* **2020**, *27*, 22. [CrossRef] [PubMed]
18. Sommergruber, W.; Ahorn, H.; Klump, H.; Seipelt, J.; Zoephel, A.; Fessl, F.; Krystek, E.; Blaas, D.; Kuechler, E.; Liebig, H.D.; et al. 2A Proteinases of Coxsackie- and Rhinovirus Cleave Peptides Derived from eIF-4γ via a Common Recognition Motif. *Virology* **1994**, *198*, 741–745. [CrossRef]
19. de Breyne, S.; Bonderoff, J.M.; Chumakov, K.M.; Lloyd, R.E.; Hellen, C.U.T. Cleavage of Eukaryotic Initiation Factor eIF5B by Enterovirus 3C Proteases. *Virology* **2008**, *378*, 118–122. [CrossRef] [PubMed]
20. Kuyumcu-Martinez, N.M.; Van Eden, M.E.; Younan, P.; Lloyd, R.E. Cleavage of Poly(A)-Binding Protein by Poliovirus 3C Protease Inhibits Host Cell Translation: A Novel Mechanism for Host Translation Shutoff. *Mol. Cell. Biol.* **2004**, *24*, 1779–1790. [CrossRef]
21. White, J.P.; Reineke, L.C.; Lloyd, R.E. Poliovirus Switches to an eIF2-Independent Mode of Translation during Infection. *J. Virol.* **2011**, *85*, 8884–8893. [CrossRef]
22. Weng, K.-F.; Li, M.-L.; Hung, C.-T.; Shih, S.-R. Enterovirus 71 3C Protease Cleaves a Novel Target CstF-64 and Inhibits Cellular Polyadenylation. *PLoS Pathog.* **2009**, *5*, e1000593. [CrossRef]

23. Lei, X.; Xiao, X.; Xue, Q.; Jin, Q.; He, B.; Wang, J. Cleavage of Interferon Regulatory Factor 7 by Enterovirus 71 3C Suppresses Cellular Responses. *J. Virol.* **2013**, *87*, 1690–1698. [CrossRef] [PubMed]
24. Lei, X.; Han, N.; Xiao, X.; Jin, Q.; He, B.; Wang, J. Enterovirus 71 3C Inhibits Cytokine Expression through Cleavage of the TAK1/TAB1/TAB2/TAB3 Complex. *J. Virol.* **2014**, *88*, 9830–9841. [CrossRef] [PubMed]
25. Li, M.-L.; Lin, J.-Y.; Chen, B.-S.; Weng, K.-F.; Shih, S.-R.; Calderon, J.D.; Tolbert, B.S.; Brewer, G. EV71 3C Protease Induces Apoptosis by Cleavage of hnRNP A1 to Promote Apaf-1 Translation. *PLoS ONE* **2019**, *14*, e0221048. [CrossRef]
26. Sweeney, T.R.; Abaeva, I.S.; Pestova, T.V.; Hellen, C.U.T. The Mechanism of Translation Initiation on Type 1 Picornavirus IRESs. *EMBO J.* **2014**, *33*, 76–92. [CrossRef] [PubMed]
27. Abdullah, S.W.; Wu, J.; Wang, X.; Guo, H.; Sun, S. Advances and Breakthroughs in IRES-Directed Translation and Replication of Picornaviruses. *mBio* **2023**, *14*, e0035823. [CrossRef]
28. López-Ulloa, B.; Fuentes, Y.; Pizarro-Ortega, M.S.; López-Lastra, M. RNA-Binding Proteins as Regulators of Internal Initiation of Viral mRNA Translation. *Viruses* **2022**, *14*, 188. [CrossRef]
29. Martínez-Salas, E.; Francisco-Velilla, R.; Fernandez-Chamorro, J.; Lozano, G.; Diaz-Toledano, R. Picornavirus IRES Elements: RNA Structure and Host Protein Interactions. *Virus Res.* **2015**, *206*, 62–73. [CrossRef] [PubMed]
30. Martínez-Salas, E. The Impact of RNA Structure on Picornavirus IRES Activity. *Trends Microbiol.* **2008**, *16*, 230–237. [CrossRef]
31. Martinez-Salas, E.; Francisco-Velilla, R.; Fernandez-Chamorro, J.; Embarek, A.M. Insights into Structural and Mechanistic Features of Viral IRES Elements. *Front. Microbiol.* **2018**, *8*, 2629. [CrossRef]
32. Pestova, T.V.; Hellen, C.U.; Wimmer, E. A Conserved AUG Triplet in the 5′ Nontranslated Region of Poliovirus Can Function as an Initiation Codon in Vitro and in Vivo. *Virology* **1994**, *204*, 729–737. [CrossRef]
33. Lulla, V.; Dinan, A.M.; Hosmillo, M.; Chaudhry, Y.; Sherry, L.; Irigoyen, N.; Nayak, K.M.; Stonehouse, N.J.; Zilbauer, M.; Goodfellow, I.; et al. An Upstream Protein-Coding Region in Enteroviruses Modulates Virus Infection in Gut Epithelial Cells. *Nat. Microbiol.* **2019**, *4*, 280–292. [CrossRef] [PubMed]
34. Guo, H.; Li, Y.; Liu, G.; Jiang, Y.; Shen, S.; Bi, R.; Huang, H.; Cheng, T.; Wang, C.; Wei, W. A Second Open Reading Frame in Human Enterovirus Determines Viral Replication in Intestinal Epithelial Cells. *Nat. Commun.* **2019**, *10*, 4066. [CrossRef]
35. Lizcano-Perret, B.; Michiels, T. Nucleocytoplasmic Trafficking Perturbation Induced by Picornaviruses. *Viruses* **2021**, *13*, 1210. [CrossRef]
36. Wang, C.; Sun, M.; Yuan, X.; Ji, L.; Jin, Y.; Cardona, C.J.; Xing, Z. Enterovirus 71 Suppresses Interferon Responses by Blocking Janus Kinase (JAK)/Signal Transducer and Activator of Transcription (STAT) Signaling through Inducing Karyopherin-A1 Degradation. *J. Biol. Chem.* **2017**, *292*, 10262–10274. [CrossRef] [PubMed]
37. Shiroki, K.; Isoyama, T.; Kuge, S.; Ishii, T.; Ohmi, S.; Hata, S.; Suzuki, K.; Takasaki, Y.; Nomoto, A. Intracellular Redistribution of Truncated La Protein Produced by Poliovirus 3Cpro-Mediated Cleavage. *J. Virol.* **1999**, *73*, 2193–2200. [CrossRef] [PubMed]
38. Wang, J.; Sun, D.; Wang, M.; Cheng, A.; Zhu, Y.; Mao, S.; Ou, X.; Zhao, X.; Huang, J.; Gao, Q.; et al. Multiple Functions of Heterogeneous Nuclear Ribonucleoproteins in the Positive Single-Stranded RNA Virus Life Cycle. *Front. Immunol.* **2022**, *13*, 989298. [CrossRef]
39. Mayeda, A.; Krainer, A.R. Regulation of Alternative Pre-mRNA Splicing by hnRNP A1 and Splicing Factor SF2. *Cell* **1992**, *68*, 365–375. [CrossRef]
40. Jean-Philippe, J.; Paz, S.; Caputi, M. hnRNP A1: The Swiss Army Knife of Gene Expression. *Int. J. Mol. Sci.* **2013**, *14*, 18999–19024. [CrossRef]
41. Michael, W.M.; Choi, M.; Dreyfuss, G. A Nuclear Export Signal in hnRNP A1: A Signal-Mediated, Temperature-Dependent Nuclear Protein Export Pathway. *Cell* **1995**, *83*, 415–422. [CrossRef]
42. Henics, T.; Sanfridson, A.; Hamilton, B.J.; Nagy, E.; Rigby, W.F. Enhanced Stability of Interleukin-2 mRNA in MLA 144 Cells. Possible Role of Cytoplasmic AU-Rich Sequence-Binding Proteins. *J. Biol. Chem.* **1994**, *269*, 5377–5383. [CrossRef]
43. Hamilton, B.J.; Burns, C.M.; Nichols, R.C.; Rigby, W.F. Modulation of AUUUA Response Element Binding by Heterogeneous Nuclear Ribonucleoprotein A1 in Human T Lymphocytes. The Roles of Cytoplasmic Location, Transcription, and Phosphorylation. *J. Biol. Chem.* **1997**, *272*, 28732–28741. [CrossRef] [PubMed]
44. Zhang, Q.-S.; Manche, L.; Xu, R.-M.; Krainer, A.R. hnRNP A1 Associates with Telomere Ends and Stimulates Telomerase Activity. *RNA* **2006**, *12*, 1116–1128. [CrossRef] [PubMed]
45. Guil, S.; Cáceres, J.F. The Multifunctional RNA-Binding Protein hnRNP A1 Is Required for Processing of miR-18a. *Nat. Struct. Mol. Biol.* **2007**, *14*, 591–596. [CrossRef] [PubMed]
46. Mili, S.; Shu, H.J.; Zhao, Y.; Piñol-Roma, S. Distinct RNP Complexes of Shuttling hnRNP Proteins with Pre-mRNA and mRNA: Candidate Intermediates in Formation and Export of mRNA. *Mol. Cell. Biol.* **2001**, *21*, 7307–7319. [CrossRef] [PubMed]
47. Piñol-Roma, S.; Dreyfuss, G. Shuttling of Pre-mRNA Binding Proteins between Nucleus and Cytoplasm. *Nature* **1992**, *355*, 730–732. [CrossRef]
48. Levengood, J.D.; Tolbert, B.S. Idiosyncrasies of hnRNP A1-RNA Recognition: Can Binding Mode Influence Function. *Semin. Cell Dev. Biol.* **2019**, *86*, 150–161. [CrossRef] [PubMed]
49. Morgan, C.E.; Meagher, J.L.; Levengood, J.D.; Delproposto, J.; Rollins, C.; Stuckey, J.A.; Tolbert, B.S. The First Crystal Structure of the UP1 Domain of hnRNP A1 Bound to RNA Reveals a New Look for an Old RNA Binding Protein. *J. Mol. Biol.* **2015**, *427*, 3241–3257. [CrossRef] [PubMed]

50. Kooshapur, H.; Choudhury, N.R.; Simon, B.; Mühlbauer, M.; Jussupow, A.; Fernandez, N.; Jones, A.N.; Dallmann, A.; Gabel, F.; Camilloni, C.; et al. Structural Basis for Terminal Loop Recognition and Stimulation of Pri-miRNA-18a Processing by hnRNP A1. *Nat. Commun.* **2018**, *9*, 2479. [CrossRef]
51. Barraud, P.; Allain, F.H.-T. Solution Structure of the Two RNA Recognition Motifs of hnRNP A1 Using Segmental Isotope Labeling: How the Relative Orientation between RRMs Influences the Nucleic Acid Binding Topology. *J. Biomol. NMR* **2013**, *55*, 119–138. [CrossRef]
52. Lin, J.-Y.; Shih, S.-R.; Pan, M.; Li, C.; Lue, C.-F.; Stollar, V.; Li, M.-L. hnRNP A1 Interacts with the 5′ Untranslated Regions of Enterovirus 71 and Sindbis Virus RNA and Is Required for Viral Replication. *J. Virol.* **2009**, *83*, 6106–6114. [CrossRef]
53. Levengood, J.D.; Tolbert, M.; Li, M.-L.; Tolbert, B.S. High-Affinity Interaction of hnRNP A1 with Conserved RNA Structural Elements Is Required for Translation and Replication of Enterovirus 71. *RNA Biol.* **2013**, *10*, 1136–1145. [CrossRef] [PubMed]
54. Lin, J.-Y.; Li, M.-L.; Huang, P.-N.; Chien, K.-Y.; Horng, J.-T.; Shih, S.-R. Heterogeneous Nuclear Ribonuclear Protein K Interacts with the Enterovirus 71 5′ Untranslated Region and Participates in Virus Replication. *J. Gen. Virol.* **2008**, *89 Pt 10*, 2540–2549. [CrossRef]
55. Lin, J.-Y.; Li, M.-L.; Shih, S.-R. Far Upstream Element Binding Protein 2 Interacts with Enterovirus 71 Internal Ribosomal Entry Site and Negatively Regulates Viral Translation. *Nucleic Acids Res.* **2009**, *37*, 47–59. [CrossRef] [PubMed]
56. Davila-Calderon, J.; Li, M.-L.; Penumutchu, S.R.; Haddad, C.; Malcolm, L.; Hargrove, A.E.; Brewer, G.; Tolbert, B.S. Enterovirus Evolution Reveals the Mechanism of an RNA-Targeted Antiviral and Determinants of Viral Replication. *bioRxiv* **2023**. [CrossRef]
57. Fialcowitz, E.J.; Brewer, B.Y.; Keenan, B.P.; Wilson, G.M. A Hairpin-like Structure within an AU-Rich mRNA-Destabilizing Element Regulates Trans-Factor Binding Selectivity and mRNA Decay Kinetics. *J. Biol. Chem.* **2005**, *280*, 22406–22417. [CrossRef] [PubMed]
58. Geuens, T.; Bouhy, D.; Timmerman, V. The hnRNP Family: Insights into Their Role in Health and Disease. *Hum. Genet.* **2016**, *135*, 851–867. [CrossRef] [PubMed]
59. Wu, X.; Brewer, G. The Regulation of mRNA Stability in Mammalian Cells: 2.0. *Gene* **2012**, *500*, 10. [CrossRef]
60. Pont, A.R.; Sadri, N.; Hsiao, S.J.; Smith, S.; Schneider, R.J. mRNA Decay Factor AUF1 Maintains Normal Aging, Telomere Maintenance, and Suppression of Senescence by Activation of Telomerase Transcription. *Mol. Cell* **2012**, *47*, 5–15. [CrossRef]
61. Laroia, G.; Cuesta, R.; Brewer, G.; Schneider, R.J. Control of mRNA Decay by Heat Shock-Ubiquitin-Proteasome Pathway. *Science* **1999**, *284*, 499–502. [CrossRef]
62. Lu, J.-Y.; Bergman, N.; Sadri, N.; Schneider, R.J. Assembly of AUF1 with eIF4G–Poly(A) Binding Protein Complex Suggests a Translation Function in AU-Rich mRNA Decay. *RNA* **2006**, *12*, 883–893. [CrossRef]
63. Sinsimer, K.S.; Gratacós, F.M.; Knapinska, A.M.; Lu, J.; Krause, C.D.; Wierzbowski, A.V.; Maher, L.R.; Scrudato, S.; Rivera, Y.M.; Gupta, S.; et al. Chaperone Hsp27, a Novel Subunit of AUF1 Protein Complexes, Functions in AU-Rich Element-Mediated mRNA Decay. *Mol. Cell. Biol.* **2008**, *28*, 5223–5237. [CrossRef] [PubMed]
64. Davila-Calderon, J.; Patwardhan, N.N.; Chiu, L.-Y.; Sugarman, A.; Cai, Z.; Penutmutchu, S.R.; Li, M.-L.; Brewer, G.; Hargrove, A.E.; Tolbert, B.S. IRES-Targeting Small Molecule Inhibits Enterovirus 71 Replication via Allosteric Stabilization of a Ternary Complex. *Nat. Commun.* **2020**, *11*, 4775. [CrossRef] [PubMed]
65. Lin, J.-Y.; Li, M.-L.; Brewer, G. mRNA Decay Factor AUF1 Binds the Internal Ribosomal Entry Site of Enterovirus 71 and Inhibits Virus Replication. *PLoS ONE* **2014**, *9*, e103827. [CrossRef] [PubMed]
66. Cathcart, A.L.; Rozovics, J.M.; Semler, B.L. Cellular mRNA Decay Protein AUF1 Negatively Regulates Enterovirus and Human Rhinovirus Infections. *J. Virol.* **2013**, *87*, 10423–10434. [CrossRef] [PubMed]
67. Lin, J.-Y.; Brewer, G.; Li, M.-L. HuR and Ago2 Bind the Internal Ribosome Entry Site of Enterovirus 71 and Promote Virus Translation and Replication. *PLoS ONE* **2015**, *10*, e0140291. [CrossRef] [PubMed]
68. Stains, J.P.; Lecanda, F.; Towler, D.A.; Civitelli, R. Heterogeneous Nuclear Ribonucleoprotein K Represses Transcription from a Cytosine/Thymidine-Rich Element in the Osteocalcin Promoter. *Biochem. J.* **2005**, *385 Pt 2*, 613–623. [CrossRef] [PubMed]
69. Cao, W.; Razanau, A.; Feng, D.; Lobo, V.G.; Xie, J. Control of Alternative Splicing by Forskolin through hnRNP K during Neuronal Differentiation. *Nucleic Acids Res.* **2012**, *40*, 8059–8071. [CrossRef]
70. Fan, X.; Xiong, H.; Wei, J.; Gao, X.; Feng, Y.; Liu, X.; Zhang, G.; He, Q.-Y.; Xu, J.; Liu, L. Cytoplasmic hnRNPK Interacts with GSK3β and Is Essential for the Osteoclast Differentiation. *Sci. Rep.* **2015**, *5*, 17732. [CrossRef]
71. Fukuda, T.; Naiki, T.; Saito, M.; Irie, K. hnRNP K Interacts with RNA Binding Motif Protein 42 and Functions in the Maintenance of Cellular ATP Level during Stress Conditions. *Genes Cells* **2009**, *14*, 113–128. [CrossRef]
72. Habelhah, H.; Shah, K.; Huang, L.; Ostareck-Lederer, A.; Burlingame, A.L.; Shokat, K.M.; Hentze, M.W.; Ronai, Z. ERK Phosphorylation Drives Cytoplasmic Accumulation of hnRNP-K and Inhibition of mRNA Translation. *Nat. Cell Biol.* **2001**, *3*, 325–330. [CrossRef]
73. Bomsztyk, K.; Denisenko, O.; Ostrowski, J. hnRNP K: One Protein Multiple Processes. *Bioessays* **2004**, *26*, 629–638. [CrossRef] [PubMed]
74. Leffers, H.; Dejgaard, K.; Celis, J.E. Characterisation of Two Major Cellular Poly(rC)-Binding Human Proteins, Each Containing Three K-Homologous (KH) Domains. *Eur. J. Biochem.* **1995**, *230*, 447–453. [CrossRef] [PubMed]
75. Luo, Z.; Dong, X.; Li, Y.; Zhang, Q.; Kim, C.; Song, Y.; Kang, L.; Liu, Y.; Wu, K.; Wu, J. PolyC-Binding Protein 1 Interacts with 5′-Untranslated Region of Enterovirus 71 RNA in Membrane-Associated Complex to Facilitate Viral Replication. *PLoS ONE* **2014**, *9*, e87491. [CrossRef] [PubMed]

76. Thyagarajan, A.; Szaro, B.G. Phylogenetically Conserved Binding of Specific K Homology Domain Proteins to the 3′-Untranslated Region of the Vertebrate Middle Neurofilament mRNA. *J. Biol. Chem.* **2004**, *279*, 49680–49688. [CrossRef] [PubMed]
77. Meng, Q.; Rayala, S.K.; Gururaj, A.E.; Talukder, A.H.; O'Malley, B.W.; Kumar, R. Signaling-Dependent and Coordinated Regulation of Transcription, Splicing, and Translation Resides in a Single Coregulator, PCBP1. *Proc. Natl. Acad. Sci. USA* **2007**, *104*, 5866–5871. [CrossRef] [PubMed]
78. Lian, W.-X.; Yin, R.-H.; Kong, X.-Z.; Zhang, T.; Huang, X.-H.; Zheng, W.-W.; Yang, Y.; Zhan, Y.-Q.; Xu, W.-X.; Yu, M.; et al. THAP11, a Novel Binding Protein of PCBP1, Negatively Regulates CD44 Alternative Splicing and Cell Invasion in a Human Hepatoma Cell Line. *FEBS Lett.* **2012**, *586*, 1431–1438. [CrossRef] [PubMed]
79. Zhang, T.; Huang, X.-H.; Dong, L.; Hu, D.; Ge, C.; Zhan, Y.-Q.; Xu, W.-X.; Yu, M.; Li, W.; Wang, X.; et al. PCBP-1 Regulates Alternative Splicing of the CD44 Gene and Inhibits Invasion in Human Hepatoma Cell Line HepG2 Cells. *Mol. Cancer* **2010**, *9*, 72. [CrossRef]
80. Malik, A.K.; Flock, K.E.; Godavarthi, C.L.; Loh, H.H.; Ko, J.L. Molecular Basis Underlying the Poly C Binding Protein 1 as a Regulator of the Proximal Promoter of Mouse Mu-Opioid Receptor Gene. *Brain Res.* **2006**, *1112*, 33–45. [CrossRef]
81. Han, S.P.; Tang, Y.H.; Smith, R. Functional Diversity of the hnRNPs: Past, Present and Perspectives. *Biochem. J.* **2010**, *430*, 379–392. [CrossRef]
82. Patton, J.G.; Mayer, S.A.; Tempst, P.; Nadal-Ginard, B. Characterization and Molecular Cloning of Polypyrimidine Tract-Binding Protein: A Component of a Complex Necessary for Pre-mRNA Splicing. *Genes Dev.* **1991**, *5*, 1237–1251. [CrossRef]
83. Hamid, F.M.; Makeyev, E.V. Regulation of mRNA Abundance by Polypyrimidine Tract-Binding Protein-Controlled Alternate 5′ Splice Site Choice. *PLoS Genet.* **2014**, *10*, e1004771. [CrossRef] [PubMed]
84. Rahman, M.A.; Masuda, A.; Ohe, K.; Ito, M.; Hutchinson, D.O.; Mayeda, A.; Engel, A.G.; Ohno, K. HnRNP L and hnRNP LL Antagonistically Modulate PTB-Mediated Splicing Suppression of CHRNA1 Pre-mRNA. *Sci. Rep.* **2013**, *3*, 2931. [CrossRef] [PubMed]
85. Söderberg, M.; Raffalli-Mathieu, F.; Lang, M.A. Identification of a Regulatory Cis-Element within the 3′-Untranslated Region of the Murine Inducible Nitric Oxide Synthase (iNOS) mRNA; Interaction with Heterogeneous Nuclear Ribonucleoproteins I and L and Role in the iNOS Gene Expression. *Mol. Immunol.* **2007**, *44*, 434–442. [CrossRef] [PubMed]
86. Majumder, M.; Yaman, I.; Gaccioli, F.; Zeenko, V.V.; Wang, C.; Caprara, M.G.; Venema, R.C.; Komar, A.A.; Snider, M.D.; Hatzoglou, M. The hnRNA-Binding Proteins hnRNP L and PTB Are Required for Efficient Translation of the Cat-1 Arginine/Lysine Transporter mRNA during Amino Acid Starvation. *Mol. Cell. Biol.* **2009**, *29*, 2899–2912. [CrossRef] [PubMed]
87. Jafarifar, F.; Yao, P.; Eswarappa, S.M.; Fox, P.L. Repression of VEGFA by CA-Rich Element-Binding microRNAs Is Modulated by hnRNP L. *EMBO J.* **2011**, *30*, 1324–1334. [CrossRef]
88. Xi, J.; Ye, F.; Wang, G.; Han, W.; Wei, Z.; Yin, B.; Yuan, J.; Qiang, B.; Peng, X. Polypyrimidine Tract-Binding Protein Regulates Enterovirus 71 Translation Through Interaction with the Internal Ribosomal Entry Site. *Virol. Sin.* **2019**, *34*, 66–77. [CrossRef] [PubMed]
89. Min, H.; Turck, C.W.; Nikolic, J.M.; Black, D.L. A New Regulatory Protein, KSRP, Mediates Exon Inclusion through an Intronic Splicing Enhancer. *Genes Dev.* **1997**, *11*, 1023–1036. [CrossRef] [PubMed]
90. Davis-Smyth, T.; Duncan, R.C.; Zheng, T.; Michelotti, G.; Levens, D. The Far Upstream Element-Binding Proteins Comprise an Ancient Family of Single-Strand DNA-Binding Transactivators. *J. Biol. Chem.* **1996**, *271*, 31679–31687. [CrossRef]
91. Lellek, H.; Kirsten, R.; Diehl, I.; Apostel, F.; Buck, F.; Greeve, J. Purification and Molecular Cloning of a Novel Essential Component of the Apolipoprotein B mRNA Editing Enzyme-Complex. *J. Biol. Chem.* **2000**, *275*, 19848–19856. [CrossRef]
92. Gherzi, R.; Chen, C.-Y.; Trabucchi, M.; Ramos, A.; Briata, P. The Role of KSRP in mRNA Decay and microRNA Precursor Maturation. *Wiley Interdiscip. Rev. RNA* **2010**, *1*, 230–239. [CrossRef]
93. Trabucchi, M.; Briata, P.; Filipowicz, W.; Rosenfeld, M.G.; Ramos, A.; Gherzi, R. How to Control miRNA Maturation? *RNA Biol.* **2009**, *6*, 536–540. [CrossRef] [PubMed]
94. Briata, P.; Chen, C.-Y.; Ramos, A.; Gherzi, R. Functional and Molecular Insights into KSRP Function in mRNA Decay. *Biochim. Biophys. Acta* **2013**, *1829*, 689–694. [CrossRef] [PubMed]
95. Chen, L.-L.; Kung, Y.-A.; Weng, K.-F.; Lin, J.-Y.; Horng, J.-T.; Shih, S.-R. Enterovirus 71 Infection Cleaves a Negative Regulator for Viral Internal Ribosomal Entry Site-Driven Translation. *J. Virol.* **2013**, *87*, 3828–3838. [CrossRef] [PubMed]
96. Zhang, J.; Chen, Q.M. Far Upstream Element Binding Protein 1: A Commander of Transcription, Translation and Beyond. *Oncogene* **2013**, *32*, 2907–2916. [CrossRef] [PubMed]
97. Duncan, R.; Bazar, L.; Michelotti, G.; Tomonaga, T.; Krutzsch, H.; Avigan, M.; Levens, D. A Sequence-Specific, Single-Strand Binding Protein Activates the Far Upstream Element of c-Myc and Defines a New DNA-Binding Motif. *Genes Dev.* **1994**, *8*, 465–480. [CrossRef] [PubMed]
98. Olanich, M.E.; Moss, B.L.; Piwnica-Worms, D.; Townsend, R.R.; Weber, J.D. Identification of FUSE-Binding Protein 1 as a Regulatory mRNA-Binding Protein That Represses Nucleophosmin Translation. *Oncogene* **2011**, *30*, 77–86. [CrossRef] [PubMed]
99. Zhang, Z.; Harris, D.; Pandey, V.N. The FUSE Binding Protein Is a Cellular Factor Required for Efficient Replication of Hepatitis C Virus. *J. Virol.* **2008**, *82*, 5761–5773. [CrossRef]
100. Chien, H.-L.; Liao, C.-L.; Lin, Y.-L. FUSE Binding Protein 1 Interacts with Untranslated Regions of Japanese Encephalitis Virus RNA and Negatively Regulates Viral Replication. *J. Virol.* **2011**, *85*, 4698–4706. [CrossRef]

101. Huang, P.-N.; Lin, J.-Y.; Locker, N.; Kung, Y.-A.; Hung, C.-T.; Lin, J.-Y.; Huang, H.-I.; Li, M.-L.; Shih, S.-R. Far Upstream Element Binding Protein 1 Binds the Internal Ribosomal Entry Site of Enterovirus 71 and Enhances Viral Translation and Viral Growth. *Nucleic Acids Res.* **2011**, *39*, 9633–9648. [CrossRef]
102. Hung, C.-T.; Kung, Y.-A.; Li, M.-L.; Brewer, G.; Lee, K.-M.; Liu, S.-T.; Shih, S.-R. Additive Promotion of Viral Internal Ribosome Entry Site-Mediated Translation by Far Upstream Element-Binding Protein 1 and an Enterovirus 71-Induced Cleavage Product. *PLoS Pathog.* **2016**, *12*, e1005959. [CrossRef]
103. Höck, J.; Meister, G. The Argonaute Protein Family. *Genome Biol.* **2008**, *9*, 210. [CrossRef] [PubMed]
104. Schirle, N.T.; MacRae, I.J. The Crystal Structure of Human Argonaute2. *Science* **2012**, *336*, 1037–1040. [CrossRef] [PubMed]
105. Pantazopoulou, V.I.; Delis, A.D.; Georgiou, S.; Pagakis, S.N.; Filippa, V.; Dragona, E.; Kloukina, I.; Chatzitheodoridis, E.; Trebicka, J.; Velentzas, A.D.; et al. AGO2 Localizes to Cytokinetic Protrusions in a P38-Dependent Manner and Is Needed for Accurate Cell Division. *Commun. Biol.* **2021**, *4*, 726. [CrossRef] [PubMed]
106. Vasudevan, S.; Steitz, J.A. AU-Rich-Element-Mediated Upregulation of Translation by FXR1 and Argonaute 2. *Cell* **2007**, *128*, 1105–1118. [CrossRef] [PubMed]
107. Brennan, C.M.; Steitz, J.A. HuR and mRNA Stability. *Cell. Mol. Life Sci.* **2001**, *58*, 266–277. [CrossRef] [PubMed]
108. Lebedeva, S.; Jens, M.; Theil, K.; Schwanhäusser, B.; Selbach, M.; Landthaler, M.; Rajewsky, N. Transcriptome-Wide Analysis of Regulatory Interactions of the RNA-Binding Protein HuR. *Mol. Cell* **2011**, *43*, 340–352. [CrossRef] [PubMed]
109. Fan, X.C.; Steitz, J.A. Overexpression of HuR, a Nuclear-Cytoplasmic Shuttling Protein, Increases the in Vivo Stability of ARE-Containing mRNAs. *EMBO J.* **1998**, *17*, 3448–3460. [CrossRef] [PubMed]
110. Uren, P.J.; Burns, S.C.; Ruan, J.; Singh, K.K.; Smith, A.D.; Penalva, L.O.F. Genomic Analyses of the RNA-Binding Protein Hu Antigen R (HuR) Identify a Complex Network of Target Genes and Novel Characteristics of Its Binding Sites. *J. Biol. Chem.* **2011**, *286*, 37063–37066. [CrossRef]
111. Liu, Y.; Zhang, Y.; Zhang, J.; Ma, J.; Xu, X.; Wang, Y.; Zhou, Z.; Jiang, D.; Shen, S.; Ding, Y.; et al. Silencing of HuR Inhibits Osteosarcoma Cell Epithelial-Mesenchymal Transition via AGO2 in Association With Long Non-Coding RNA XIST. *Front. Oncol.* **2021**, *11*, 601982. [CrossRef]
112. Zhu, Y.; Wang, B.; Huang, H.; Zhao, Z. Enterovirus 71 Induces Anti-Viral Stress Granule-like Structures in RD Cells. *Biochem. Biophys. Res. Commun.* **2016**, *476*, 212–217. [CrossRef]
113. Wang, H.; Chang, L.; Wang, X.; Su, A.; Feng, C.; Fu, Y.; Chen, D.; Zheng, N.; Wu, Z. MOV10 Interacts with Enterovirus 71 Genomic 5′UTR and Modulates Viral Replication. *Biochem. Biophys. Res. Commun.* **2016**, *479*, 571–577. [CrossRef] [PubMed]
114. Gregersen, L.H.; Schueler, M.; Munschauer, M.; Mastrobuoni, G.; Chen, W.; Kempa, S.; Dieterich, C.; Landthaler, M. MOV10 Is a 5′ to 3′ RNA Helicase Contributing to UPF1 mRNA Target Degradation by Translocation along 3′ UTRs. *Mol. Cell* **2014**, *54*, 573–585. [CrossRef] [PubMed]
115. Meister, G.; Landthaler, M.; Peters, L.; Chen, P.Y.; Urlaub, H.; Lührmann, R.; Tuschl, T. Identification of Novel Argonaute-Associated Proteins. *Curr. Biol.* **2005**, *15*, 2149–2155. [CrossRef] [PubMed]
116. Abudu, A.; Wang, X.; Dang, Y.; Zhou, T.; Xiang, S.-H.; Zheng, Y.-H. Identification of Molecular Determinants from Moloney Leukemia Virus 10 Homolog (MOV10) Protein for Virion Packaging and Anti-HIV-1 Activity. *J. Biol. Chem.* **2012**, *287*, 1220–1228. [CrossRef] [PubMed]
117. Rahman, S.; Islam, R. Mammalian Sirt1: Insights on Its Biological Functions. *Cell Commun. Signal.* **2011**, *9*, 11. [CrossRef] [PubMed]
118. Yamamoto, H.; Schoonjans, K.; Auwerx, J. Sirtuin Functions in Health and Disease. *Mol. Endocrinol.* **2007**, *21*, 1745–1755. [CrossRef] [PubMed]
119. Kwon, H.-S.; Ott, M. The Ups and Downs of SIRT1. *Trends Biochem. Sci.* **2008**, *33*, 517–525. [CrossRef]
120. Pagans, S.; Pedal, A.; North, B.J.; Kaehlcke, K.; Marshall, B.L.; Dorr, A.; Hetzer-Egger, C.; Henklein, P.; Frye, R.; McBurney, M.W.; et al. SIRT1 Regulates HIV Transcription via Tat Deacetylation. *PLoS Biol.* **2005**, *3*, e41. [CrossRef]
121. Ren, J.-H.; Tao, Y.; Zhang, Z.-Z.; Chen, W.-X.; Cai, X.-F.; Chen, K.; Ko, B.C.B.; Song, C.-L.; Ran, L.-K.; Li, W.-Y.; et al. Sirtuin 1 Regulates Hepatitis B Virus Transcription and Replication by Targeting Transcription Factor AP-1. *J. Virol.* **2014**, *88*, 2442–2451. [CrossRef]
122. Han, Y.; Wang, L.; Cui, J.; Song, Y.; Luo, Z.; Chen, J.; Xiong, Y.; Zhang, Q.; Liu, F.; Ho, W.; et al. SIRT1 Inhibits EV71 Genome Replication and RNA Translation by Interfering with the Viral Polymerase and 5′UTR RNA. *J. Cell Sci.* **2016**, *129*, 4534–4547. [CrossRef]
123. Matter, N.; Herrlich, P.; König, H. Signal-Dependent Regulation of Splicing via Phosphorylation of Sam68. *Nature* **2002**, *420*, 691–695. [CrossRef] [PubMed]
124. Paronetto, M.P.; Cappellari, M.; Busà, R.; Pedrotti, S.; Vitali, R.; Comstock, C.; Hyslop, T.; Knudsen, K.E.; Sette, C. Alternative Splicing of the Cyclin D1 Proto-Oncogene Is Regulated by the RNA-Binding Protein Sam68. *Cancer Res.* **2010**, *70*, 229–239. [CrossRef] [PubMed]
125. Sánchez-Jiménez, F.; Sánchez-Margalet, V. Role of Sam68 in Post-Transcriptional Gene Regulation. *Int. J. Mol. Sci.* **2013**, *14*, 23402–23419. [CrossRef] [PubMed]
126. Fumagalli, S.; Totty, N.F.; Hsuan, J.J.; Courtneidge, S.A. A Target for Src in Mitosis. *Nature* **1994**, *368*, 871–874. [CrossRef]
127. Macias, M.J.; Wiesner, S.; Sudol, M. WW and SH3 Domains, Two Different Scaffolds to Recognize Proline-Rich Ligands. *FEBS Lett.* **2002**, *513*, 30–37. [CrossRef]

128. Galarneau, A.; Richard, S. The STAR RNA Binding Proteins GLD-1, QKI, SAM68 and SLM-2 Bind Bipartite RNA Motifs. *BMC Mol. Biol.* **2009**, *10*, 47. [CrossRef]
129. Lin, Q.; Taylor, S.J.; Shalloway, D. Specificity and Determinants of Sam68 RNA Binding: Implications for the biological function of k homology domains. *J. Biol. Chem.* **1997**, *272*, 27274–27280. [CrossRef]
130. Lawrence, P.; Schafer, E.A.; Rieder, E. The Nuclear Protein Sam68 Is Cleaved by the FMDV 3C Protease Redistributing Sam68 to the Cytoplasm during FMDV Infection of Host Cells. *Virology* **2012**, *425*, 40–52. [CrossRef]
131. McBride, A.E.; Schlegel, A.; Kirkegaard, K. Human Protein Sam68 Relocalization and Interaction with Poliovirus RNA Polymerase in Infected Cells. *Proc. Natl. Acad. Sci. USA* **1996**, *93*, 2296–2301. [CrossRef]
132. Zhang, H.; Song, L.; Cong, H.; Tien, P. Nuclear Protein Sam68 Interacts with the Enterovirus 71 Internal Ribosome Entry Site and Positively Regulates Viral Protein Translation. *J. Virol.* **2015**, *89*, 10031–10043. [CrossRef]
133. Xu, P.; Tong, W.; Chen, Y.-M. FUSE Binding Protein FUBP3 Is a Potent Regulator in Japanese Encephalitis Virus Infection. *Virol. J.* **2021**, *18*, 224. [CrossRef] [PubMed]
134. Sharma, M.; Anandram, S.; Ross, C.; Srivastava, S. FUBP3 Regulates Chronic Myeloid Leukaemia Progression through PRC2 Complex Regulated PAK1-ERK Signalling. *J. Cell. Mol. Med.* **2023**, *27*, 15–29. [CrossRef] [PubMed]
135. Tsai, F.-J.; Lin, C.-W.; Lai, C.-C.; Lan, Y.-C.; Lai, C.-H.; Hung, C.-H.; Hsueh, K.-C.; Lin, T.-H.; Chang, H.C.; Wan, L.; et al. Kaempferol Inhibits Enterovirus 71 Replication and Internal Ribosome Entry Site (IRES) Activity through FUBP and HNRP Proteins. *Food Chem.* **2011**, *128*, 312–322. [CrossRef] [PubMed]
136. Huang, H.-I.; Chang, Y.-Y.; Lin, J.-Y.; Kuo, R.-L.; Liu, H.-P.; Shih, S.-R.; Wu, C.-C. Interactome Analysis of the EV71 5′ Untranslated Region in Differentiated Neuronal Cells SH-SY5Y and Regulatory Role of FBP3 in Viral Replication. *Proteomics* **2016**, *16*, 2351–2362. [CrossRef] [PubMed]
137. Brush, M.H.; Weiser, D.C.; Shenolikar, S. Growth Arrest and DNA Damage-Inducible Protein GADD34 Targets Protein Phosphatase 1α to the Endoplasmic Reticulum and Promotes Dephosphorylation of the α Subunit of Eukaryotic Translation Initiation Factor 2. *Mol. Cell. Biol.* **2003**, *23*, 1292–1303. [CrossRef] [PubMed]
138. Zhu, T.; Jiang, X.; Xin, H.; Zheng, X.; Xue, X.; Chen, J.-L.; Qi, B. GADD34-Mediated Dephosphorylation of eIF2α Facilitates Pseudorabies Virus Replication by Maintaining de Novo Protein Synthesis. *Vet. Res.* **2021**, *52*, 148. [CrossRef] [PubMed]
139. Pakos-Zebrucka, K.; Koryga, I.; Mnich, K.; Ljujic, M.; Samali, A.; Gorman, A.M. The Integrated Stress Response. *EMBO Rep.* **2016**, *17*, 1374–1395. [CrossRef] [PubMed]
140. Dalet, A.; Argüello, R.J.; Combes, A.; Spinelli, L.; Jaeger, S.; Fallet, M.; Vu Manh, T.-P.; Mendes, A.; Perego, J.; Reverendo, M.; et al. Protein Synthesis Inhibition and GADD34 Control IFN-β Heterogeneous Expression in Response to dsRNA. *EMBO J.* **2017**, *36*, 761–782. [CrossRef]
141. Clavarino, G.; Cláudio, N.; Couderc, T.; Dalet, A.; Judith, D.; Camosseto, V.; Schmidt, E.K.; Wenger, T.; Lecuit, M.; Gatti, E.; et al. Induction of GADD34 Is Necessary for dsRNA-Dependent Interferon-β Production and Participates in the Control of Chikungunya Virus Infection. *PLoS Pathog.* **2012**, *8*, e1002708. [CrossRef]
142. Minami, K.; Tambe, Y.; Watanabe, R.; Isono, T.; Haneda, M.; Isobe, K.-I.; Kobayashi, T.; Hino, O.; Okabe, H.; Chano, T.; et al. Suppression of Viral Replication by Stress-Inducible GADD34 Protein via the Mammalian Serine/Threonine Protein Kinase mTOR Pathway. *J. Virol.* **2007**, *81*, 11106–11115. [CrossRef]
143. Wang, X.; Liao, Y.; Yap, P.L.; Png, K.J.; Tam, J.P.; Liu, D.X. Inhibition of Protein Kinase R Activation and Upregulation of GADD34 Expression Play a Synergistic Role in Facilitating Coronavirus Replication by Maintaining De Novo Protein Synthesis in Virus-Infected Cells. *J. Virol.* **2009**, *83*, 12462–12472. [CrossRef] [PubMed]
144. Ishaq, M.; Marshall, H.; Natarajan, V. GADD34 Attenuates HIV-1 Replication by Viral 5′-UTR TAR RNA-Mediated Translational Inhibition. *Virology* **2020**, *540*, 119–131. [CrossRef]
145. Brush, M.H.; Shenolikar, S. Control of Cellular GADD34 Levels by the 26S Proteasome. *Mol. Cell. Biol.* **2008**, *28*, 6989–7000. [CrossRef] [PubMed]
146. Li, H.; Li, W.; Zhang, S.; Qiu, M.; Li, Z.; Lin, Y.; Tan, J.; Qiao, W. Enterovirus 71 Activates GADD34 via Precursor 3CD to Promote IRES-Mediated Viral Translation. *Microbiol. Spectr.* **2022**, *10*, e01388-21. [CrossRef] [PubMed]
147. Venkataramanan, S.; Gadek, M.; Calviello, L.; Wilkins, K.; Floor, S.N. DDX3X and DDX3Y Are Redundant in Protein Synthesis. *RNA* **2021**, *27*, 1577–1588. [CrossRef] [PubMed]
148. Schröder, M. Human DEAD-Box Protein 3 Has Multiple Functions in Gene Regulation and Cell Cycle Control and Is a Prime Target for Viral Manipulation. *Biochem. Pharmacol.* **2010**, *79*, 297–306. [CrossRef]
149. Soto-Rifo, R.; Rubilar, P.S.; Limousin, T.; de Breyne, S.; Décimo, D.; Ohlmann, T. DEAD-Box Protein DDX3 Associates with eIF4F to Promote Translation of Selected mRNAs. *EMBO J.* **2012**, *31*, 3745–3756. [CrossRef]
150. Su, Y.-S.; Tsai, A.-H.; Ho, Y.-F.; Huang, S.-Y.; Liu, Y.-C.; Hwang, L.-H. Stimulation of the Internal Ribosome Entry Site (IRES)-Dependent Translation of Enterovirus 71 by DDX3X RNA Helicase and Viral 2A and 3C Proteases. *Front. Microbiol.* **2018**, *9*, 1324. [CrossRef]
151. Gao, J.; Choudhry, H.; Cao, W. Apolipoprotein B mRNA Editing Enzyme Catalytic Polypeptide-like Family Genes Activation and Regulation during Tumorigenesis. *Cancer Sci.* **2018**, *109*, 2375–2382. [CrossRef]
152. Wang, H.; Zhong, M.; Li, Y.; Li, K.; Wu, S.; Guo, T.; Cen, S.; Jiang, J.; Li, Z.; Li, Y. APOBEC3G Is a Restriction Factor of EV71 and Mediator of IMB-Z Antiviral Activity. *Antivir. Res.* **2019**, *165*, 23–33. [CrossRef]

153. Fehrholz, M.; Kendl, S.; Prifert, C.; Weissbrich, B.; Lemon, K.; Rennick, L.; Duprex, P.W.; Rima, B.K.; Koning, F.A.; Holmes, R.K.; et al. The Innate Antiviral Factor APOBEC3G Targets Replication of Measles, Mumps and Respiratory Syncytial Viruses. *J. Gen. Virol.* **2012**, *93*, 565–576. [CrossRef] [PubMed]
154. Nguyen, D.H.; Hu, J. Reverse Transcriptase- and RNA Packaging Signal-Dependent Incorporation of APOBEC3G into Hepatitis B Virus Nucleocapsids. *J. Virol.* **2008**, *82*, 6852–6861. [CrossRef] [PubMed]
155. Okada, A.; Iwatani, Y. APOBEC3G-Mediated G-to-A Hypermutation of the HIV-1 Genome: The Missing Link in Antiviral Molecular Mechanisms. *Front. Microbiol.* **2016**, *7*, 2027. [CrossRef] [PubMed]
156. Li, Z.; Ning, S.; Su, X.; Liu, X.; Wang, H.; Liu, Y.; Zheng, W.; Zheng, B.; Yu, X.-F.; Zhang, W. Enterovirus 71 Antagonizes the Inhibition of the Host Intrinsic Antiviral Factor A3G. *Nucleic Acids Res.* **2018**, *46*, 11514–11527. [CrossRef] [PubMed]
157. Wickham, L.; Duchaîne, T.; Luo, M.; Nabi, I.R.; DesGroseillers, L. Mammalian Staufen Is a Double-Stranded-RNA- and Tubulin-Binding Protein Which Localizes to the Rough Endoplasmic Reticulum. *Mol. Cell. Biol.* **1999**, *19*, 2220–2230. [CrossRef]
158. Kim, Y.K.; Furic, L.; Desgroseillers, L.; Maquat, L.E. Mammalian Staufen1 Recruits Upf1 to Specific mRNA 3′UTRs so as to Elicit mRNA Decay. *Cell* **2005**, *120*, 195–208. [CrossRef]
159. Furic, L.; Maher-Laporte, M.; DesGroseillers, L. A Genome-Wide Approach Identifies Distinct but Overlapping Subsets of Cellular mRNAs Associated with Staufen1- and Staufen2-Containing Ribonucleoprotein Complexes. *RNA* **2008**, *14*, 324–335. [CrossRef]
160. Chen, Y.-M.; Ou, B.-T.; Chen, C.-Y.; Chan, H.-H.; Chen, C.-J.; Wang, R.Y. Staufen1 Protein Participates Positively in the Viral RNA Replication of Enterovirus 71. *Viruses* **2019**, *11*, 142. [CrossRef]
161. Blackham, S.L.; McGarvey, M.J. A Host Cell RNA-Binding Protein, Staufen1, Has a Role in Hepatitis C Virus Replication before Virus Assembly. *J. Gen. Virol.* **2013**, *94 Pt 11*, 2429–2436. [CrossRef]
162. Abrahamyan, L.G.; Chatel-Chaix, L.; Ajamian, L.; Milev, M.P.; Monette, A.; Clément, J.-F.; Song, R.; Lehmann, M.; DesGroseillers, L.; Laughrea, M.; et al. Novel Staufen1 Ribonucleoproteins Prevent Formation of Stress Granules but Favour Encapsidation of HIV-1 Genomic RNA. *J. Cell Sci.* **2010**, *123 Pt 3*, 369–383. [CrossRef]
163. de Lucas, S.; Peredo, J.; Marión, R.M.; Sánchez, C.; Ortín, J. Human Staufen1 Protein Interacts with Influenza Virus Ribonucleoproteins and Is Required for Efficient Virus Multiplication. *J. Virol.* **2010**, *84*, 7603–7612. [CrossRef] [PubMed]
164. Mauger, D.M.; Lin, C.; Garcia-Blanco, M.A. hnRNP H and hnRNP F Complex with Fox2 To Silence Fibroblast Growth Factor Receptor 2 Exon IIIc. *Mol. Cell. Biol.* **2008**, *28*, 5403–5419. [CrossRef] [PubMed]
165. Xu, C.; Xie, N.; Su, Y.; Sun, Z.; Liang, Y.; Zhang, N.; Liu, D.; Jia, S.; Xing, X.; Han, L.; et al. HnRNP F/H Associate with hTERC and Telomerase Holoenzyme to Modulate Telomerase Function and Promote Cell Proliferation. *Cell Death Differ.* **2020**, *27*, 1998–2013. [CrossRef] [PubMed]
166. Thiel, G.; Cibelli, G. Regulation of Life and Death by the Zinc Finger Transcription Factor Egr-1. *J. Cell. Physiol.* **2002**, *193*, 287–292. [CrossRef] [PubMed]
167. Ho, B.-C.; Yu, S.-L.; Chen, J.J.W.; Chang, S.-Y.; Yan, B.-S.; Hong, Q.-S.; Singh, S.; Kao, C.-L.; Chen, H.-Y.; Su, K.-Y.; et al. Enterovirus-Induced miR-141 Contributes to Shutoff of Host Protein Translation by Targeting the Translation Initiation Factor eIF4E. *Cell Host Microbe* **2011**, *9*, 58–69. [CrossRef] [PubMed]
168. Song, Y.; Cheng, X.; Yang, X.; Zhao, R.; Wang, P.; Han, Y.; Luo, Z.; Cao, Y.; Zhu, C.; Xiong, Y.; et al. Early Growth Response-1 Facilitates Enterovirus 71 Replication by Direct Binding to the Viral Genome RNA. *Int. J. Biochem. Cell Biol.* **2015**, *62*, 36–46. [CrossRef]
169. Kedersha, N.L.; Gupta, M.; Li, W.; Miller, I.; Anderson, P. RNA-Binding Proteins TIA-1 and TIAR Link the Phosphorylation of eIF-2 Alpha to the Assembly of Mammalian Stress Granules. *J. Cell Biol.* **1999**, *147*, 1431–1442. [CrossRef]
170. Gottschald, O.R.; Malec, V.; Krasteva, G.; Hasan, D.; Kamlah, F.; Herold, S.; Rose, F.; Seeger, W.; Hänze, J. TIAR and TIA-1 mRNA-Binding Proteins Co-Aggregate under Conditions of Rapid Oxygen Decline and Extreme Hypoxia and Suppress the HIF-1α Pathway. *J. Mol. Cell Biol.* **2010**, *2*, 345–356. [CrossRef]
171. White, J.P.; Lloyd, R.E. Poliovirus Unlinks TIA1 Aggregation and mRNA Stress Granule Formation. *J. Virol.* **2011**, *85*, 12442–12454. [CrossRef]
172. Wu, S.; Wang, Y.; Lin, L.; Si, X.; Wang, T.; Zhong, X.; Tong, L.; Luan, Y.; Chen, Y.; Li, X.; et al. Protease 2A Induces Stress Granule Formation during Coxsackievirus B3 and Enterovirus 71 Infections. *Virol. J.* **2014**, *11*, 192. [CrossRef]
173. Wang, X.; Wang, H.; Li, Y.; Jin, Y.; Chu, Y.; Su, A.; Wu, Z. TIA-1 and TIAR Interact with 5′-UTR of Enterovirus 71 Genome and Facilitate Viral Replication. *Biochem. Biophys. Res. Commun.* **2015**, *466*, 254–259. [CrossRef]
174. Lin, J.-Y.; Kung, Y.-A.; Shih, S.-R. Antivirals and Vaccines for Enterovirus A71. *J. Biomed. Sci.* **2019**, *26*, 65. [CrossRef]
175. de Breyne, S.; Yu, Y.; Unbehaun, A.; Pestova, T.V.; Hellen, C.U.T. Direct Functional Interaction of Initiation Factor eIF4G with Type 1 Internal Ribosomal Entry Sites. *Proc. Natl. Acad. Sci. USA* **2009**, *106*, 9197–9202. [CrossRef] [PubMed]
176. Beckham, S.A.; Matak, M.Y.; Belousoff, M.J.; Venugopal, H.; Shah, N.; Vankadari, N.; Elmlund, H.; Nguyen, J.H.C.; Semler, B.L.; Wilce, M.C.J.; et al. Structure of the PCBP2/Stem-Loop IV Complex Underlying Translation Initiation Mediated by the Poliovirus Type I IRES. *Nucleic Acids Res.* **2020**, *48*, 8006–8021. [CrossRef] [PubMed]
177. Walter, B.L.; Parsley, T.B.; Ehrenfeld, E.; Semler, B.L. Distinct Poly(rC) Binding Protein KH Domain Determinants for Poliovirus Translation Initiation and Viral RNA Replication. *J. Virol.* **2002**, *76*, 12008–12022. [CrossRef] [PubMed]
178. Blyn, L.B.; Swiderek, K.M.; Richards, O.; Stahl, D.C.; Semler, B.L.; Ehrenfeld, E. Poly(rC) Binding Protein 2 Binds to Stem-Loop IV of the Poliovirus RNA 5′ Noncoding Region: Identification by Automated Liquid Chromatography-Tandem Mass Spectrometry. *Proc. Natl. Acad. Sci. USA* **1996**, *93*, 11115–11120. [CrossRef]

179. Andreev, D.E.; Hirnet, J.; Terenin, I.M.; Dmitriev, S.E.; Niepmann, M.; Shatsky, I.N. Glycyl-tRNA Synthetase Specifically Binds to the Poliovirus IRES to Activate Translation Initiation. *Nucleic Acids Res.* **2012**, *40*, 5602–5614. [CrossRef]
180. Meerovitch, K.; Svitkin, Y.V.; Lee, H.S.; Lejbkowicz, F.; Kenan, D.J.; Chan, E.K.; Agol, V.I.; Keene, J.D.; Sonenberg, N. La Autoantigen Enhances and Corrects Aberrant Translation of Poliovirus RNA in Reticulocyte Lysate. *J. Virol.* **1993**, *67*, 3798–3807. [CrossRef]
181. Bedard, K.M.; Daijogo, S.; Semler, B.L. A Nucleo-Cytoplasmic SR Protein Functions in Viral IRES-Mediated Translation Initiation. *EMBO J.* **2007**, *26*, 459–467. [CrossRef]
182. Hunt, S.L.; Hsuan, J.J.; Totty, N.; Jackson, R.J. Unr, a Cellular Cytoplasmic RNA-Binding Protein with Five Cold-Shock Domains, Is Required for Internal Initiation of Translation of Human Rhinovirus RNA. *Genes Dev.* **1999**, *13*, 437–448. [CrossRef]
183. Davila-Calderon, J.; Penumutchu, S.; Tolbert, B. A Regulatory Interplay: hnRNP A1 and AUF1 Compete for the Same IRES Domain to Regulate Viral Translation in EV71. *FASEB J.* **2019**, *33*, 625.5. [CrossRef]
184. Courtney, D.G. Post-Transcriptional Regulation of Viral RNA through Epitranscriptional Modification. *Cells* **2021**, *10*, 1129. [CrossRef] [PubMed]

viruses

MDPI

Article

Translation of Overlapping Open Reading Frames Promoted by Type 2 IRESs in Avian Calicivirus Genomes

Yani Arhab, Tatyana V. Pestova and Christopher U. T. Hellen *

Department of Cell Biology, SUNY Downstate Health Sciences University, Brooklyn, NY 11203, USA; yani.arhab@downstate.edu (Y.A.); tatyana.pestova@downstate.edu (T.V.P.)
* Correspondence: christopher.hellen@downstate.edu

Abstract: Caliciviruses have positive-sense RNA genomes, typically with short 5′-untranslated regions (5′UTRs) that precede the long open reading frame 1 (ORF1). Exceptionally, some avian caliciviruses have long 5′UTRs containing a picornavirus-like internal ribosomal entry site (IRES), which was likely acquired by horizontal gene transfer. Here, we identified numerous additional avian calicivirus genomes with IRESs, predominantly type 2, and determined that many of these genomes contain a ~200–300 codon-long ORF (designated ORF1*) that overlaps the 5′-terminal region of ORF1. The activity of representative type 2 IRESs from grey teal calicivirus (GTCV) and *Caliciviridae* sp. isolate yc-13 (RaCV1) was confirmed by in vitro translation. Toeprinting showed that in cell-free extracts and in vitro reconstituted reactions, ribosomal initiation complexes assembled on the ORF1* initiation codon and at one or two AUG codons in ORF1 at the 3′-border and/or downstream of the IRES. Initiation at all three sites required eIF4A and eIF4G, which bound to a conserved region of the IRES; initiation on the ORF1* and principal ORF1 initiation codons involved eIF1/eIF1A-dependent scanning from the IRES's 3′-border. Initiation on these IRESs was enhanced by the IRES *trans*-acting factors (ITAFs) Ebp1/$ITAF_{45}$, which bound to the apical subdomain Id of the IRES, and PTB (GTCV) or PCBP2 (RaCV1).

Keywords: calicivirus; IRES; translation initiation; overlapping reading frame; Ebp1

1. Introduction

Viruses depend on the cellular translation apparatus. They have evolved strategies to usurp it for the synthesis of viral proteins, to exploit its capabilities to maximize utilization of the coding capacity of viral mRNAs, and to suppress the translation of cellular mRNAs during infection, particularly of mRNAs that are induced as part of the innate immune response [1,2].

The *Caliciviridae* family comprises at least eleven genera of viruses and has a broad host range, including birds, fish, mammals and reptiles [3]. Caliciviruses have single-stranded, positive-sense RNA genomes of 6.4–8.5 kb with a VPg (viral protein genome-linked) covalently bound to the 5′-terminus, a short 5′-UTR and a 3′-terminal poly(A) tail [4]. A subgenomic mRNA that is 3′-coterminal to the genomic RNA is synthesized during infection, and it is also VPg-linked. Calicivirus genomes generally contain two or three open reading frames (ORFs): the 5′-proximal ORF1 encodes a large polyprotein that is cleaved by a virus-encoded protease to yield non-structural proteins. The structural proteins VP1 and VP2 are encoded by ORF2 and ORF3 and are expressed from the subgenomic mRNA. These ORFs overlap, and RNA elements upstream of the ORF2 stop codon subvert post-termination recycling and promote reinitiation on the ORF3 initiation codon [5–7]. Exceptionally, murine norovirus subgenomic mRNAs encode an overlapping ORF4 within the VP1 coding region [8,9].

The canonical initiation process [10] begins with binding of the eukaryotic initiation factor (eIF)2-GTP/Met-$tRNA_i^{Met}$ complex, eIF1, eIF1A and eIF3 to the ribosomal 40S subunit to form a 43S preinitiation complex. Cellular mRNAs have a 5′-terminal m^7GpppG cap that is recognized by the eIF4E cap-binding subunit of eIF4F, which also contains eIF4A (helicase) and eIF4G subunits. It recruits the 43S complex to the capped end of the mRNA. The 43S complex scans downstream to locate the initiation codon in a process that requires eIF1 and eIF1A and that involves the unwinding of mRNA by eIF4F, which is aided by eIF4B. Scanning on highly structured 5′-UTRs also requires the DExH-box protein DHX29. The establishment of base pairing between the initiation codon and the anticodon of Met-tRNA$_i^{Met}$ arrests the scanning 43S complex and forms a 48S complex after the eIF5-induced hydrolysis of eIF2-bound GTP. The subsequent eIF5B-mediated reorientation of Met-tRNA$_i^{Met}$, joining of a 60S ribosomal subunit and release of factors forms an 80S ribosome that is competent to begin translation.

The 5′-ends of calicivirus mRNAs are linked to VPg rather than to a m^7Gppp cap, so the initiation of translation is independent of the cap-eIF4E interaction. Instead, VPg recruits eIF4F and associated factors such as eIF3 via interactions with eIF4E and eIF4G and exploits the stress granule assembly factor G3BP1 to promote the binding of the 40S subunit to the VPg-linked mRNA [11–13]. This distinct mechanism of ribosomal recruitment to calicivirus mRNAs may account for their preferential translation in infected cells in which the translation of cellular mRNAs is impaired [14]. Subsequent steps in this initiation process remain uncharacterized. Caliciviruses therefore utilize strategies such as the synthesis of subgenomic mRNA, synthesis of a polyprotein that is proteolytically cleaved to yield mature products, translation of linked ORFs and overlapping ORFs to maximize utilization of their genomic coding capacity, and the non-canonical initiation and termination–reinitiation mechanisms for the translation of viral mRNA.

The 5′UTRs of a subset of calicivirus genomes contain an IRES and can therefore exploit alternative non-canonical mechanisms for the initiation of translation. IRESs are complex, highly structured RNA elements that form structural classes, each of which uses a different subset of initiation factors to mediate initiation by a distinct mechanism (Table 1) [15]. Type 1 and type 2 IRESs, exemplified by poliovirus and encephalomyocarditis virus (EMCV), respectively, are ~450 nt long and consist of five domains, designated II–VI in type 1 and H–L in type 2 IRESs. Both types of IRES contain highly conserved, functionally important motifs. In type 2 IRESs, they include a C-rich loop, the GNRA tetraloop and an AAA motif in apical arms of domain I [16–19], an A-rich stem-loop wedged between J and K domains [20], and an apical bipartite motif in domain J [21–23]. Type 1 and type 2 IRESs both have a 3′-terminal Yn-Xm-AUG motif, in which a Yn pyrimidine tract (n = 8–10 nt) is separated by a spacer (m = 18–20 nt) from an AUG triplet [24]. This triplet is the initiation codon for the viral polyprotein in type 2 IRESs, although initiation can also occur downstream of it [25–27]. Type 5 IRESs are also ~450 nt long and are chimeric, containing a Yn-Xm-AUG motif and constituent elements that resemble domains from type 1 and type 2 IRESs [28,29].

Initiation on type 1 and type 2 IRESs relies on their specific interaction with the central eIF4A-binding domain of eIF4G [20–23,30–33]. These IRESs can consequently function without eIF4E and the N-terminal region of eIF4G to which it binds, which is commonly cleaved from eIF4G by virus-encoded proteases during infection, shutting off cellular translation. 48S complex formation on type 2 IRESs requires eIF2, eIF3, the central domain of eIF4G and eIF4A, and is enhanced by eIF4B [30,32,33]; scanning to AUG codons downstream of the Yn-Xm-AUG motif additionally requires eIF1 and eIF1A [27]. These IRESs also commonly require specific IRES trans-acting factors (ITAFs). The principal ITAF for type 1 IRESs is the poly(C) binding protein 2 [31], whereas type 2 IRESs require the pyrimidine tract binding protein (PTB) and in some instances, ErbB3-binding protein 1 (Ebp1, also known as $ITAF_{45}$) [30,32–36].

Table 1. Classes of viral IRES.

IRES Class	Representative Member	Virus Family
1	Poliovirus	*Picornaviridae*
2	Encephalomyocarditis virus	*Picornaviridae*
	Foot-and-mouth disease virus	*Picornaviridae*
	Caliciviridae sp. isolate yc-13 "Red-crowned crane calicivirus"	*Caliciviridae*
3	Hepatitis A virus	*Picornaviridae*
4	Hepatitis C virus	*Flaviviridae*
	Ruddy turnstone calicivirus A	*Caliciviridae*
5	Aichivirus	*Picornaviridae*
	Caliciviridae sp. isolate hwf182cal1	*Caliciviridae*
6	Cricket paralysis virus	*Dicistroviridae*

Type 4 IRESs, exemplified by hepatitis C virus (HCV), are ~330nt long and employ a fundamentally different mechanism. They bind directly to the 40S subunit, positioning the initiation codon in the ribosomal P site so that the recruitment of eIF2-GTP/Met-tRNA$_i^{Met}$ forms a 48S complex without the involvement of group 4 eIFs [37,38]. Subunit joining on IRES types 1, 2, 4 and 5 is mediated by eIF5 and eIF5B.

Type 6 IRESs, exemplified by the Cricket paralysis virus intergenic region, are the smallest IRESs, and they function by the most streamlined mechanism, binding directly to 40S subunits and/or 80S ribosomes to bypass the entire initiation process and instead commencing directly with elongation [39,40].

Seven caliciviruses belonging to the *Nacovirus* genus of *Caliciviridae* and to another currently unclassified clade of caliciviruses were found by us to contain elements that are closely related to picornavirus type 2 IRESs (e.g., Figure 1B,C), one to have a type 4 IRES and one to have a type 5 IRES [41]. A subsequent report identified a type 2 IRES in Trumpeter swan calicivirus [42]. Some members of these two genera (e.g., Goose calicivirus strain N [43] and Turkey calicivirus isolate L11043 [44]) have short 5′ non-coding regions, whereas Duck calicivirus 2 strain DuCV-2_B6 [45] has a 275 nt long 5′UTR but lacks an IRES. It is not known whether these sequences correspond to incomplete genomes or to genomic mRNAs that naturally have short 5′UTRs. This point is addressed below.

We previously determined that the type 4 IRES of Ruddy turnstone calicivirus mediates initiation by a mechanism that involves the direct binding of the IRES to the 40S subunit [41] analogous to that used by type 4 IRESs such as that of HCV, completely bypassing the requirement for the eIF4 group of initiation factors. In the present report, we determined that several additional calicivirus genomes contain IRESs, identified conserved ORFs (designated ORF1*) that overlap ORF1 in distinct groups of caliciviruses, and characterized the mechanism of initiation of translation of ORF1 and ORF1* mediated by these type 2 IRESs. These studies add to the understanding of viral genome compaction strategies, of viral translation mechanisms and of the dissemination of genetic elements between viruses by recombination.

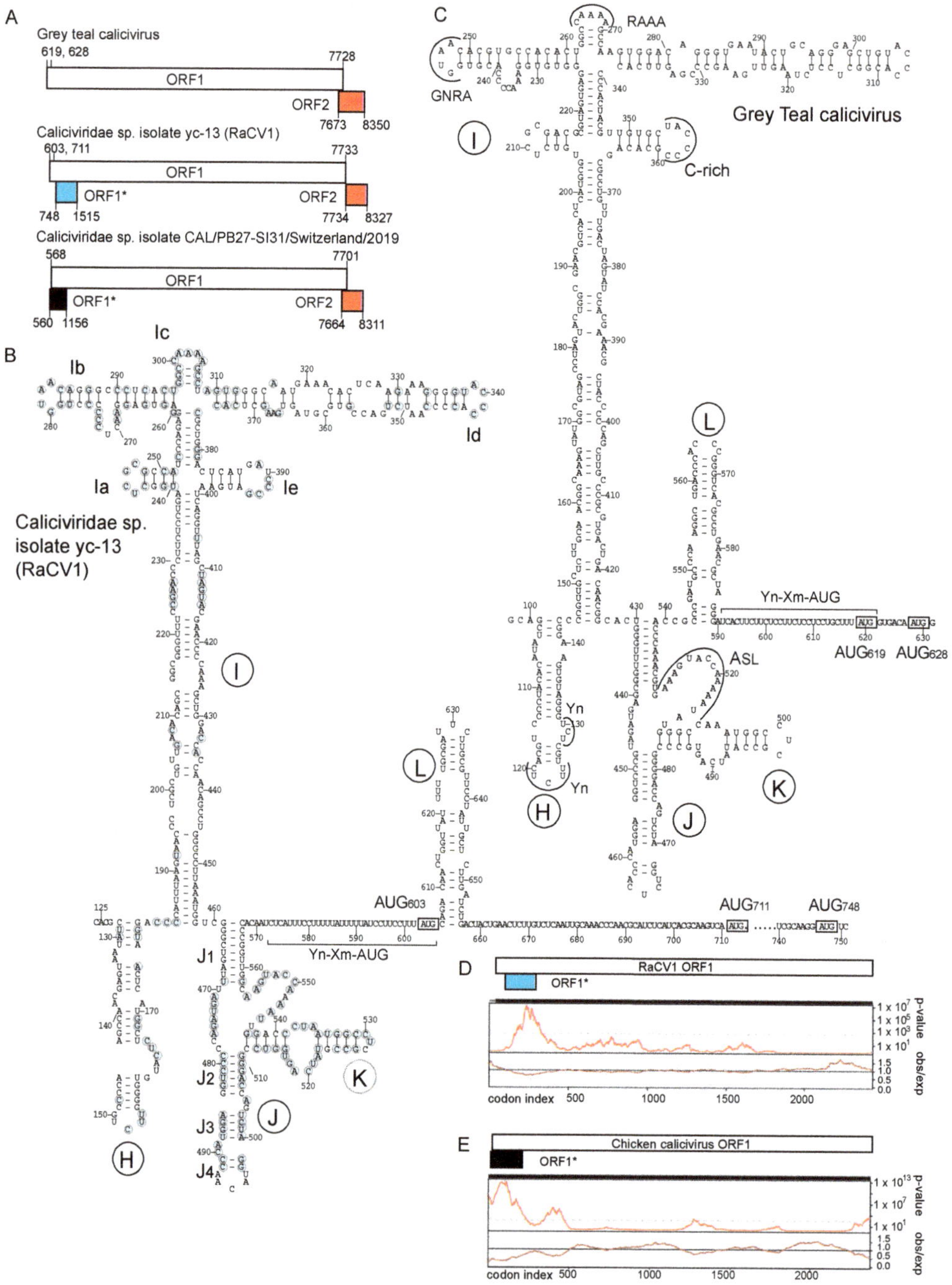

Figure 1. Models of the genomes and type 2 IRESs of representative avian caliciviruses. (**A**) Schematic representation of the coding regions of grey teal calicivirus (GTCV) isolate MW09, *Caliciviridae* sp.

isolate yc-13 (RaCV1) and *Caliciviridae* sp. Isolate CAL/PB27-SI31/Switzerland/2019. ORF1 is cleaved into mature non-structural proteins, ORF2 encodes the VP2 capsid protein. ORF1* is the overlapping open reading frame identified in this report. (**B**,**C**) Models of the secondary structures of (b) RaCV1 and (c) GTCV IRESs. IRES domains are labeled sequentially from H to L; nucleotides that are identical in 10 of the 13 sequences in Supplementary Figure S1 are indicated by blue circles. (**B**) Subdomains in domain I are labeled sequentially from Ia to Ie, and helices in domain J are labeled J_1 to J_4. (**C**) The model is annotated to show conserved sequence motifs, including oligopyrimidine (Yn) motifs in domain H, GNRA, RAAA and C-rich motifs at the apex of domain I, the ASL domain in the J-K domain and the Yn-Xm-AUG motif downstream of the J-K domain. Nucleotides are numbered, and the initiation codons AUG_{603}, AUG_{711} and AUG_{748} (RaCV1) and AUG_{619} and AUG_{628} (GTCV) are boxed. (**D**,**E**) Synonymous-site variability in a subset of avian caliciviruses. (**D**) Above: map of RaCV1 ORF1 and the overlapping ORF1*. Below: analysis of synonymous-site variability in an alignment of eight avian calicivirus sequences. (**E**) Above: map of Chicken calicivirus isolate Chicken/NLD/2019/V_M_030_calici_8 ORF1 and the overlapping ORF1*. Below: analysis of synonymous-site variability in an alignment of eleven chicken calicivirus sequences. (**D**,**E**) The lower panel (obs/exp) (darker red lines) indicates the relative amount of synonymous-site variability as represented by the ratio of the observed number of synonymous substitutions to the expected number, (**D**) in a 101-codon and (**E**) in a 75-codon sliding window. The upper panels (brighter red lines) show the corresponding *p*-values. (**A**,**D**,**E**) ORF2 coding regions are colored red, and the two classes of ORF1* are colored blue and black, respectively.

2. Materials and Methods

2.1. Sequences

Sequences were analyzed from the following caliciviruses (name followed by Genbank accession number): Avocet calicivirus isolate MW21 (MH453804.1), Bird calicivirus isolate Cormcali01 (OR062614.1), *Caliciviridae* sp. isolate 361bR-k141_109560 (MZ679042.1), *Caliciviridae* sp. isolate hwf182cal1 (MT138020.1), *Caliciviridae* sp. isolate xftoti59cal1 (MT138028.1), Calicivirus chicken/V0021/Bayern/2004 (NC_075411.1), Chicken calicivirus ChCV-052 (MH992118.1), Chicken calicivirus isolate GA/1478/2003 (MN810874.1), Chicken calicivirus isolate GA/1479/2004 (MN810875.1), Chicken calicivirus isolate CaliciD62/2013 (KM254170.1), Chicken calicivirus strain RS/BR/15/1R-1 (MG846434.1), Chicken calicivirus strain RS/BR/15/3S-1 (MG846429.1), Chicken calicivirus strain RS/BR/15/4S-1 (MG846430.1), Chicken calicivirus strain RS/BR/15/6R (MG846433.1), Chicken calicivirus strain RS/BR/15/1R-1 (MG846434.1), Chicken calicivirus strain RS/BR/2015 (KY120883.1), Duck calicivirus 2 strain DuCV-2_B6 (MN175552.1), Duck calicivirus 2 strain DuCV-2_B76 (MN175556.1), Duck calicivirus isolate MW20 (MH453811.1), Goose calicivirus strain N (KJ473715.1), Goose calicivirus isolate MA/CHN/2018 (MN068022.1), Goose calicivirus isolate H146 (KY399947.1), Grey teal calicivirus (MK204392.1), Duck calicivirus isolate MW20 (MH453811.1), Mute swan feces associated calicivirus strain Abbotsbury/A/2016 (MW588067.1), Norovirus G1 (NC_001959.2), Pink-eared duck calicivirus I (MK204416.1), *Caliciviridae* sp. isolate yc-13 red-crowned crane virus (KY312552.1), Ruddy Turnstone calicivirus A isolate MW19 (MH453861.1), Temminck's stint calicivirus isolate SIB_91 (ON815296.1), Trumpeter swan calicivirus (OP271827.1), Turkey calicivirus isolate L11043 (JQ347522.1), Wilkes virus isolate Antarctic11 (MT025075.1) from the following caliciviruses (name followed by GenBase accession number): Goose calicivirus isolate NC-19BV-37-k77-711556 (C_AA057594.1) and Duck calicivirus isolate NC-19BV-45-k77-747816 (C_AA057597.1), and from the following calicivirus *metagenome*-assembled genomes (MAGs) (name followed by Genbank accession number): Bavaria virus isolate A5_2S5_SA_2022 (PP228889.1), *Bavovirus* sp. isolate Chicken/NLD/2019/V_M_032_calici_1 (MW684839.1), *Bavovirus* sp. isolate Chicken/NLD/2019/V_M_021_calici_5 (MW684836.1), *Bavovirus* sp. isolate Chicken/NLD/2019/V_M_029_calici_1 (MW684837.1), Chicken calicivirus isolate Environment/NLD/2019/VE_8_calici_66 (MW684834.1), Chicken calicivirus isolate Chicken/NLD/2019/V_M_019_calici_3 (MW684835.1), Chicken calicivirus isolate Chicken/NLD/2019/

V_M_030_calici_8 (MW684838.1), Chicken calicivirus isolate Chicken/NLD/2019/V_M_036_calici_2 (MW684840.1), Chicken calicivirus isolate Chicken/NLD/2019/V_M_041_calici_11 (MW684841.1), Chicken calicivirus isolate Chicken/NLD/2019/V_M_050_calici_2 (MW684842.1), Chicken calicivirus isolate Chicken/NLD/2019/V_M_051_calici_5 (MW684843.1), Chicken calicivirus isolate Chicken/NLD/2019/V_M_052_calici_7 (MW684844.1), Chicken calicivirus isolate Chicken/NLD/2019/V_M_053_calici_1 (MW684845.1), Chicken calicivirus isolate Chicken/NLD/2019/V_M_056_calici_1 (MW684847.1) *Caliciviridae* sp. isolate CAL/PB30-SII34/Switzerland/2019 (OM469263.1), *Caliciviridae* sp. isolate CAL/PB10-HII34X/Switzerland/2019 (OM469260.1), Caliciviridae sp. isolate CAL/PB27-SI31/ Switzerland/2019 (OM469262.1), *Caliciviridae* sp. isolate bbr034cal1 (MT138025.1), *Caliciviridae* sp. isolate ybb044cal01nc (MT138022.1), *Caliciviridae* sp. isolate mag166cal1 (MT138021.1), *Caliciviridae* sp. isolate cftwhg06cal1 (MT138017.1), *Caliciviridae* sp. isolate fmg067cal1nc (MT138023.1), *Caliciviridae* sp. isolate zftwig05sapo1nc (MT138019.1), Duck calicivirus isolate PI202116/k141_160805 (PP512781.1), Shelduck calicivirus isolate MW17 (MH453874.1).

Sequences were analyzed from the following picornaviruses (name followed by Genbank accession number): Anativirus A (AV; formerly Duck picornavirus TW90A) (AY563023.1), Mute swan feces associated sapelovirus 1 strain Abbotsbury/B/2016 (MW588168.1) and Red-necked stint Picornavirus B-like (MK204387.1).

2.2. Nucleotide Sequence Alignment and Modelling of RNA Structures

Candidate IRES sequences were identified in the NCBI database (accessed 1 June 2024) using BLASTN (http://www.ncbi.nlm.nih.gov/BLAST/) to search nucleotide collection (nr/nt) sequences, using the following parameters: E, 1000; word size, 11; match/mismatch scores, 1/1; gap costs, 2/1. Hits were characterized by six-frame translation (https://www.ncbi.nlm.nih.gov/orffinder/ and http://molbiol.ru/eng/scripts/01_13.html) to identify potential initiation codons and open reading frames.

Sequence alignments of IRESs made using Clustal Omega (http://www.ebi.ac.uk/Tools/msa/clustalo/) were adjusted manually using established IRES structures as guides [41,46]. Pairwise comparisons were made using EMBOSS matcher (https://www.ebi.ac.uk/jdispatcher/psa/emboss_matcher) with default parameters. Structural elements were modeled using Mfold (http://unafold.rna.albany.edu/?q=mfold) [47] with default parameters. ORF1* sequences were aligned using Clustal Omega with default parameters.

Synonymous site conservation was analyzed with SYNPLOT2 [48], using codon-respecting alignments of (a) complete ORF1 sequences from *Caliciviridae* sp. isolate yc-13 (RaCV1), Duck calicivirus 2 strain DuCV-2_B6, Goose calicivirus isolate NC-19BV-37-k77-711556, Goose calicivirus strain N, Pink-eared duck calicivirus I, Turkey calicivirus isolate L11043, Wilkes virus isolate Antarctic11 and Caliciviridae sp. isolate xftoti59cal1 and a 101 codon sliding window and (b) ORF1* and ORF1 sequences from calicivirus *metagenome*-assembled genomes (MAGs) with Genbank accession numbers: MW684838.1, MW684840.1, MW684841.1, MW684842.1, MW684843.1, MW684844.1, MW684845.1, MW684847.1, OM469263.1, OM469260.1 and OM469262.1 and a 75 codon sliding window.

2.3. Phylogenetic Analysis

Sequences were aligned using Clustal Omega, and gaps and ambiguously aligned regions were removed using BMGE [49], in both instances using default settings, and maximum- likelihood phylogenetic trees were inferred using PhyML with Smart Model Selection (LG + G + I + F model), implemented in NGPhylogeny [50,51]. The phylogenetic tree with branch supports was visualized using iTOL [52] and annotated using Adobe Illustrator.

2.4. Plasmids

Expression vectors for His_6-tagged eIF1, eIF1A [53], eIF4A, eIF4B [33], eIF4A (R362Q) [54], $eIF4GI_{653–1115}$, $eIF4GI_{736–1008}$ and $eIF4GI_{736–988}$ [55], cysteine-less and T830C $eIF4GI_{736–1115}$ [21], $eIF4GI_{653–1600}$ [30], eIF5 [56], DHX29 [57], *Escherichia coli* methionyl tRNA synthetase [58] and nPTB [59] have been described. The vector pET15b-His$_6$-Ebp1 for expression in *E. coli* of N-terminally His$_6$-tagged Ebp1 was made by inserting the appropriate DNA sequence between Nde1/BamH1 sites of pET15b (GenScript, Piscataway NJ, USA). Ebp1 cysteine substitution mutants were made (Genscript) by introducing individual surface-exposed cysteine residues at positions 22, 37, 55, 65, 94, 131, 165, 208, 221, 254, 271, 311 and 360.

The transcription vectors for dicistronic DC Aichivirus IRES-containing mRNA [28] and monocistronic EMCV nt 373–1656 mRNA [41] have been described. pUC57-FMDV nt 278–891 was made by inserting FMDV O$_1$K nt 278–891 (GenBank: X00871.1) flanked by SnaBI and NdeI sites and an upstream T7 promoter into the EcoRV site of pUC57 (Genscript).

pUC57[T7-Stem-GTCV] was prepared (Genscript) by synthesizing a DNA fragment with flanking HindIII and BamH1 sites that corresponded to a T7 promoter, a stable hairpin ($\Delta G = -32.40$ kcal/mol) and GTCV nt 1–1494) with tandem UAA-UGA stop codons inserted after nt 1461 and substitutions to replace codons at positions 216, 225, 241, 254 and 263 in the ORF initiating at AUG$_{528}$ with AUG codons to increase the methionine content of the 278 amino acid (a.a.)-long (31.7 kDa) coding sequence). This DNA fragment was inserted into pUC57. This plasmid was used by GenScript to generate [GUAA-UACG], [GUAA-CCUU], [destab. Dom L], [AUG$_{619}$AAA], [AUG$_{628}$AAA], [AUG$_{723}$AAA] and [AUG$_{619}$AUC] substitution mutants and the [ΔIb] and [ΔStem 1] deletion mutants.

The transcription vector pUC57[T7-Stem-XL-GTCV] was prepared (GenScript) by inserting a 1533 nt long BstEII-BamHI DNA fragment of pUC57[T7-Stem-GTCV] corresponding to the entire modified GTCV sequence (i.e., with tandem stop codons and AUG codon substitutions) into DC-Stem-ΔXL-HalV [60] to replace the intergenic region and ORF2 by GTCV nt 1–1494.

pUC57-T7-Stem-RaCV1[nt 1–1386] was prepared (GenScript) by inserting DNA corresponding to a T7 promotor and a stable hairpin ($\Delta G = -32.40$ kcal/mol) linked to nt 1–1386 of the RaCV1 genome or a stable hairpin ($\Delta G = -33.40$ kcal/mol) linked to RaCV1 nt 1–1538 between HindIII and BamHI sites of pUC57. The RaCV1 sequence was modified (Genscript) to generate [UAA$_{612}$], [UAA$_{612}$UAA$_{720}$], [GAA366-8CUC],[GUG$_{357–9}$CUC], [ACC$_{343–5}$UAG], [CCC$_{340–2}$AAA], substitution mutants, [ΔC341], [Δnt 334–347] and [Δnt 317–366] deletion mutants and [Stem1 + AUG$_{630}$], [Stem2 + AUG$_{630}$] and [Stem3 + AUG$_{630}$] insertion mutants.

The transcription vector pUC57-T7-Stem-RaCV1[nt 1–1538] was prepared (GenScript) by inserting DNA corresponding to a T7 promotor and a stable hairpin ($\Delta G = -33.40$ kcal/mol) linked to RaCV1 nt 1–1538 with flanking ApaI and BamHI sites into pUC57. The RaCV1 sequence was modified to place a UAA stop codon at nt 1497–1499 and to replace codons at positions 245 and 289 in the ORF initiating at AUG$_{603}$ (ORF1) and at positions 173, 180 and 182 in the ORF initiating at AUG$_{748}$ (ORF1*) with AUG codons to increase the methionine content of the 298 amino acid (a.a.)-long (33.5 kDa) ORF1 and 182 a.a.-long (19.8 kDa) ORF1* coding sequences. This plasmid was used (GenScript) to generate [GUAA-UACG] and [GUAA-CCUU] substitution mutants and the [Δ1b] deletion mutant.

pUC57[T7-DC RaCV1 wt] and pUC57[T7-DC RaCV1(GUAA$_{280–283}$-CCUU)] were prepared by inserting DNA corresponding to RaCV1 nt 124–797 flanked by 5′-XhoI and 3′-BamHI sites into DC Aichivirus [28], replacing the Aichivirus IRES by the RaCV1 IRES, and a fragment of the adjacent 287 a.a.-long (31.3 kDa) ORF. pUC57[T7-DC RaCV1 wt] was used (Genscript) to generate [AUG$_{603}$-context], [AUG$_{603}$-AAG], [AUG$_{711}$AAA], [AUG$_{630}$], [AUG$_{669}$], [J$_2$-dest] and [J$_3$-dest] substitution mutants and [Stem1 + AUG$_{630}$], [Stem2 + AUG$_{630}$] and [Stem3 + AUG$_{630}$] insertion mutants.

pUC57 plasmids were linearized by EcoRV. mRNAs were transcribed with T7 RNA polymerase.

2.5. *Purification of Initiation Factors, Elongation Factors and Ribosomal Subunits*

Ribosomal 40S and 60S subunits, native eIF2, eIF3, eIF5B, eEF1H, eEF2 and total aminoacyl-tRNA synthetases were purified from RRL (Green Hectares, Oregon, WI), and recombinant eIF1, eIF1A, eIF4A, eIF4A^{R362Q}, eIF4B, eIF4GI$_{653–1115}$, eIF4GI$_{736–988}$ eIF4GI$_{736–1008}$, cysteine-less and T830C eIF4GI$_{736–1115}$, eIF4GI$_{653–1600}$, nPTB, PCBP2 and *E. coli* methionyl tRNA-synthetase were expressed in *E. coli* BL21(DE3) and purified as described [31,40,61]. Recombinant Ebp1 was expressed in 4 L *E. coli* BL21 (DE3) for 18–20 h at 16 °C after induction with 1 mM IPTG and purified by Ni-NTA-agarose affinity chromatography. Ebp1 was then purified by FPLC on a monoQ 5/50 GL column with fractions being collected across a linear 100–500 mM KCl gradient: Ebp1 eluted at 200 mM KCl. Native total (Σ) calf liver tRNA (Promega, Madison WI) was aminoacylated with all amino acids using native aminoacyl-tRNA synthetases [61] or with methionine and recombinant *E. coli* methionyl-tRNA synthetase [58].

2.6. *Assembly and Analysis of Ribosomal Complexes*

The 48S complexes were assembled by incubating 50 nM RNA with 100 nM 40S subunits, 125 nM 60S, 125 nM total tRNAs, 0.5 μM eIF2, 0.2 μM eIF3, 1 μM eIF1 (unless otherwise indicated), 1 μM eIF1A, 0.8 μM eIF4A, 0.1 μM eIF4B, 0.8 μM eIF4G$_{653–1600}$, 0.25 μM Ebp1 and 0.25 μM PCBP2 or 0.25 μM PTB for 15 min at 37 °C in 40 μL buffer A (20 mM Tris, pH 7.5, 2 mM dithiothreitol, 0.25 mM spermidine, 2.5 mM $MgCl_2$) supplemented with 1 mM ATP, 0.5 mM GTP, 100 mM KCl and 1.5 U/μL RNAse inhibitor. The elongation competence of the resulting 48S complexes was assayed by incubation with 0.25 μM eIF5, 125 nM eIF5B and 120 nM 60S subunits for 10 min, which was followed by the addition of 100 nM eEF1H and 110 nM eEF2 and incubation for another 10 min. The positions of 48S and 80S complexes were determined by primer extension inhibition using AMV reverse transcriptase (Promega) and γ^{32}P-end-labeled primers 5′-GAGATGCGTTGGGTTTGCAA-3′ (complementary to RaCV1 nt 678–697), 5′-GGCACATTCACGACAATTG-3′ (complementary to RaCV1 nt 788–806), 5′-ATTGTAGGCAGTGTAGGCGT-3′ (complementary to RaCV1 nt 808–827) or 5′-GAATCGACGGGTGTAACA-3′ (complementary to GTCV nt 693–710).

2.7. *In Vitro Translation*

Monocistronic mRNAs (4 pmol) or dicistronic mRNAs (2 pmol) were translated for 1 h at 37 °C using the Flexi RRL system (Promega) in 20 μL reaction mixtures with 70 mM KCl, 1.3 mM $MgCl_2$, 0.4 mM complete amino acids (minus methionine) and 0.5 mCi/mL [^{35}S]methionine (43.5 TBq/mmol) supplemented with ~20 pmol eIF4A^{wt} or dominant-negative eIF4A^{R362Q} as indicated. Translation products were analyzed by electrophoresis using Nu-PAGE 4–12% Bis-Tris-Gel (Invitrogen, Carlsbad, CA), which was followed by autoradiography.

2.8. *Fe(II)-BABE Modification of eIF4G$_{726–1115}$ and Ebp1*

Ebp1 and eIF4G$_{726–1115}$ mutant proteins were derivatized with Fe(II)-BABE (Dojindo, Rockville, MD, USA) [62,63] by incubating 30 μM eIF4G$_{726–1115}$ or Ebp1 with 1 mM Fe(II)-BABE in 100 mL buffer B (80 mM HEPES, 260 mM KCl, 10% glycerol) for 30 min at 37 °C. Derivatized proteins were separated from unincorporated reagent by buffer exchange on Amicon 10k filter units and stored at −80 °C.

2.9. *Directed Hydroxyl Radical Cleavage*

RNA/[Fe(II)-BABE]-Ebp1 complexes were formed by incubating 2 pmol RNA and 5 pmol [Fe(II)-BABE]-Ebp1 in 50 μL buffer C (20 mM Tris, pH 7.5, 0.25 mM spermidine, 2.5 mM $MgCl_2$, 0.1 mM dithiothreitol) supplemented with 1 mM ATP, 0.5 mM GTP and 1.5 U/μL RNAse inhibitor for 15 min at 37 °C. IRES/eIF4G$_{726–1115}$ complexes with or without eIF4A were prepared as described [21]. To generate hydroxyl radicals, reaction mixtures were incubated on ice for 10 min, supplemented with 0.06% H_2O_2 and 5 mM ascorbic acid [62] and incubated on ice for 10 min. Reactions were quenched by adding 20 mM thiourea. RNAs were phenol-extracted, ethanol-precipitated and analyzed by toeprinting using AMV reverse transcriptase and the γ^{32}P-end-labeled primers 5′-GTCTACTAAATCAGCCCGAC-3′ (complementary to RaCV1 nt 458–477), 5′-GCAATAGGAACGCAAGAATCGC-3′ (complementary to RaCV1 nt 625–646), 5′-GAGATG

CGTTGGGTTTGCAA-3′ (complementary to RaCV1 nt 678–697), 5′-TTTACCGGAGCGGTA TAGTC-3′ (complementary to GTCV nt 488–507), 5′-GGAGAAGAAGTGSTCCCTAG-3′ (complementary to GTCV nt 584–603), 5′-GAGTGTCGCTTGTTACC-3′ (complementary to FMDV nt 603–597) or 5′-GCAAGTCTCTTGTTCCATGG-3′ (complementary to EMCV nt 863–844), as appropriate.

2.10. Quantification and Statistical Analysis

All in vitro experiments were repeated at least three times, and they included technical and biological replicates. Representative gel images are shown.

3. Results

3.1. Type 2 and Type 4 IRESs in Avian Caliciviruses

Our initial analysis identified picornavirus-like type 2, type 4 and type 5 IRESs in seven caliciviruses from the *Nacovirus* genus of *Caliciviridae* and another currently unclassified clade of caliciviruses [41]. We now identified type 2 IRESs in *Caliciviridae* sp. isolate 361bR-k141_109560 [64], Goose calicivirus isolate NC-19BV-37-k77-711556 [65], Duck calicivirus isolate PI202116/k141_160805 and four closely related chicken caliciviruses [66,67] as well as partial type 2 IRESs in Duck calicivirus isolate NC-19BV-45-k77-747816 and some chicken caliciviruses (Supplementary Figure S1). Several recently described genomes contain short, possibly incomplete 5′UTRs that have high levels of nucleotide sequence identity to 3′ terminal regions of the full-length IRES-containing 5′UTRs. Pairwise sequence identity between these putative calicivirus type 2 IRESs from the 5′ border of domain H to the 3′ border of domain J is high (44–100%). They contain elements that are important for the function of picornavirus type 2 IRESs, including a C-rich loop, a GNRA tetraloop and an AAA motif in apical arms of domain I, an A-rich stem-loop that wedges between the minor grooves of J and K domains, a bipartite sequence/structural motif at the apex of domain J and a 3′-terminal Yn-Xm-AUG motif. These elements coincide with the most highly conserved nucleotides in calicivirus type 2 IRESs (Figure 1B), suggesting that these motifs are also important for their function.

Several structural properties distinguish calicivirus type 2 IRESs from the best-characterized picornavirus counterparts, such as those of EMCV and FMDV. Stemloop Id emerges from the four-way junction at the apex of domain I and is 10–15 nt long in EMCV and FMDV IRESs [46], whereas the equivalent domain in calicivirus IRESs is hyperextended (~67 nt long). The sequence downstream of the J-K domain is highly variable: some calicivirus type 2 IRESs such as that of grey teal calicivirus (GTCV) contain a hairpin (domain L) between the J-K domain and the Yn-Xm-AUG motif (Figure 1C), whereas in others, such as *Caliciviridae* sp. isolate yc-13 ("Red-crowned crane calicivirus" (RaCV1) [68], domain L overlaps the AUG triplet of the Yn-Xm-AUG motif (Figure 1B). In picornavirus type 2 IRESs, domain L is located upstream of the Yn-Xm-AUG motif [46,69], and mutational analysis of the FMDV IRES suggested that it makes a minor contribution to IRES activity [18,70]. The separation between the Yn-Xm-AUG motif and the ORF1 initiation codon ranges from 6 nt in GTCV to 216 nt in Duck calicivirus isolate MW20. The sequence divergence between calicivirus IRESs, structural differences at their 3′ borders and the variable separation between the Yn-Xm-AUG motif and ORF1 together suggest that the recombinational exchange of IRESs between caliciviruses and picornaviruses may have occurred on multiple occasions. This conclusion is supported by our analysis of other recently reported calicivirus genomes, which revealed that Temminck's stint calicivirus [71] has a type 4 IRES that is related to that of Red-necked stint Picornavirus B-like [72] (74% nucleotide identity), and that Bird calicivirus isolate Cormcali01 from cormorants [73] contains a type 4 IRES that is related to those of the picornaviruses Anativirus A1 isolate TW90A [74,75] and Mute swan feces associated sapelovirus 1 [76] (>60% nucleotide identity).

3.2. An Overlapping Open Reading Frame in Avian Calicivirus Genomes

In most of these type 2 IRESs, the AUG of the Yn-Xm-AUG motif is in poor nucleotide context, suggesting that it will be bypassed and that translation of the principal open reading frame (ORF1) initiates at a downstream location. In addition to multiple potential initiation codons downstream of the motif and in frame with ORF1, many of these IRES-containing genomes contain an overlapping ORF, which is designated here as ORF1* (Figure 1A). The ORF1* amino acid sequences form two principal conserved groups, which are epitomized by RaCV1 and *Caliciviridae* sp. Isolate CAL/PB27-SI31/Switzerland/2019, respectively (Supplementary Figures S2 and S3). The former group is divergent (33–100% a.a. sequence identity and, with the exception of incomplete sequences, 213–317 amino acids (a.a.) long), but similarity is high over the C-terminal ~70 a.a.-long region, whereas the latter group is highly homologous (197–198 a.a.-long; 90–100% a.a. sequence identity). The ORF1* initiation codons in both groups almost all have suboptimal nucleotide context with pyrimidine residues at −3 and/or at +4 positions. Notably, some calicivirus genomes that lack 5′-terminal IRESs nevertheless encode ORF1* sequences. Thus, Turkey calicivirus isolate L11043, Goose calicivirus N and Duck calicivirus 2 strain DuCV-2_B6, which have 5′UTRs that are 8, 10 and 275 nt long, respectively, all encode an RaCV1-like ORF1* sequence. Only four of the fifteen chicken caliciviruses contain complete type 2 IRESs, but they all contain ORF1* sequences. The presence in caliciviruses that have short 5′UTRs of ORF1* sequences that are closely related to ORF1* sequences in IRES-containing genomes suggests that ORF1* is conserved because it has an important function. It could possibly be translated following end-dependent and IRES-mediated modes of initiation, although alternatively, ORF1*-containing genomes with short 5′UTRs may simply be incomplete. A few avian caliciviruses, such as GTCV, lack ORF1*.

Phylogenetic analysis indicates that the two major ORF1*-encoding classes of avian calicivirus genomes form coherent groups, and that the GTCV and Trumpeter swan calicivirus genomes constitute a separate cluster (Figure 2). These observations are consistent with the results of previously reported phylogenetic analyses [42,65–67,71,73,74,77]. Notably, numerous other calicivirus genomes also encode ORF1* sequences, forming additional groups that include (a) Duck calicivirus isolates MW20 and NC-19BV-45-k77-747816, (b) *Caliciviridae* sp. isolate cftwhg06cal1 and Goose calicivirus isolates H146 and MA/CHN/2028, (c) *Caliciviridae* sp.isolate ybb044cal01nc (Supplementary Figure S4), (d) numerous members of the *Bavovirus* genus (Supplementary Figure S5) and (e) Ruddy turnstone calicivirus A isolate MW19, Bird calicivirus isolate Cormacali01 and Trumpter swan calicivirus. The first two of these groups of ORFs show limited similarity to the major group epitomized by RaCV1, and like them, they initiate downstream of the ORF1 initiation codon. The disposition of coding sequences at the 5′-end of *Bavovirus* genomes is more complex: (i) many genomic sequences appear to be incomplete, so that open reading frame(s) commonly begin at the 5′-terminal nucleotide, encoding an N-terminal extension of the annotated ORF, and (ii) a second conserved ORF (ORF1′) either overlaps ORF1 or is in the same reading frame and is separated from it by a short intergenic region (Supplementary Figure S5). ORF1* elements therefore appear to be characteristic of most avian calicivirus genomes and to form several distinct groups.

Overlapping functional elements in mRNAs such as coding regions constrain sequence evolution and can be identified on the basis of reduced synonymous site variation [48]. Analysis of the aligned ORF1 coding sequences of (a) eleven chicken calicivirus and (b) nine other avian calicivirus genomes revealed a region of enhanced synonymous site conservation in each group that ended at the conserved location of the ORF1* stop codon, which is consistent with the presence of functionally important, translated overlapping ORFs (Figure 1D,E).

To locate ORF1 and ORF1* initiation codons, and to ascertain whether ORF1* is accessible to initiating ribosomes, we assayed the translation of mRNAs containing GTCV and RaCV1 5′UTRs in rabbit reticulocyte lysate (RRL) and analyzed initiation in reactions reconstituted in vitro from the individual purified components of the translation apparatus.

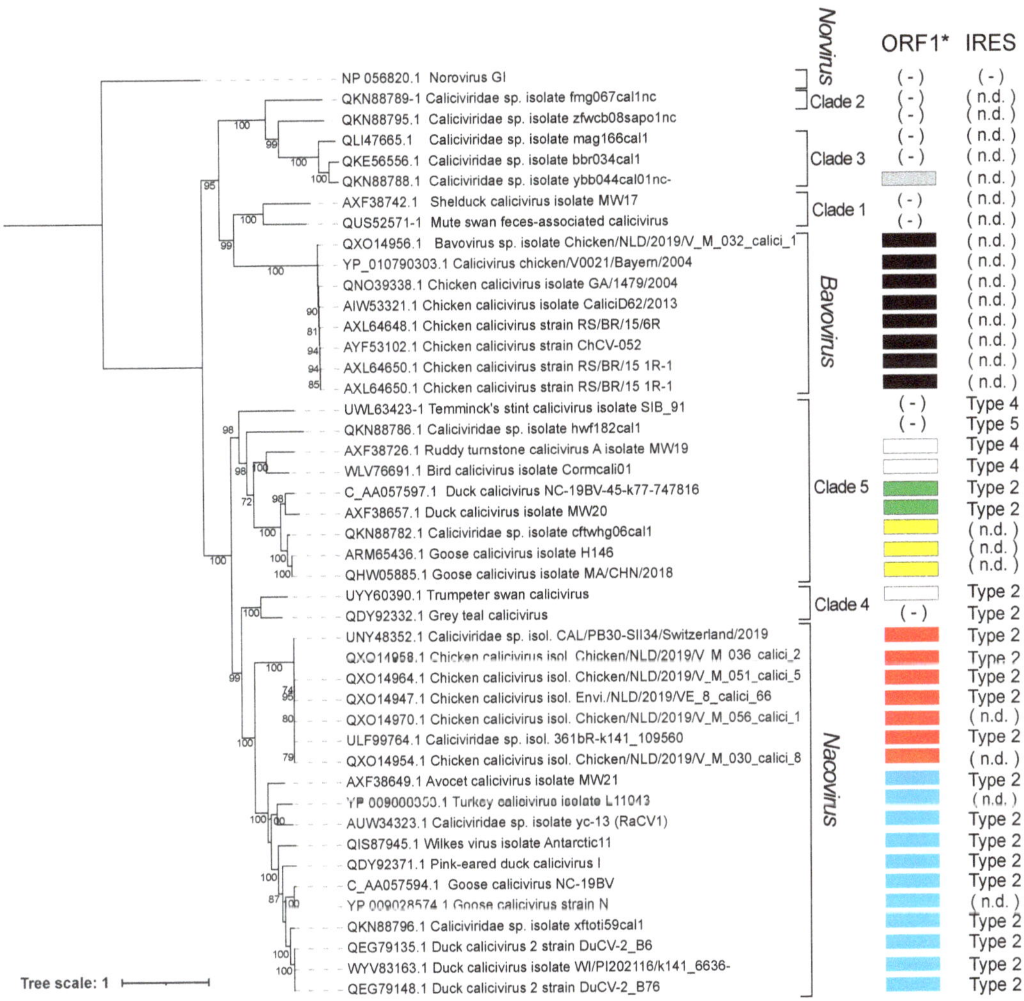

Figure 2. Maximum likelihood phylogenetic tree of the ORF1 polyprotein of avian members of the virus family *Caliciviridae*, which where analyzed using amino acid sequences deposited in Genbank or derived computationally from nucleotide sequence deposited in Genbase. Viruses are identified by name and Genbank protein or Genbase nucleotide accession number, as appropriate. Bootstrap values over 70% are shown. The scale bar represents the number of amino acid substitutions per site. Viruses are assigned to established genera and to clades proposed by Canuti et al. [42]. The key to the right indicates that IRES type in the viral genome, and the occurrence of overlapping ORF1* sequences, color-coded to indicate the different conserved groups (light blue: members listed in Supplementary Figure S2; red: members listed in Supplementary Figure S3; yellow, green and grey: members listed in Supplementary Figures S4A, S4B and S4C, respectively; and black: members listed in Supplementary Figure S5). White rectangles indicate potential ORF1* sequences in Ruddy turnstone calicivirus, Bird calicivirus isolate Cormacali01 and Trumpeter swan calicivirus genomes. (n.d.) indicates genomes that may be incomplete so that definitive conclusions regarding the presence of IRESs and/or ORF1* sequences cannot be made. (-) indicates genomes that appear to lack ORF1* or recognized IRES sequences. This analysis excluded ORF sequences that are <50 a.a long. Bavovirus 5′NTRs are long and we cannot exclude that they might have IRES activity.

3.3. Initiation of Translation on the Grey Teal Calicivirus (GTCV) IRES

To test the activity of the GTCV 5′UTR as an IRES, we assayed the translation in RRL of a dicistronic mRNA with a stable 5′-terminal hairpin followed by part of the cyclin B2 ('XL') coding sequence, the GTCV 5′UTR and part of the adjacent ORF (Figure 3A). The GTCV ORF was robustly translated, whereas XL translation was very weak, which is consistent with the inhibitory effect of the 5′-terminal hairpin on end-dependent initiation [78] (Figure 3B). Translation of both ORFs in this mRNA and in a control dicistronic mRNA containing the Aichivirus IRES [28] was almost completely inhibited by dominant-negative $eIF4A^{R362Q}$, which acts by sequestering eIF4G. The GTCV 5′UTR therefore contains an IRES, and initiation on it requires eIF4A and eIF4G.

The principal ORF in the GTCV genome contains in-frame AUG_{523} and AUG_{547} triplets upstream of the Yn tract and in-frame AUG_{619}, AUG_{628} and AUG_{723} codons downstream of it (Figure 1C). AUG_{619} is part of the Yn-Xm-AUG motif, and it has poor nucleotide context (UUU<u>AUG</u>G), like the AUG codons in the Yn-Xm-AUG motif of other type 2 IRESs, which are consequently weak initiation codons and are frequently bypassed by ribosomes during initiation [25,27,79,80]. The in vitro translation of monocistronic GTCV mRNA with a 5′-terminal hairpin ($\Delta G = -32.4$ kcal/mol) to block end-mediated ribosomal attachment yielded a single 32 kDa product, which is consistent with initiation at AUG_{619} or AUG_{628} (Figure 3C, lane 2). Translation was strongly impaired by $AUG_{619}AAA$ substitutions, whereas $AUG_{628}AAA$ substitutions eliminated translation of the 32 kDa product but yielded low levels of a 28 kDa product, which is consistent with initiation at the next downstream AUG triplet, AUG_{723}. Notably, an $AUG_{619}AUC$ substitution reduced but did not abrogate translation of the 32 kDa product (Figure 3C, lane 8), whereas $AUG_{723}AAA$ substitutions had no effect on its synthesis (Figure 3C, lane 5). These results are consistent with a model in which AUG_{619} is part of the Yn-Xm-AUG motif which acts to promote internal ribosomal entry, so that mutation of it reduces initiation at all IRES-dependent sites, whereas AUG_{628} is the start codon for the polyprotein, and its substitution leads to initiation downstream. The identity of the initiation codon was confirmed by toeprinting ribosomal complexes formed on the IRES in RRL in the presence of GMP-PNP (which arrests initiation at the stage of 48S complex formation), and by toeprinting 80S ribosomes formed on the IRES in the presence of cycloheximide (which inhibits elongation) (Figure 3D). Ribosomes and ribosomal initiation complexes arrest primer extension $^{+}15$–$^{+}17$ nt from the $^{+}1$ nucleotide of the codon in the P site, yielding characteristic toeprints. The appearance of toeprints $^{+}15$–$^{+}17$ from AUG_{628} confirmed that this triplet is the primary initiation codon for GTCV ORF1.

The GTCV IRES contains a GNRA tetraloop at the apex of subdomain Ib (Figure 1E) that is characteristic of picornavirus type 2 IRESs and is important for their function. Consistently, CCUU and UACG substitutions in the tetraloop and deletion of the apical half of subdomain Ib all abrogated IRES function (Figure 3F, lanes 3–5). Whereas the tetraloop is a conserved feature of all type 2 IRESs, domain L is variable in terms of sequence, structure, and position relative to the JK domain and the Yn-Xm-AUG motif [46,81,82]. The effect of deleting part or all of domain L (Figure 3E; Figure 3F, lanes 9, 10) indicated that it does not have an essential function. However, destabilization of the apical half of this domain almost abolished IRES activity (Figure 3F, lane 6), which is possibly due to the disruption of the optimal spatial orientation of elements such as the JK domain and the Yn-Xm-AUG motif.

In vitro reconstitution showed that 40S subunits, $Met\text{-}tRNA^{Met}{}_{i}$, eIFs 1, 1A, 2, 3, 4A, 4B and 4F were sufficient for 48S complex formation on the GTCV and confirmed AUG_{628} as the initiation codon (Figure 3G, lane 2). Initiation required initiator tRNA (Figure 3G, lane 5) and was enhanced by PTB and Ebp1 (Figure 3G, lanes 2, 3); eIF4F could be substituted by eIF4A and $eIF4G_{653\text{–}1600}$ and less effectively by eIF4A and $eIF4G_{736\text{–}1115}$ (Figure 3H, lanes 2–4). DHX29 enhances initiation on type 5 IRESs but not on type 2 (EMCV), type 4 or type 6 IRESs [28,29], and consistently, it did not enhance initiation on the GTCV IRES at AUG_{628} (Figure 3G, lanes 3, 4). A reverse transcriptase (RT) stop appeared at C_{626} independently of 48S complex formation, PTB and Ebp1 (Figure 3G, lanes 2–5; Figure 3H, lanes 2–4), indicating that it is likely induced by canonical initiation factors, potentially eIF4G/eIF4A, which induces conformational changes in the EMCV IRES that lead to the appearance of RT stops downstream of the Yn-Xm-AUG motif [21].

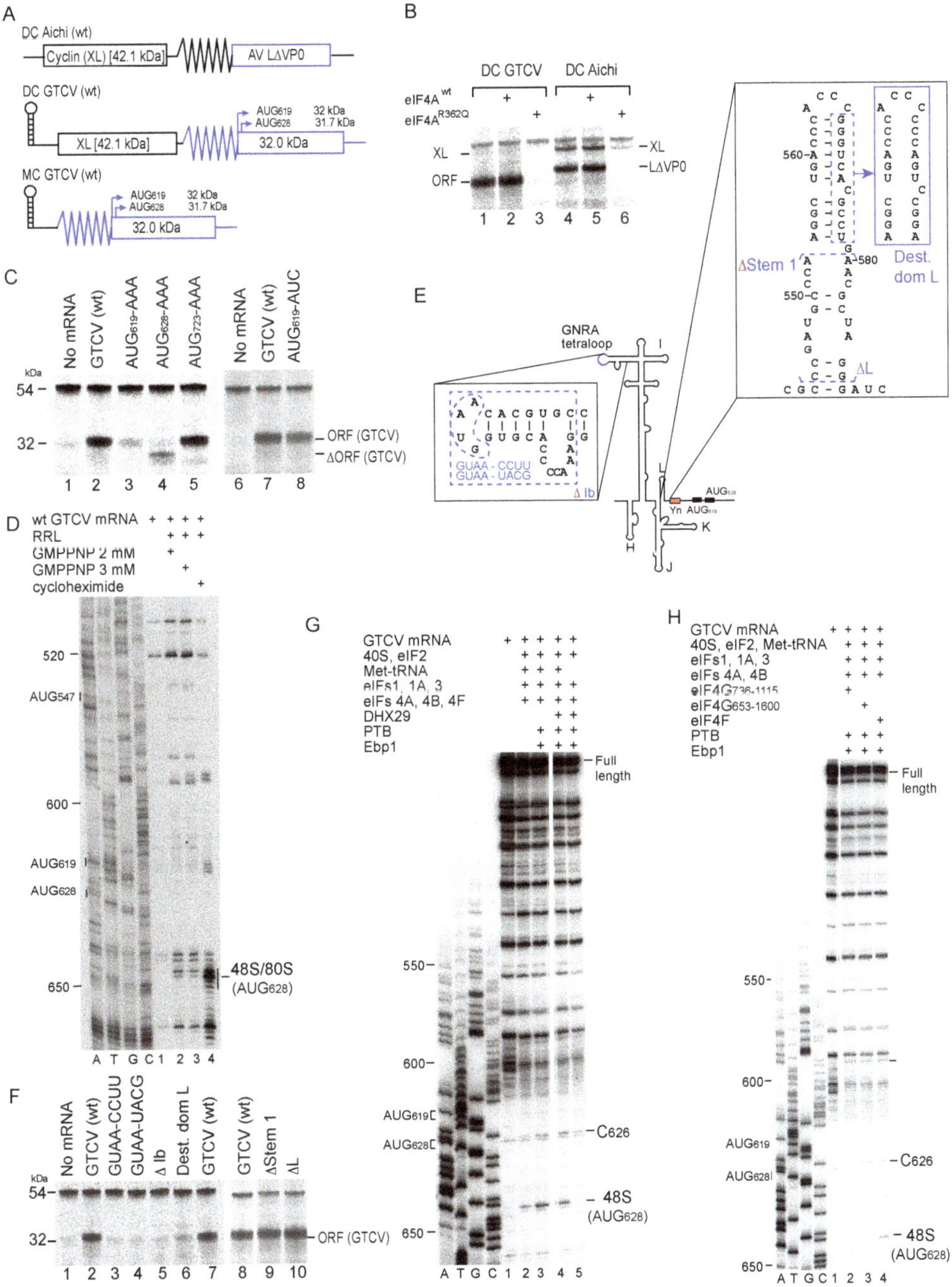

Figure 3. The mechanism of initiation on the grey teal calicivirus (GTCV) IRES. (**A**) Schematic representations of monocistronic (MC) GTCV and dicistronic (DC) GTCV and Aichivirus mRNAs,

with IRES sequences represented by zigzag lines, a stable hairpin placed upstream of the first cistron (the cyclin B2 (XL) ORF), and potential initiation codons and corresponding translation products indicated on the downstream GTCV ORF. (**B**,**C**,**F**) Translational activity of the GTCV and Aichivirus IRESs in RRL, showing (**B**) their sensitivity to inhibition by dominant-negative eIF4A^{R362Q} and the effect (**C**) of substitution of potential GTCV initiation codons and (**F**) of substitutions and deletion mutations in subdomains Ib and domain L. The XL translation product and products resulting from initiation at GTCV and Aichivirus initiation codons resolved by SDS-PAGE are indicated. (**D**) Toeprint analysis of 48S complexes and 80S ribosomes formed on wt GTCV MC mRNA in RRL in the presence of GMPPNP and cycloheximide, as indicated. (**E**) Model of the IRES, annotated to show mutations in subdomains Ib and domain L. (**G**,**H**) Toeprinting analysis of 48S complex formation on GTCV MC mRNA in the presence of 40S subunits, unfractionated Met-tRNA$_i^{Met}$, eIFs, PTB1, Ebp1, DHX29, eIF4F, eIF4G$_{736–1115}$ or eIF4G$_{653–1600}$, as indicated. Toeprints caused by 48S complexes assembled on AUG$_{628}$ and a factor-induced stop at C$_{626}$ are indicated on the right. A dideoxynucleotide sequence generated with the same primer was run in parallel on each gel (lanes A, T, G, C) to show the corresponding viral sequence. Lanes in panel (**H**) separated by a white line were juxtaposed from the same gel. Lanes 6–8 of panel (**C**) and lanes 8–10 of panel (**F**) were derived from the same gel.

3.4. Initiation of Translation on the Red-Capped Crane Calicivirus 1 (RaCV1) IRES

The RaCV1 IRES has properties that are distinct from those of the GTCV IRES, and we therefore selected it for a series of complementary experiments. Domain L of the RaCV1 IRES is located immediately downstream of the Yn-Xm-AUG motif rather than between domain J and this motif (as in GTCV) and unlike GTCV, the RaCV1 genome contains an overlapping ORF (ORF1*) within ORF1. The translation of ORF1 might start at UUU$\underline{\text{AUG}_{603}}$C, which is part of the Yn- Xm-AUG motif, or downstream at UCA$\underline{\text{AUG}_{711}}$G. In the wild-type genome and in the monocistronic mRNAs used here, the translation of ORF1* could start at AGG$\underline{\text{AUG}_{748}}$U, but dicistronic mRNAs containing the RaCV1 5′UTR and proximal coding sequence linked to a reporter ORF contained an adventitious C$_{729}$G substitution that introduced an upstream AUG codon (AGC$\underline{\text{AUG}_{729}}$A), and initiation at this codon would yield an N-terminally extended 22.1 kDa form of ORF1* (Figure 4A). The translation of monocistronic mRNA in RRL yielded a weak 28 kDa product, a prominent 24 kDa product and a weak 19 kDa product, predicted to initiate at AUG$_{603}$, AUG$_{711}$ and AUG$_{748}$, respectively (Figure 5B, lane 2). In vitro translation of dicistronic mRNAs in RRL yielded a prominent 42 kDa product that corresponded to the XL product of the upstream ORF, weak 35 and 22 kDa translation products and a prominent 31 kDa product (Figure 4B, lane 2). The last three polypeptides correspond to the predicted products of translation of RaCV1 ORF1 and ORF1*. The RaCV1 5′UTR therefore functions as an IRES.

Substitutions that improved the context of AUG$_{603}$ from UUU$\underline{\text{AUG}}$C to ACC$\underline{\text{AUG}}$G in dicistronic mRNAs significantly increased the prominence of the 35 kDa translation product while reducing that of the 31 and 23 kDa products (Figure 4B, lane 3). The substitution of AUG$_{603}$ abolished translation of the 35 kDa product, whereas the elimination of AUG$_{711}$ abolished translation of the 31 kDa polypeptide and strongly enhanced translation of the 23 kDa ORF1* product that resulted from initiation at AUG$_{729}$ (Figure 4B, lanes 4, 5). In in vitro reconstitution experiments, 48S complexes assembled on monocistronic RaCV1 mRNA at AUG$_{603}$, AUG$_{711}$ and AUG$_{748}$ (Figure 4D, lanes 1–5), which is consistent with the assignment of these triplets as initiation codons. Changes in the prominence of the translation products resulting from the improvement of AUG$_{603}$ context, insertion of AUG$_{729}$ or elimination of AUG$_{603}$ or AUG$_{711}$ are consistent with an initiation mechanism involving scanning after ribosomal attachment at or upstream of AUG$_{603}$.

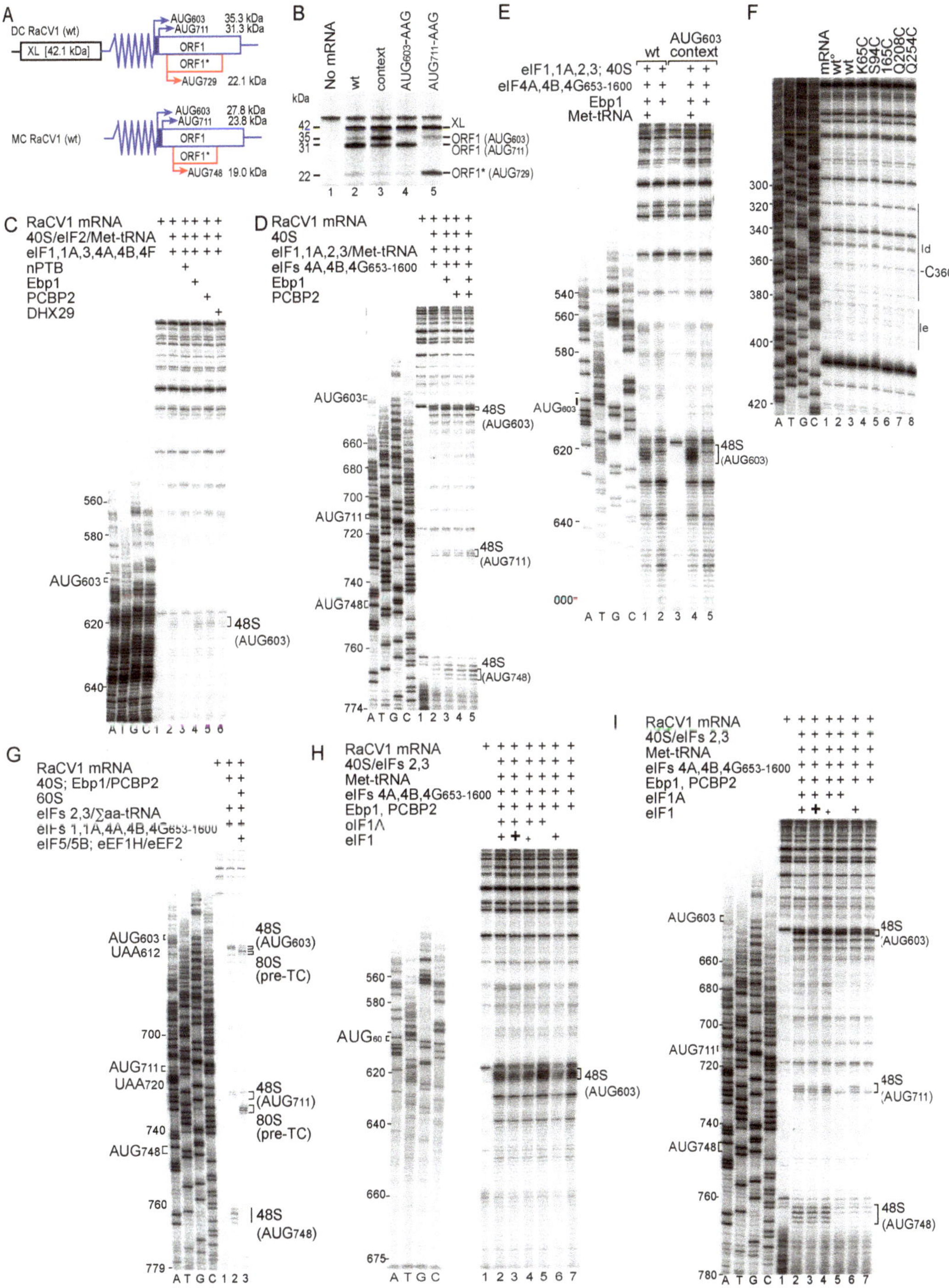

Figure 4. Initiation factor requirements for initiation on the RaCV1 IRES. (**A**) Schematic representation of monocistronic (MC) and dicistronic (DC) RaCV1 mRNAs showing IRESs represented by zigzag

lines, the upstream cyclin B2 (XL) ORF and the downstream RaCV1 ORF1 and ORF1* with potential initiation codons and corresponding translation products. (**B**) Translation activity of the RaCV1 IRES and initiation codon selection in RRL mediated by the wt RaCV1 IRES and variants with improved nucleotide context for AUG_{603} or substitution of AUG_{603} or AUG_{711} by AAG codons. The XL translation product and products resulting from initiation at AUG_{603} or AUG_{711} in ORF1 and at AUG_{729} upstream of and in-frame with ORF1* resolved by SDS-PAGE are indicated. (**C–E,G–I**) Toeprint analysis of formation of 48S initiation complexes and 80S pre-termination complexes (pre-TC) on MC RaCV1 mRNA in the presence of combinations of 40S subunits, 60S subunits, Met-$tRNA_i^{Met}$ or total aminoacylated tRNAs (Σaa-tRNAs), eIFs, Ebp1, PCBP2, DHX29, eEF1H and eEF2, as indicated. In panels (**H,I**), lanes 2, 3 and 4 included 1 μM eIF1, 3 μM eIF1 and 0.3 μM eIF1, respectively. Toeprints caused by 48S complexes on AUG_{603}, AUG_{711} and AUG_{748}, and by pre-TCs assembled on UAA_{612} and UAA_{720} are indicated on the right. A dideoxynucleotide sequence generated with the same primer was run in parallel on each gel (lanes A, T, G, C) to show the corresponding viral sequence. (**F**) Primer extension analysis conducted using the wt RaCV1 IRES and unmodified wt Ebp1 (wt°; lane 2) or Fe(II)BABE-modified wt or mutant Ebp1 with substitutions as indicated (lanes 3–8). The cDNA product labeled C_{360} terminated at this nucleotide. Reference lanes A, T, G, C show the wt RaCV1 sequence generated with the same primer.

3.5. *The Mechanism of Initiation on the RaCV1 IRES*

To characterize the mechanism of initiation on the RaCV1 IRES in detail, the process was reconstituted in vitro from individual purified components of the translation apparatus. The activity of the RaCV1 IRES in mediating initiation in RRL (Figure 4B) indicates that the canonical eIFs and the ITAFs present in RRL are sufficient to support initiation on the IRES.

Initiation on different picornavirus type 2 IRESs is enhanced by Ebp1 and/or the pyrimidine tract-binding protein PTB [32–36,59]. We tested the influence of these and other ITAFs individually and in combination on initiation on the RaCV1 IRES. DHX29, which is required for initiation on type 5 IRESs (e.g., [28,29]), had no effect, and nPTB inhibited initiation complex formation, whereas initiation at AUG_{603}, AUG_{711} and AUG_{748} was enhanced by Ebp1, by PCBP2 (Figure 4C, lanes 2–5) and synergistically by Ebp1 and PCBP2 (Figure 4D, lanes 2–5). Incubation of the IRES with unmodified wt Ebp1 and with a panel of Fe(II)BABE-modified wt and mutant forms of Ebp1 yielded a toeprint at nt 360 (Figure 4F), indicating that Ebp1 bound specifically and independently to subdomain Id. 48S complex formation at AUG_{603} and AUG_{711} of ORF1 and at AUG_{748} of ORF1* was dependent on the presence of Met-$tRNA^{Met}_i$ (e.g., Figure 4E, lane 5). AUG_{603} has poor nucleotide context (UUU<u>AUG</u>C), and its improvement (to ACC<u>AUG</u>G) strongly enhanced initiation at this codon (Figure 4E, lanes 1, 4). The translational competence of the in vitro assembled 48S complexes was assayed using RaCV1 IRES-linked mRNA with substitutions that placed UAA termination codons 9nt downstream of AUG_{603} and AUG_{711}. The addition of 60S subunits, eIF5, eIF5B, eEF1H, eEF2 and total aminoacylated tRNAs led to the formation of elongation-competent 80S ribosomes that stalled on the UAA_{612} and UAA_{720} stop codons, forming pre-termination complexes (Figure 4G). AUG_{748} was not followed by a nearby in-frame stop codon, so that although 48S complexes that had assembled on this codon were able to undergo subunit joining and elongation in these conditions, as evident by the disappearance of corresponding stops downstream of AUG_{748} (Figure 4G, lane 3), these elongation complexes did not form pre-TCs that were detectable by toeprinting.

The assembly of 48S complexes at AUG_{603} was impaired by the omission of eIF1A, whereas the omission of eIF1 with or without eIF1A enhanced their formation at this codon (Figure 4H). The formation of 48S complexes on AUG_{711} and particularly at AUG_{748} was strongly reduced by the omission of either or both factors (Figure 4I). These data implicate eIF1A in promoting the recruitment of 43S complexes to the IRES and eIF1 in their movement to initiation codons downstream of the attachment site, by promotion of their adoption of a scanning-competent conformation and by discrimination against initiation at AUG_{603} because of its unfavorable nucleotide context [53,63,83].

The presence of domain L between AUG_{603} and the downstream AUG_{711} and AUG_{748} raises the question of whether 43S complexes unwind this domain completely to scan downstream or if they can also bypass this domain and be loaded downstream of it. To investigate this, AUG codons with the context AUCAUGU were inserted at the apex of domain L (AUG_{630}) or downstream of it (AUG_{669}) (Figure 5A). Both codons were recognized and used efficiently in RRL (Figure 5C, lanes 3, 6) and in in vitro reconstituted reactions (Figure 5D, lane 8; Figure 5E, lanes 8, 11). Notably, initiation at these and downstream codons (AUG_{711}, AUG_{748}) was strongly dependent on eIF1, and its omission led to the accumulation of 48S complexes at AUG_{603} (Figure 5D, lanes 8, 9, 11, 12; Figure 5E, lanes 8, 9, 11, 12).

The stabilization of domain L ($\Delta G = -24.4$ kcal/mol) (Figure 5A) almost completely abrogated IRES-mediated translation in RRL (Figure 5B, lane 3) and initiation in in vitro reconstituted reactions (Figure 5D, lanes 1, 2). Note that in these experiments, the stabilized domain L arrested reverse transcriptase, yielding stops at the base of this domain (Figure 5D, lanes 1, 2; Figure 5E, lanes 1, 2). The loss of function could be due to a failure of 48S complexes to bind to the IRES or to an inability to unwind the hairpin and scan downstream. An AUG codon placed at the apex of a stabilized domain L ("Stem1 + AUG_{630}") (Figure 5A) was not utilized in RRL or in in vitro reconstituted reactions (Figure 5C, lanes 2–4; Figure 5F, lanes 1–4), but a low level of initiation occurred at AUG_{711}, suggesting that a small proportion of ribosomes bypasses domain L. The interruption of base-pairing in the helix in [Stem2 + AUG_{630}] mRNA enabled initiation to occur at the AUG_{630}-equivalent codon (Figure 5C, lanes 3, 5; Figure 5F, lanes 1, 2, 5, 6), but its utilization was almost completely eliminated if it was sequestered in a stable hairpin placed on top of a stem with alternating bulges and helical elements ([Stem3 + AUG_{630}]) (Figure 5A). Initiation on AUG_{603} in this mRNA was markedly higher than on [Stem2 + AUG_{630}] mRNA containing AUG_{630} without the stable apical hairpin (Figure 5F, lanes 8), indicating that 43S complexes can bind to the Yn-Xm-AUG motif when the proximal stem of domain L contains discontinuities, but subsequent scanning is blocked by the stable apical hairpin. It is well established that initiation by scanning ribosomal complexes on weak initiation codons is enhanced by stable structures optimally placed ~14 nucleotides downstream [84].

3.6. Conserved Structures and Interactions in RaCV1 IRES Function

The sensitivity of RaCV1 IRES-mediated translation to inhibition by dominant-negative $eIF4A^{R362Q}$ (Figure 6A) implicates the eIF4G/eIF4A complex in this process as for initiation on GTCV and other type 2 IRESs (Figure 3B). The 48S complexes formed on the RaCV1 IRES in in vitro reconstituted reactions that included $eIF4G_{653-1600}$ and eIF4A in place of eIF4F (Figure 5E), indicating that initiation is independent of eIF4E and of the N-terminal region of eIF4G that includes the eIF4E binding site. eIF4G also contains an eIF3-binding domain and two eIF4A-binding domains (Figure 6B).

To establish whether they are required to support initiation on the IRES, $eIF4G_{653-1600}$ was substituted by recombinant fragments of eIF4G. The 48S complex formation at AUG_{603} was unaffected by deletion of the C-terminal eIF4A-binding domain and of the eIF3-binding site (a.a. 1015–1104) but was reduced by further C-terminal truncation to a.a. 989, which was likely due to the weakened binding of eIF4G to eIF4A (Figure 6C, lanes 2–5). Initiation at AUG_{711} was more strongly affected than at AUG_{603} by these deletions. These elements of eIF4G therefore contribute to the scanning competence of 43S complexes following their recruitment to this IRES. The omission of eIF4B impaired initiation, particularly at AUG_{711} and AUG_{748} (Figure 5E, lanes 4, 5). eIF4B therefore also contributes to the scanning competence of 43S complexes.

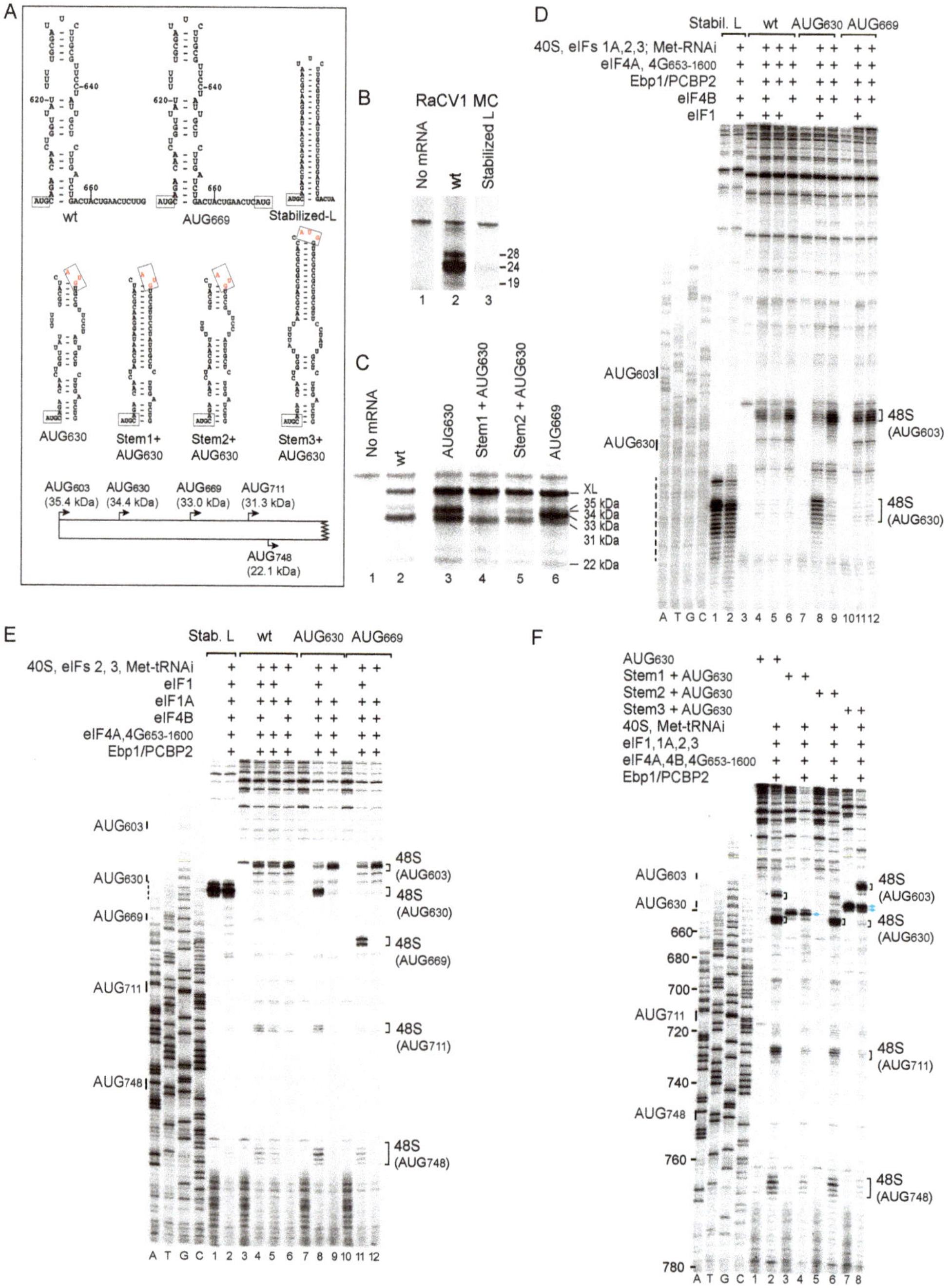

Figure 5. Influence of AUG codons and structural elements inserted downstream of the Yn-Xm-AUG motif on RaCV1 IRES-mediated initiation of translation. (**A**) Upper and middle panels: sequence and structure of IRES domain L and variants thereof, showing base-pairing and the positions of inserted codons AUG_{630} and AUG_{699}. Lower panel: schematic representation of the coding potential of RaCV1 mRNAs, showing initiation codons present in different mutants and the molecular weights

of corresponding translation products. (**B**,**C**) In vitro translation of (**B**) monocistronic and (**C**) dicistronic mRNAs containing the indicated wild-type and mutant IRESs. Panels are annotated to show translation products resolved by SDS-PAGE. (**D–F**) Toeprint analysis of 48S complexes formed on (**D**,**E**) wild-type and (**D–F**) mutant IRES-containing mRNAs by incubation with 40S subunits, Met-tRNA$_i^{Met}$, eIFs, Ebp1 and PCBP2 as indicated. Panels are annotated to show the positions of potential initiation codons on the left and of toeprints corresponding to 48S complexes on the right, and (**D**,**E**) with dashed lines to indicate RT stops induced on mRNA in the absence of factors by the stabilization of domain L. A dideoxynucleotide sequence generated with the same primer was run in parallel on each gel (lanes A, T, G, C) to show the corresponding wild-type sequence. Separation of lanes in panels (**B**,**C**) by white lines indicates that they were juxtaposed from the same gel.

Initiation on type 2 IRESs relies on the specific binding of eIF4G's central domain to the J-K domain, which is enhanced by eIF4A [30,55]. We used directed hydroxyl radical cleavage (DHRC) to determine whether eIF4G binds to the RaCV1 IRES in the same manner. In this approach, Fe(II) is site-specifically tethered by the linker 1-(*p*-bromoacetamidobenzyl)-EDTA (BABE) to a cysteine residue on the surface of an RNA-binding protein. Hydroxyl radicals generated by ascorbic acid/H_2O_2 treatment cleave mRNA with an intensity that is distance dependent. Cleavage is detected by primer extension inhibition and is strong within 22 Å, has medium intensity between 12 and 36 Å and is weak within 20–44 Å of the tethering position [85]. We used a BABE-linked variant of eIF4G$_{736–11116}$ containing a single cysteine residue (in place of Thr-830) that binds type 2 IRESs in a known orientation [21,23]. Primer extension revealed cleavage at GG$_{483–4}$, CA$_{507–8}$ and GG$_{558–9}$ that was strongly enhanced by eIF4A (Figure 6D) and that mapped to positions in the J-K domain that coincide directly with the strongest cleavages induced by this BABE-modified eIF4G variant in the EMCV IRES [21]. They overlap a bipartite motif near the apex of domain J that constitutes part of the eIF4G binding site in type 2 IRESs and is important for their function [21–23]. Consistently, destabilizing substitutions in the apical helices of domain J (Figure 6E) moderately impaired RaCV1 IRES-mediated translation (Figure 6F).

Analysis of another conserved element in the IRES showed that substitutions in the apical GNRA tetraloop of subdomain Ib in monocistronic and dicistronic mRNAs and deletion of Ib in monocistronic mRNA strongly impaired initiation at AUG$_{603}$, AUG$_{711}$ and AUG$_{748}$ in RRL (Figure 7A, lanes 1, 2, 4, 5; Figure 7B, lane 3). Consistently, these mutations strongly impaired 48S complex formation at these codons in in vitro reconstituted initiation reactions (Figure 7C,D)

3.7. Specific Binding of Ebp1 to Calicivirus Type 2 IRESs

Initiation on GTCV and RaCV1 IRESs uses overlapping sets of ITAFs: PTB enhances GTCV IRES function but is not required by the RaCV1 IRES, whereas initiation on the RaCV1 IRES is enhanced by PCBP2 (Figures 3G and 4C,D). Initiation on both IRESs requires Ebp1, which was first identified as an ITAF that enhances the activity of the type 2 FMDV IRES [32]. An important step in elucidating how Ebp1 promotes initiation on GTCV and RaCV1 IRESs is to determine how it interacts with them. We employed directed hydroxyl radical cleavage to determine the binding site and the orientation of bound Ebp1 on this and other IRESs. If the structure of a Fe(II)BABE-modified protein is known, then cleavage at multiple sites enables the position and orientation of the protein relative to the RNA ligand to be established. Ebp1 consists of two anti-parallel β-sheets, each flanked by a pair of α-helices (Figure 8A) [35], and it adopts different conformations on the ribosome (e.g., [86–88]). It binds to the tunnel exit on the 60S subunit, and in one of the several reported structures [86], it is held in place by interaction with the ribosomal RNA expansion segment ES27L. However, it is important to note that Ebp1's role in promoting 48S complex formation does not involve the 60S subunit.

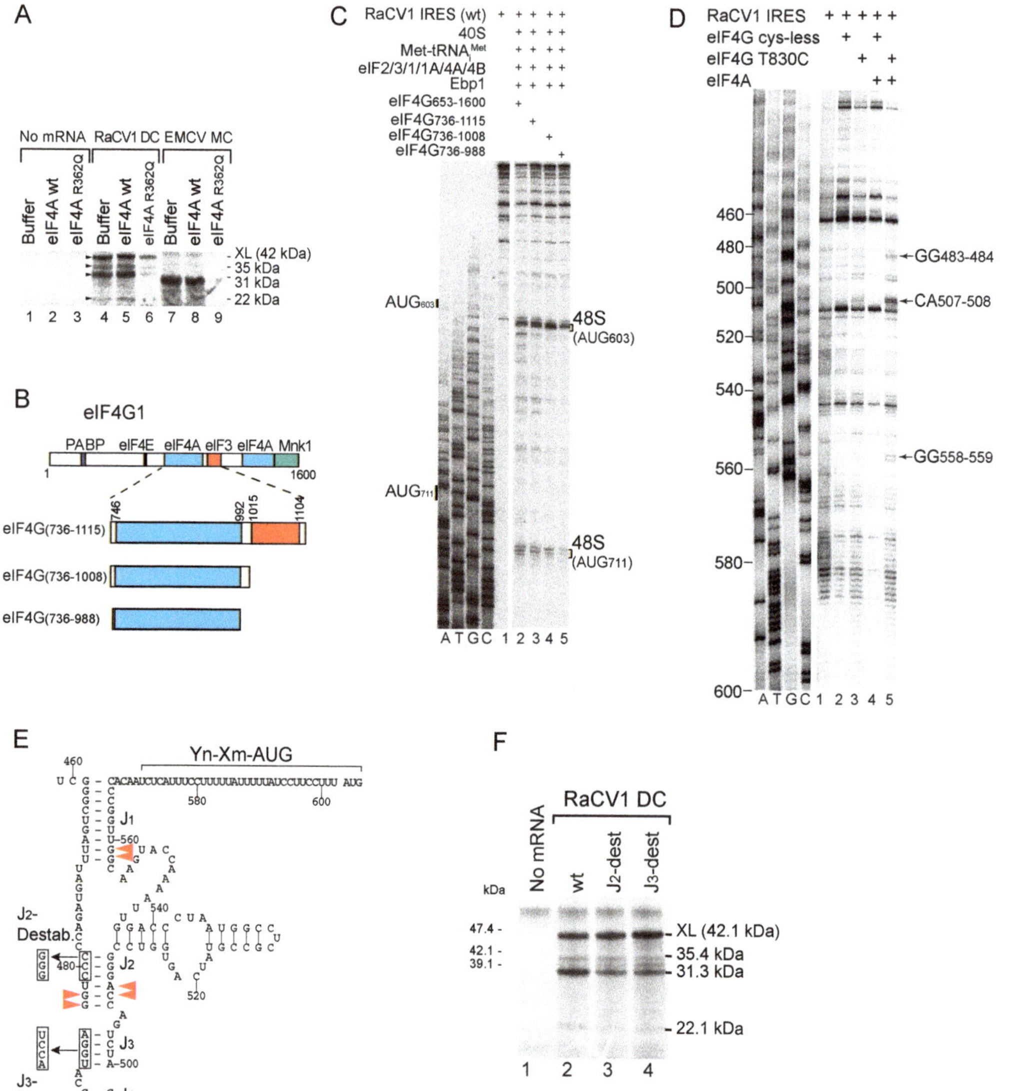

Figure 6. Requirement for subunits of eIF4F for initiation on the RaCV1 IRES. (**A**) Influence of recombinant eIF4A (wt) and eIF4A^{R362Q} on translation in RRL initiated on RaCV1 (lanes 4–6) and EMCV IRESs (lanes 7–9), (**B**) Schematic representation of full-length and truncated fragments of eIF4G1, showing factor binding sites. (**C**) Toeprinting analysis of 48S complex formation on RaCV1 MC mRNA in the in vitro reconstituted system in the presence of 40S subunits, Met-tRNA$_i^{Met}$, Ebp1, eIFs 1, 1A, 2, 3, 4A, 4B, and eIF4G$_{736–1599}$, eIF4G$_{736–1115}$, eIF4G$_{736–1108}$ or eIF4G$_{736–988}$, as indicated. Toeprints from 48S complexes assembled on AUG$_{603}$ and AUG$_{711}$ codons are indicated on the right. Separation of lanes by a white line indicates that they were juxtaposed from the same gel. (**D**) Primer extension analysis of directed hydroxyl radical cleavage of the wt RaCV1 IRES from Fe(II)-tethered eIF4G$_{736–1115}$-[T830C], in the presence/absence of eIF4A, as indicated. A dideoxynucleotide sequence generated with the same primer was run in parallel on gels shown in panels (**C**,**D**) (lanes A, T, G, C)

to show the corresponding wild-type sequence. The white lines in panel (**D**) indicate that the exposure of DHRC lanes 1–5 differed from that of the sequencing lanes A, T, G, C, but that sets of lanes were juxtaposed from the same gel. (**E**) Model of the J-K domain and adjacent Yn-Xm-AUG motif of the RaCV1 IRES, annotated to show structural elements (as in Figure 1B), sites of hydroxyl radical cleavage from Cys_{830} on $eIF4G_{736–1115}$ (indicated by blue arrowheads) and destabilizing substitutions in mutant J_2-dest and J_3-dest RaCV1 IRESs. (**F**) In vitro translation of dicistronic mRNA containing the wt RaCV1 IRES (lane 2), J_2-dest and J_3-dest mutant IRESs (lanes 3, 4). The panel is annotated to show translation products resolved by SDS-PAGE.

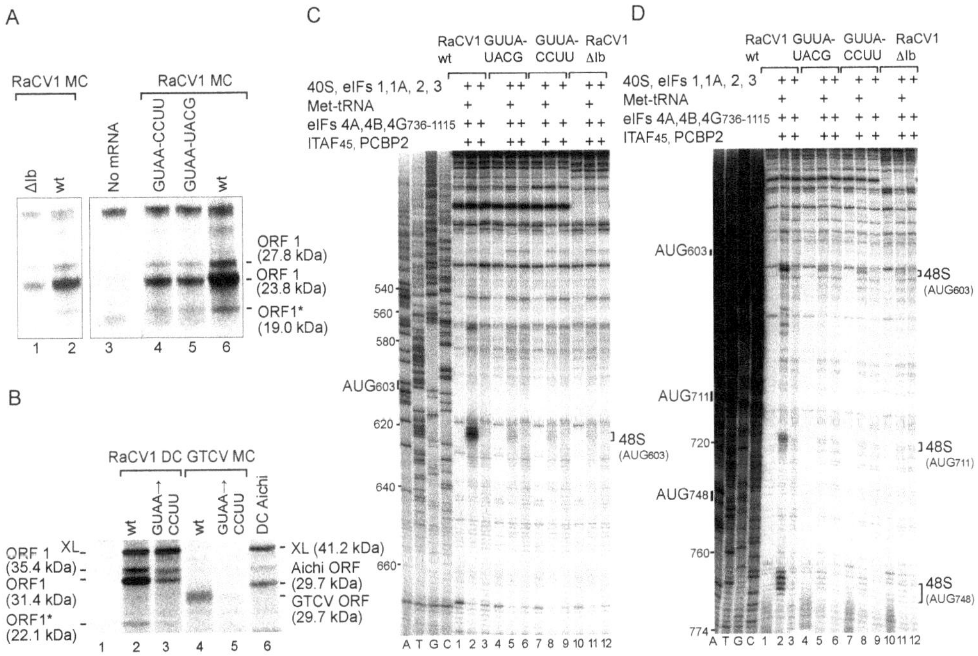

Figure 7. Influence of the GNRA tetraloop at the apex of subdomain Ib on RaCV1 IRES function. (**A**,**B**) In vitro translation of monocistronic and dicistronic mRNAs containing the indicated wild-type and mutant GTCV and RaCV1 IRESs. Panels are annotated to show translation products resolved by SDS-PAGE. Separation of lanes in panel (**A**) by a white line indicates that they were juxtaposed from the same gel. (**C**,**D**) Toeprint analysis of 48S complexes formed on wild-type and mutant IRES-containing mRNAs by incubation with 40S subunits, $Met\text{-}tRNA_i^{Met}$, initiation factors, Ebp1 and PCBP2 as indicated. Panels are annotated to show the positions of initiation codons on the left of toeprints corresponding to 48S complexes on the right. A dideoxynucleotide sequence generated with the same primer was run in parallel on each gel (lanes A, T, G, C) to show the corresponding wild-type sequence.

The native cysteine residues in Ebp1 could not be substituted to yield a cysteine-less variant without loss of activity, but the incubation of wt Ebp1 with Fe(II)BABE) did not induce the cleavage of type 2 IRESs (Figure 8B, lane 3), suggesting that native cysteine residues in it are not surface-exposed and thus that cysteines could be placed at surface-exposed locations for DHRC experiments. We generated a panel of Ebp1 moieties with additional cysteine residues at positions 22, 37, 55, 65, 94, 131, 165, 208, 22, 254, 271, 311 or 360 (Figure 8A). Fe(II)BABE-linked cysteine residues at positions 37, 55, 65, 131, 165, 271 and 360 induced strong cleavage in the central and apical elements of subdomain Id of the RaCV1 IRES (Figure 8B,C; summarized in Figure 8E). Weaker, more dispersed cleavages mapped to the central stem of domain I.

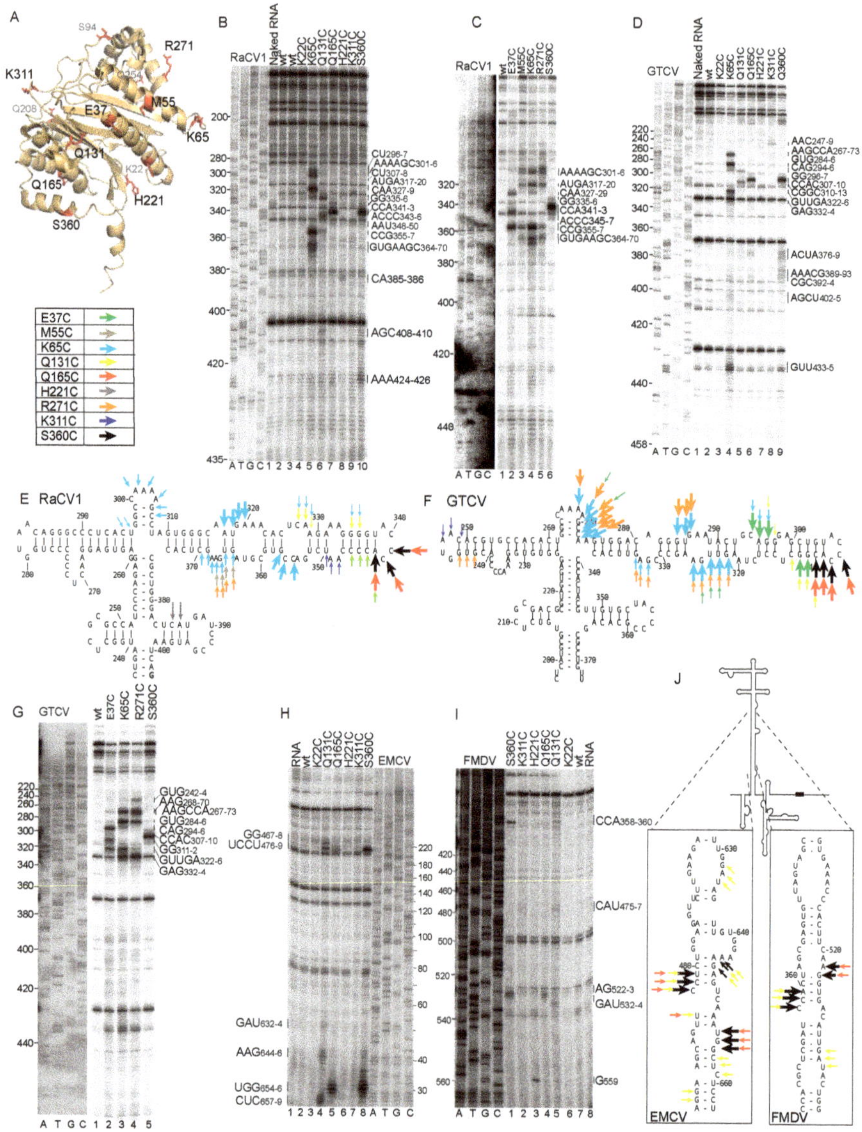

Figure 8. Interaction of Ebp1 with type 2 IRESs. (**A**) Upper panel: schematic representation of Ebp1 showing the positions of residues substituted by cysteine residues in individual mutants. Residues labeled in large black font induced cleavage of the bound IRES; residues labeled in smaller gray font did not. Lower panel: the color scheme used to indicate residues that induced cleavage on models of (**E**) the RaCV1 IRES, (**F**) the GTCV IRES and (**J**) EMCV and FMDV IRESs. (**B–D**,**G–I**) primer extension analysis of DHRC of (**B**,**C**) the RaCV1 IRES, (**D**,**G**) the GTCV IRES, (**H**) the EMCV IRES and (**I**) the FMDV IRES from Fe(II)BABE-linked Ebp1 mutants. Sites of cleavage are indicated (**B–D**,**G**,**I**) on the right, (**H**) on the left and (**E**,**F**,**J**) mapped onto IRES models. Reference lanes A, T, G, C on each panel show the appropriate viral sequence generated with the same primer, except in (**H**), which shows the FMDV sequence and is numbered to indicate the number of residues from the 3′-end of the primer. Separation of lanes by white lines in panels (**C**,**G**) indicates that they were juxtaposed from the same gels; the intensity of sequence lanes in (**C**) was increased.

The principal Ebp1 binding site on the RaCV1 IRES corresponds to an element that occurs in all avian calicivirus type 2 IRESs. The patterns of cleavage of GTCV and RaCV1 IRESs by Fe(II)-BABE-derivatized Ebp1 were very similar (compare Figure 8B–D,G), indicating that it binds to them at equivalent locations and with an identical orientation (Figure 8E,F). However, this binding site is not present in EMCV or FMDV IRESs even though they both bind to Ebp1 [35]. DHRC showed that Fe(II)-BABE-derivatized Ebp1 induced cleavage in the irregular stem of domain I of EMCV and FMDV IRESs with some weaker cleavages mapping to the junction of I, J-K and L domains (Figure 8H,I; summarized in Figure 8J). Notably, the cleavage of EMCV and FMDV IRESs was induced most effectively by residues that induced the strong cleavage of RaCV1 and GTCV IRESs, indicating that the same surface of Ebp1 binds to these IRESs.

The similarity in patterns of Ebp1-directed hydroxyl radical cleavage of the RaCV1 and GTCV IRESs is evidence for the specificity of binding. To characterize the functional importance of binding, we compared DHRC by Ebp1 and 48S complex formation on a panel of mutant RaCV1 IRESs with substitutions or deletions in subdomain Id (Figure 9A).

Levels of DHRC and initiation activity on these mutants correlated well: mutations 4 and 5 had little or no effect on cleavage (Figure 9D, lanes 5–12) or on the level of 48S complex formation (Figure 9B, lanes 6, 7), mutations 3, 6 and 7 abrogated cleavage (and therefore, binding) (Figure 9D, lanes 1–4; Figure 9E, lanes 1–8), and they supported 48S complex formation at up to only 50% of the wild-type level (Figure 9B, lanes 5, 8, 9), whereas mutations 1 and 2 exhibited site-specific defects in cleavage (Figure 9C, lanes 5–12) and mild reductions in the level of 48S complex formation (Figure 9B, lanes 3, 4).

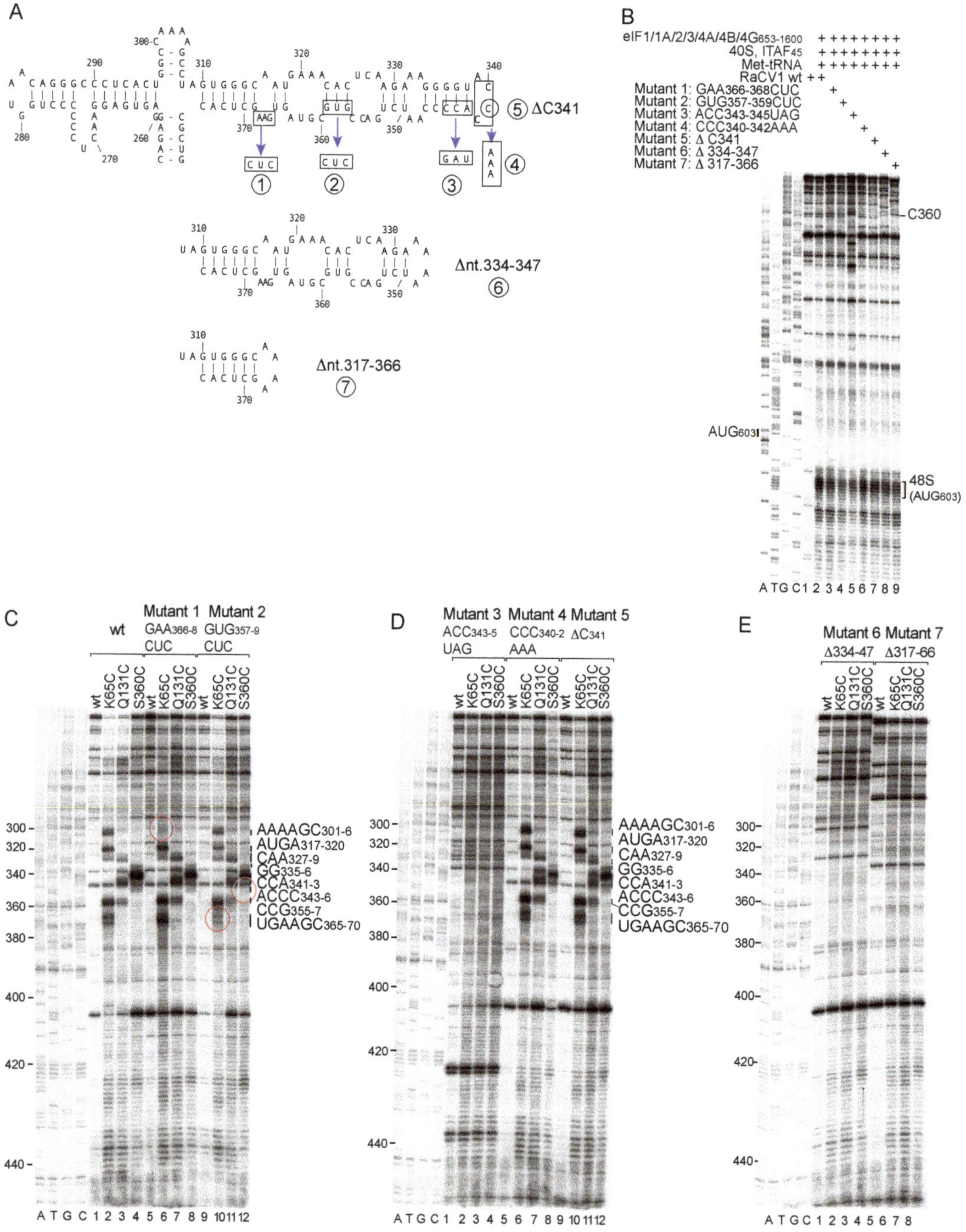

Figure 9. Influence of mutations in RaCV1 IRES subdomain Id on binding of Ebp1 assessed by directed hydroxyl radical probing and IRES activity. (**A**) Schematic model of RaCV1 IRES subdomain Id showing the substitutions/deletions in mutants 1–5 (upper panel) and residual predicted structure in deletion mutants 6 (middle panel) and 7 (lower panel). (**B**) Toeprint analysis of 48S complexes formed

on wild-type and mutant IRES-containing mRNAs on incubation with 40S subunits, Met-tRNA$_i^{Met}$, initiation factors, Ebp1 and PCBP2 as indicated. Panels are annotated to show the positions of initiation codons on the left and of toeprints corresponding to 48S complexes on the right. (**C–E**) Primer extension analysis of directed hydroxyl radical cleavage of ((**C**) lanes 1–4) wt and (**C–E**) mutant RaCV1 IRESs from the indicated Fe(II)-tethered Ebp1 variants. Sites of cleavage are indicated on the right. (**C**) Circles show sites of defective cleavage associated with specific IRES mutations. A dideoxynucleotide sequence generated with the same primer and (**B**,**D**,**E**) wild-type and (**C**) [ACC$_{343-5}$UAG] mutant IRESs was run in parallel on each gel.

4. Discussion

Caliciviruses use various mechanisms to maximize the utilization of their coding capacity, including (i) synthesis of a polyprotein that is cleaved to yield individual functional proteins, (ii) synthesis of subgenomic mRNA to allow the independent expression of structural proteins, and (iii) substoichiometric translation of a downstream ORF by a termination–reinitiation mechanism promoted by elements of the viral mRNA. We have now identified examples of another mechanism: translation of overlapping ORFs. This mechanism had previously been reported for an alternative coding region within ORF2 in the subgenomic mRNA of noroviruses [8,9]. The examples identified here are exemplified by translation of the ORF1 and ORF1* moieties of RaCV1 and *Caliciviridae* sp. Isolate CAL/PB27-SI31/Switzerland/2019.

The translation of ORF1* in the first group initiates downstream of the ORF1 initiation codon and yields a 239–317 a.a.-long polypeptide, whereas ORF1* in the second group initiates upstream of the ORF1 initiation codon and encodes a ~200 a.a.-long polypeptide. There are several precedents for IRES-mediated initiation at multiple initiation codons, either in-frame (as in FMDV), leading to the synthesis of an N-terminally extended polyprotein [25,79], or in overlapping reading frames, e.g., mediated by the type 1 poliovirus IRES [89,90], by the type 2 IRES of Theiler's murine encephalomyelitis virus (leading to synthesis of the L* protein) [26] and likely by Cosavirus, Hunnivirus, Rosavirus and Sicinivirus type 2 IRESs [90]. Initiation at AUG codons downstream of the RaCV1 IRES involves scanning after IRES-mediated recruitment of the 43S complex and is dependent on eIF1 and eIF1A (Figure 4H). The nucleotide context of the ORF1 initiation codon in the RaCV1 group is in many instances suboptimal, so that it would be bypassed by an appreciable proportion of scanning ribosomes, ensuring the translation of ORF1*. The function of ORF1* is unknown, and it does not have significant sequence or structural homology with proteins of known function. Consistent with other polypeptides encoded by alternative reading frames [91], ORF1* moieties are predicted to be largely unstructured and might therefore function by binding to viral or cellular proteins. For example, the TMEV L* protein binds to RNAse L, an effector in the interferon antiviral response, and prevents its activation [92,93].

The exploitation of overlapping ORFs described above contributes to maximizing utilization of the coding capacity of calicivirus genomes. The presence of an IRES to promote translation of the overlapping ORFs potentially enables the virus to sustain the translation of both ORFs in circumstances that would impair cap-dependent translation. Calicivirus infections lead to the shut-off of cellular translation that is associated with changes in the translation apparatus, including cleavage of the poly(A) binding protein PABP, dissociation of eIF3 from eIF4G, and phosphorylation of eIF4E [14,94,95]. Notably, the EMCV IRES is less dependent on PABP for translation than cellular mRNAs [96], and its function is enhanced by dephosphorylation/activation of the regulatory eIF4E-binding protein (4E-BP) and by its overexpression [97], whereas the translation of cellular mRNAs and several caliciviral mRNAs is inhibited by 4E-BP [98]. Interestingly, the translation of calicivirus mRNAs that depend on the VPg-eIF4E interaction is also inhibited by 4E-BP [11,99], suggesting that the acquisition of a type 2 IRES by a calicivirus genome may alter its ability to sustain translation in specific stress conditions, resulting in a gain of function. As noted here for the RaCV1 IRES and elsewhere for the EMCV IRES [55], initiation on type

2 IRESs is not dependent on the eIF4G–eIF3 interaction. Taken together, these observations suggest that the presence of IRESs in calicivirus genomes, likely the result of horizontal gene transfer, may enhance their ability to usurp the cellular translation apparatus and to target cellular translation in a manner that does not impair viral translation.

Biochemical analysis revealed similarities and differences between the mechanism of initiation on calicivirus type 2 IRESs and on the canonical EMCV, FMDV and TMEV type 2 IRESs. The calicivirus IRESs require the same set of canonical initiation factors for 48S complex formation as these picornavirus IRESs [27,30,32,33,53,59]. The process similarly involves the binding of eIF4G to the JK domain, and it also does not depend on the eIF4G–eIF3 interaction ([55]; Figure 6), suggesting that another interaction must be responsible for directing association of the IRES with the 43S complex. However, eIF4G's eIF3-binding domain enhanced initiation at RaCV1 codons downstream of the Yn-Xm-AUG motif, possibly indicating that the eIF3-eIF4G interaction is important during subsequent scanning, for example to promote processivity. Scanning also required eIF1, as noted previously for scanning after the recruitment of 43S complexes to type 1 IRESs and the type 2 FMDV IRES [27,31]. Our analysis shows that as in picornavirus type 2 IRESs, the eIF4G-binding elements in domain J [21,22], the conserved apical region of domain I (including the GNRA tetraloop) [16–19] and the Yn-Xm-AUG motif are all important for the function of calicivirus type 2 IRESs. Notably, despite its conservation and importance, the function of the tetraloop remains unknown.

Differences in the ITAF requirements of different type 2 IRESs have been noted previously, exemplified by the Ebp1-dependence of the FMDV IRES but not of the EMCV and TMEV IRESs [32] and the conditional PTB-dependence of EMCV IRESs, the latter being determined either by mutations in the IRES or by its linkage to different heterologous reporters [100]. Ebp1 dependence is likely also conditional, because the affinity of its binding to EMCV and FMDV IRESs is comparable despite their different requirements for this factor [35]. ITAF requirements on calicivirus type 2 IRESs also varied: the activities of both GTCV and RaCV1 IRESs were enhanced by Ebp1, but the GTCV IRES was stimulated by PTB, whereas initiation on the RaCV1 IRES was enhanced by PCBP2. A requirement for PCBP2 by type 2 IRESs has not previously been reported, although it binds specifically to the EMCV and FMDV IRESs without contributing to their function [101]. It may be that its function is also conditional, compensating for an activity or structural characteristic that is suboptimal in calicivirus type 2 IRESs but not in EMCV or FMDV IRESs. Whether PCBP2 binds to the apex of domain I in an analogous manner to its interaction with the apical tetraloop-containing region of domain IV in type 1 IRESs [31,100,102–104] remains to be determined. The function of Ebp1 in IRES-mediated initiation also remains obscure. The results of DHRC experiments showed that it binds directly to a conserved location on calicivirus type 2 IRESs that differs from the binding site on EMCV and FMDV IRESs. Mutations in the RaCV1 IRES that abrogated binding (assayed by DHRC) led to a partial loss of function. Interestingly, the silencing of Ebp1 expression in human, bovine, canine and porcine kidney cells resulted in a similar partial impairment of FMDV IRES activity [35,36]. Ebp1 is therefore important but not absolutely essential: in light of the differences in its binding site of calicivirus and picornaviruses IRESs, its function may, like other ITAFs, be to promote adoption by IRESs of an active conformation [105].

Supplementary Materials: The following supporting information can be downloaded at https://www.mdpi.com/article/10.3390/v16091413/s1. Supplementary Figure S1. Sequence alignment of the conserved core of calicivirus type 2 IRESs. Supplementary Figure S2. Aligned ORF1* sequences from *Caliciviridae* sp. isolate yc-13 and related caliciviruses. Supplementary Figure S3. Aligned ORF1* sequences from *Caliciviridae* sp. isolate CAL/PB27-SI31/Switzerland/2019 and related caliciviruses. Supplementary Figure S4. (A) Aligned ORF1* sequences from *Caliciviridae* sp. isolate cftwhg06cal1, Goose calicivirus isolate MA/CHN/2018 and Goose calicivirus isolate H146, (B) Aligned ORF1* sequences from Duck calicivirus isolates MW20 and NC-19BV-45-k77-747816, (C) ORF1* sequence from *Caliciviridae* sp. isolate ybb044cal01nc. Supplementary Figure S5. Open reading frames (ORFs) upstream of or overlapping ORF1 in avian calicivirus genomes.

Author Contributions: Conceptualization, C.U.T.H. and T.V.P.; Investigation, Y.A., T.V.P. and C.U.T.H.; Data Curation, Y.A. and C.U.T.H.; Writing—Original Draft Preparation, C.U.T.H.; Writing—Review and Editing, Y.A., T.V.P. and C.U.T.H.; Visualization, C.U.T.H. and T.V.P.; Supervision, T.V.P. and C.U.T.H.; Project Administration, C.U.T.H.; Funding Acquisition, T.V.P. and C.U.T.H. All authors have read and agreed to the published version of the manuscript.

Funding: This research was funded by National Institutes of Health (NIH) awards 5 R35 GM122602 to T.V.P. and 5 R01 GM097014 to C.H.

Institutional Review Board Statement: Not applicable.

Data Availability Statement: The original data presented in the study are openly available in Mendeley Data at https://data.mendeley.com/drafts/9mxx48n3sc.1 from 1 September 2024.

Conflicts of Interest: The authors declare no conflicts of interest.

References

1. Stern-Ginossar, N.; Thompson, S.R.; Mathews, M.B.; Mohr, I. Translational control in virus-infected cells. *Cold Spring Harb. Perspect. Biol.* **2019**, *11*, a033001. [CrossRef]
2. Ho, J.S.Y.; Zhu, Z.; Marazzi, I. Unconventional viral gene expression mechanisms as therapeutic targets. *Nature* **2021**, *593*, 362–371. [CrossRef] [PubMed]
3. Vinjé, J.; Estes, M.K.; Esteves, P.; Green, K.Y.; Katayama, K.; Knowles, N.J.; L'Homme, Y.; Martella, V.; Vennema, H.; White, P.A.; et al. ICTV virus taxonomy profile: *Caliciviridae*. *J. Gen. Virol.* **2019**, *100*, 1469–1470. [CrossRef] [PubMed]
4. Alhatlani, B.; Vashist, S.; Goodfellow, I. Functions of the 5′ and 3′ ends of calicivirus genomes. *Virus Res.* **2015**, *206*, 134–143. [CrossRef] [PubMed]
5. Meyers, G. Translation of the minor capsid protein of a calicivirus is initiated by a novel termination-dependent reinitiation mechanism. *J. Biol. Chem.* **2003**, *278*, 34051–34060. [CrossRef]
6. Pöyry, T.A.; Kaminski, A.; Connell, E.J.; Fraser, C.S.; Jackson, R.J. The mechanism of an exceptional case of reinitiation after translation of a long ORF reveals why such events do not generally occur in mammalian mRNA translation. *Genes Dev.* **2007**, *21*, 3149–3162. [CrossRef]
7. Zinoviev, A.; Hellen, C.U.T.; Pestova, T.V. Multiple mechanisms of reinitiation on bicistronic calicivirus mRNAs. *Mol. Cell* **2015**, *57*, 1059–1073. [CrossRef]
8. Thackray, L.B.; Wobus, C.E.; Chachu, K.A.; Liu, B.; Alegre, E.R.; Henderson, K.S.; Kelley, S.T.; Virgin, H.W., 4th. Murine noroviruses comprising a single genogroup exhibit biological diversity despite limited sequence divergence. *J. Virol.* **2007**, *81*, 10460–10473. [CrossRef]
9. McFadden, N.; Bailey, D.; Carrara, G.; Benson, A.; Chaudhry, Y.; Shortland, A.; Heeney, J.; Yarovinsky, F.; Simmonds, P.; Macdonald, A.; et al. Norovirus regulation of the innate immune response and apoptosis occurs via the product of the alternative open reading frame 4. *PLoS Pathog.* **2011**, *7*, e1002413. [CrossRef]
10. Jackson, R.J.; Hellen, C.U.; Pestova, T.V. The mechanism of eukaryotic translation initiation and principles of its regulation. *Nat. Rev. Mol. Cell. Biol.* **2010**, *11*, 113–127. [CrossRef]
11. Goodfellow, I.; Chaudhry, Y.; Gioldasi, I.; Gerondopoulos, A.; Natoni, A.; Labrie, L.; Laliberté, J.F.; Roberts, L. Calicivirus translation initiation requires an interaction between VPg and eIF 4E. *EMBO Rep.* **2005**, *6*, 968–972. [CrossRef]
12. Leen, E.N.; Sorgeloos, F.; Correia, S.; Chaudhry, Y.; Cannac, F.; Pastore, C.; Xu, Y.; Graham, S.C.; Matthews, S.J.; Goodfellow, I.G.; et al. A conserved interaction between a C-terminal motif in norovirus VPg and the HEAT-1 domain of eIF4G is essential for translation initiation. *PLoS Pathog.* **2016**, *12*, e1005379, Erratum in: *PLoS Pathog.* **2016**, *12*, e1005509. [CrossRef]
13. Hosmillo, M.; Lu, J.; McAllaster, M.R.; Eaglesham, J.B.; Wang, X.; Emmott, E.; Domingues, P.; Chaudhry, Y.; Fitzmaurice, T.J.; Tung, M.K.; et al. Noroviruses subvert the core stress granule component G3BP1 to promote viral VPg-dependent translation. *eLife* **2019**, *8*, e46681. [CrossRef]
14. Emmott, E.; Sorgeloos, F.; Caddy, S.L.; Vashist, S.; Sosnovtsev, S.; Lloyd, R.; Heesom, K.; Locker, N.; Goodfellow, I. Norovirus-mediated modification of the translational landscape via virus and host-induced cleavage of translation initiation factors. *Mol. Cell. Proteom.* **2017**, *16*, S215–S229. [CrossRef] [PubMed]
15. Arhab, Y.; Bulakhov, A.G.; Pestova, T.V.; Hellen, C.U.T. Dissemination of Internal Ribosomal Entry Sites (IRES) Between Viruses by Horizontal Gene Transfer. *Viruses* **2020**, *12*, 612. [CrossRef] [PubMed]
16. López de Quinto, S.; Martínez-Salas, E. Conserved structural motifs located in distal loops of aphthovirus internal ribosome entry site domain 3 are required for internal initiation of translation. *J. Virol.* **1997**, *71*, 4171–4175. [CrossRef]
17. Fernández-Miragall, O.; Martínez-Salas, E. Structural organization of a viral IRES depends on the integrity of the GNRA motif. *RNA* **2003**, *9*, 1333–1344. [CrossRef] [PubMed]
18. Martínez-Salas, E.; López de Quinto, S.; Ramos, R.; Fernández-Miragall, O. IRES elements: Features of the RNA structure contributing to their activity. *Biochimie* **2002**, *84*, 755–763. [CrossRef]

19. Robertson, M.E.; Seamons, R.A.; Belsham, G.J. A selection system for functional internal ribosome entry site (IRES) elements: Analysis of the requirement for a conserved GNRA tetraloop in the encephalomyocarditis virus IRES. *RNA* **1999**, *5*, 1167–1179. [CrossRef]
20. Imai, S.; Kumar, P.; Hellen, C.U.; D'Souza, V.M.; Wagner, G. An accurately preorganized IRES RNA structure enables eIF4G capture for initiation of viral translation. *Nat. Struct. Mol. Biol.* **2016**, *23*, 859–864. [CrossRef]
21. Kolupaeva, V.G.; Lomakin, I.B.; Pestova, T.V.; Hellen, C.U. Eukaryotic initiation factors 4G and 4A mediate conformational changes downstream of the initiation codon of the encephalomyocarditis virus internal ribosomal entry site. *Mol. Cell. Biol.* **2003**, *23*, 687–698. [CrossRef]
22. Clark, A.T.; Robertson, M.E.; Conn, G.L.; Belsham, G.J. Conserved nucleotides within the J domain of the encephalomyocarditis virus internal ribosome entry site are required for activity and for interaction with eIF4G. *J. Virol.* **2003**, *77*, 12441–12449. [CrossRef]
23. Imai, S.; Suzuki, H.; Fujiyoshi, Y.; Shimada, I. Dynamically regulated two-site interaction of viral RNA to capture host translation initiation factor. *Nat. Commun.* **2023**, *14*, 4977. [CrossRef] [PubMed]
24. Jang, S.K.; Pestova, T.V.; Hellen, C.U.; Witherell, G.W.; Wimmer, E. Cap-independent translation of picornavirus RNAs: Structure and function of the internal ribosomal entry site. *Enzyme* **1990**, *44*, 292–309. [CrossRef] [PubMed]
25. Clarke, B.E.; Sangar, D.V.; Burroughs, J.N.; Newton, S.E.; Carroll, A.R.; Rowlands, D.J. Two initiation sites for foot-and-mouth disease virus polyprotein in vivo. *J. Gen. Virol.* **1985**, *66*, 2615–2626. [CrossRef]
26. Kong, W.P.; Roos, R.P. Alternative translation initiation site in the DA strain of Theiler's murine encephalomyelitis virus. *J. Virol.* **1991**, *65*, 3395–3399. [CrossRef] [PubMed]
27. Andreev, D.E.; Fernandez-Miragall, O.; Ramajo, J.; Dmitriev, S.E.; Terenin, I.M.; Martinez-Salas, E.; Shatsky, I.N. Differential factor requirement to assemble translation initiation complexes at the alternative start codons of foot-and-mouth disease virus RNA. *RNA* **2007**, *13*, 1366–1374. [CrossRef] [PubMed]
28. Yu, Y.; Sweeney, T.R.; Kafasla, P.; Jackson, R.J.; Pestova, T.V.; Hellen, C.U. The mechanism of translation initiation on Aichivirus RNA mediated by a novel type of picornavirus IRES. *EMBO J.* **2011**, *30*, 4423–4436. [CrossRef]
29. Sweeney, T.R.; Dhote, V.; Yu, Y.; Hellen, C.U. A distinct class of internal ribosomal entry site in members of the Kobuvirus and proposed Salivirus and Paraturdivirus genera of the Picornaviridae. *J. Virol.* **2012**, *86*, 1468–1486. [CrossRef]
30. Pestova, T.V.; Shatsky, I.N.; Hellen, C.U. Functional dissection of eukaryotic initiation factor 4F: The 4A subunit and the central domain of the 4G subunit are sufficient to mediate internal entry of 43S preinitiation complexes. *Mol. Cell. Biol.* **1996**, *16*, 6870–6878. [CrossRef]
31. Sweeney, T.R.; Abaeva, I.S.; Pestova, T.V.; Hellen, C.U. The mechanism of translation initiation on Type 1 picornavirus IRESs. *EMBO J.* **2014**, *33*, 76–92. [CrossRef]
32. Pilipenko, E.V.; Pestova, T.V.; Kolupaeva, V.G.; Khitrina, E.V.; Poperechnaya, A.N.; Agol, V.I.; Hellen, C.U. A cell cycle-dependent protein serves as a template-specific translation initiation factor. *Genes Dev.* **2000**, *14*, 2028–2045. [CrossRef]
33. Pestova, T.V.; Hellen, C.U.; Shatsky, I.N. Canonical eukaryotic initiation factors determine initiation of translation by internal ribosomal entry. *Mol. Cell. Biol.* **1996**, *16*, 6859–6869. [CrossRef] [PubMed]
34. Borovjagin, A.; Pestova, T.; Shatsky, I. Pyrimidine tract binding protein strongly stimulates in vitro encephalomyocarditis virus RNA translation at the level of preinitiation complex formation. *FEBS Lett.* **1994**, *351*, 299–302. [CrossRef]
35. Monie, T.P.; Perrin, A.J.; Birtley, J.R.; Sweeney, T.R.; Karakasiliotis, I.; Chaudhry, Y.; Roberts, L.O.; Matthews, S.; Goodfellow, I.G.; Curry, S. Structural insights into the transcriptional and translational roles of Ebp1. *EMBO J.* **2007**, *26*, 3936–3944. [CrossRef] [PubMed]
36. Kanda, T.; Ozawa, M.; Tsukiyama-Kohara, K. IRES-mediated translation of foot-and-mouth disease virus (FMDV) in cultured cells derived from FMDV-susceptible and -insusceptible animals. *BMC Vet Res.* **2016**, *12*, 66. [CrossRef]
37. Pestova, T.V.; Shatsky, I.N.; Fletcher, S.P.; Jackson, R.J.; Hellen, C.U. A prokaryotic-like mode of cytoplasmic eukaryotic ribosome binding to the initiation codon during internal translation initiation of hepatitis C and classical swine fever virus RNAs. *Genes Dev.* **1998**, *12*, 67–83. [CrossRef] [PubMed]
38. Brown, Z.P.; Abaeva, I.S.; De, S.; Hellen, C.U.T.; Pestova, T.V.; Frank, J. Molecular architecture of 40S translation initiation complexes on the hepatitis C virus IRES. *EMBO J.* **2022**, *41*, e110581. [CrossRef] [PubMed]
39. Wilson, J.E.; Pestova, T.V.; Hellen, C.U.; Sarnow, P. Initiation of protein synthesis from the A site of the ribosome. *Cell* **2000**, *102*, 511–520. [CrossRef]
40. Pestova, T.V.; Hellen, C.U. Translation elongation after assembly of ribosomes on the Cricket paralysis virus internal ribosomal entry site without initiation factors or initiator tRNA. *Genes Dev.* **2003**, *17*, 181–186. [CrossRef]
41. Arhab, Y.; Miścicka, A.; Pestova, T.V.; Hellen, C.U.T. Horizontal gene transfer as a mechanism for the promiscuous acquisition of distinct classes of IRES by avian caliciviruses. *Nucleic Acids Res.* **2022**, *50*, 1052–1068. [CrossRef] [PubMed]
42. Canuti, M.; Wilson, L.; Bowes, V.; Redford, T.; Dufour, S.C.; Lang, A.S.; Verhoeven, J.T.P. A novel calicivirus discovered in trumpeter swans (*Cygnus buccinator*) expands the richness of known avian caliciviruses. *Curr. Res. Microb. Sci.* **2022**, *3*, 100169. [CrossRef] [PubMed]
43. Wang, F.; Wang, M.; Dong, Y.; Zhang, B.; Zhang, D. Genetic characterization of a novel calicivirus from a goose. *Arch. Virol.* **2017**, *162*, 2115–2118. [CrossRef] [PubMed]
44. Wolf, S.; Reetz, J.; Hoffmann, K.; Gründel, A.; Schwarz, B.A.; Hänel, I.; Otto, P.H. Discovery and genetic characterization of novel caliciviruses in German and Dutch poultry. *Arch. Virol.* **2012**, *157*, 1499–1507. [CrossRef] [PubMed]

45. Canuti, M.; Kroyer, A.N.K.; Ojkic, D.; Whitney, H.G.; Robertson, G.J.; Lang, A.S. Discovery and characterization of novel RNA viruses in aquatic North American wild birds. *Viruses* **2019**, *11*, 768. [CrossRef]
46. Pilipenko, E.V.; Blinov, V.M.; Chernov, B.K.; Dmitrieva, T.M.; Agol, V.I. Conservation of the secondary structure elements of the 5′-untranslated region of cardio- and aphthovirus RNAs. *Nucleic Acids Res.* **1989**, *17*, 5701–5711. [CrossRef]
47. Zuker, M. Mfold web server for nucleic acid folding and hybridization prediction. *Nucleic Acids Res.* **2003**, *31*, 3406–3415. [CrossRef]
48. Firth, A.E. Mapping overlapping functional elements embedded within the protein-coding regions of RNA viruses. *Nucleic Acids Res.* **2014**, *42*, 12425–12439. [CrossRef]
49. Criscuolo, A.; Gribaldo, S. BMGE (Block Mapping and Gathering with Entropy): A new software for selection of phylogenetic informative regions from multiple sequence alignments. *BMC Evol. Biol.* **2010**, *10*, 210. [CrossRef]
50. Lefort, V.; Longueville, J.E.; Gascuel, O. SMS: Smart model selection in PhyML. *Mol. Biol. Evol.* **2017**, *34*, 2422–2424. [CrossRef]
51. Lemoine, F.; Correia, D.; Lefort, V.; Doppelt-Azeroual, O.; Mareuil, F.; Cohen-Boulakia, S.; Gascuel, O. NGPhylogeny.fr: New generation phylogenetic services for non-specialists. *Nucleic Acids Res.* **2019**, *47*, W260–W265. [CrossRef] [PubMed]
52. Letunic, I.; Bork, P. Interactive Tree of Life (iTOL) v6: Recent updates to the phylogenetic tree display and annotation tool. *Nucleic Acids Res.* **2024**, *52*, W78–W82. [CrossRef]
53. Pestova, T.V.; Borukhov, S.I.; Hellen, C.U. Eukaryotic ribosomes require initiation factors 1 and 1A to locate initiation codons. *Nature* **1988**, *394*, 854–859. [CrossRef] [PubMed]
54. Pause, A.; Méthot, N.; Svitkin, Y.; Merrick, W.C.; Sonenberg, N. Dominant negative mutants of mammalian translation initiation factor eIF-4A define a critical role for eIF-4F in cap-dependent and cap-independent initiation of translation. *EMBO J.* **1994**, *13*, 1205–1215. [CrossRef] [PubMed]
55. Lomakin, I.B.; Hellen, C.U.; Pestova, T.V. Physical association of eukaryotic initiation factor 4G (eIF4G) with eIF4A strongly enhances binding of eIF4G to the internal ribosomal entry site of encephalomyocarditis virus and is required for internal initiation of translation. *Mol. Cell. Biol.* **2000**, *20*, 6019–6029. [CrossRef]
56. Pestova, T.V.; Lomakin, I.B.; Lee, J.H.; Choi, S.K.; Dever, T.E.; Hellen, C.U. The joining of ribosomal subunits in eukaryotes requires eIF5B. *Nature* **2000**, *403*, 332–335. [CrossRef]
57. Skabkin, M.A.; Skabkina, O.V.; Dhote, V.; Komar, A.A.; Hellen, C.U.; Pestova, T.V. Activities of Ligatin and MCT-1/DENR in eukaryotic translation initiation and ribosomal recycling. *Genes Dev.* **2010**, *24*, 1787–1801. [CrossRef]
58. Lomakin, I.B.; Shirokikh, N.E.; Yusupov, M.M.; Hellen, C.U.; Pestova, T.V. The fidelity of translation initiation: Reciprocal activities of eIF1, IF3 and YciH. *EMBO J.* **2006**, *25*, 196–210. [CrossRef]
59. Pilipenko, E.V.; Viktorova, E.G.; Guest, S.T.; Agol, V.I.; Roos, R.P. Cell-specific proteins regulate viral RNA translation and virus-induced disease. *EMBO J.* **2001**, *20*, 6899–6908. [CrossRef]
60. Abaeva, I.S.; Pestova, T.V.; Hellen, C.U. Attachment of ribosomal complexes and retrograde scanning during initiation on the Halastavi árva virus IRES. *Nucleic Acids Res.* **2016**, *44*, 2362–2377. [CrossRef]
61. Pisarev, A.V.; Unbehaun, A.; Hellen, C.U.; Pestova, T.V. (2007) Assembly and analysis of eukaryotic translation initiation complexes. *Methods Enzymol.* **2007**, *430*, 147–177. [CrossRef]
62. Culver, G.M.; Noller, H.F. Directed hydroxyl radical probing of RNA from iron(II) tethered to proteins in ribonucleoprotein complexes. *Methods Enzymol.* **2000**, *318*, 461–475. [CrossRef]
63. Lomakin, I.B.; Kolupaeva, V.G.; Marintchev, A.; Wagner, G.; Pestova, T.V. Position of eukaryotic initiation factor eIF1 on the 40S ribosomal subunit determined by directed hydroxyl radical probing. *Genes Dev.* **2003**, *17*, 2786–2797. [CrossRef] [PubMed]
64. Chen, Y.M.; Sadiq, S.; Tian, J.H.; Chen, X.; Lin, X.D.; Shen, J.J.; Chen, H.; Hao, Z.Y.; Wille, M.; Zhou, Z.C.; et al. RNA viromes from terrestrial sites across China expand environmental viral diversity. *Nat. Microbiol.* **2022**, *7*, 1312–1323. [CrossRef]
65. Liu, J.; Li, X.; Song, W.; Zeng, X.; Li, H.; Yang, L.; Wang, D. The multi-kingdom microbiome of wintering migratory birds in Poyang Lake, China. *Viruses* **2024**, *16*, 396. [CrossRef] [PubMed]
66. Kubacki, J.; Qi, W.; Fraefel, C. Differential viral genome diversity of healthy and RSS-affected broiler flocks. *Microorganisms* **2022**, *10*, 1092. [CrossRef]
67. Kwok, K.T.T.; de Rooij, M.M.T.; Messink, A.B.; Wouters, I.M.; Smit, L.A.M.; Cotton, M.; Heederik, D.J.J.; Koopmans, M.P.G.; Phan, M.V.T. Establishing farm dust as a useful viral metagenomic surveillance matrix. *Sci. Rep.* **2022**, *12*, 16308. [CrossRef] [PubMed]
68. Wang, Y.; Yang, S.; Liu, D.; Zhou, C.; Li, W.; Lin, Y.; Wang, X.; Shen, Q.; Wang, H.; Li, C.; et al. The fecal virome of red-crowned cranes. *Arch. Virol.* **2019**, *164*, 3–16. [CrossRef]
69. Drexler, J.F.; Baumgarte, S.; Luna, L.K.; Stöcker, A.; Almeida, P.S.; Ribeiro, T.C.; Petersen, N.; Herzog, P.; Pedroso, C.; Brites, C.; et al. Genomic features and evolutionary constraints in Saffold-like cardioviruses. *J. Gen. Virol.* **2010**, *91*, 1418–1427. [CrossRef]
70. López de Quinto, S.; Lafuente, E.; Martínez-Salas, E. IRES interaction with translation initiation factors: Functional characterization of novel RNA contacts with eIF3, eIF4B, and eIF4GII. *RNA* **2001**, *7*, 1213–1226. [CrossRef]
71. Matsvay, A.; Dyachkova, M.; Sai, A.; Burskaia, V.; Artyushin, I.; Shipulin, G. Complete genome sequence, molecular characterization and phylogenetic relationships of a Temminck's Stint calicivirus: Evidence for a new genus within Caliciviridae family. *Microorganisms* **2022**, *10*, 1540. [CrossRef]
72. Wille, M.; Shi, M.; Klaassen, M.; Hurt, A.C.; Holmes, E.C. Virome heterogeneity and connectivity in waterfowl and shorebird communities. *ISME J.* **2019**, *13*, 2603–2616. [CrossRef] [PubMed]

73. Wu, Y.; Lu, C.; Zhao, R.; He, Y.; Hou, J.; Sun, Y.; Yang, S.; Qin, Z.; Ji, L.; Wang, Y.; et al. A new bird calicivirus detected in feces of cormorants. *Acta Virol.* **2023**, *68*, 12515. [CrossRef]
74. Shan, T.; Yang, S.; Wang, H.; Wang, H.; Zhang, J.; Gong, G.; Xiao, Y.; Yang, J.; Wang, X.; Lu, J.; et al. Virome in the cloaca of wild and breeding birds revealed a diversity of significant viruses. *Microbiome* **2022**, *10*, 60. [CrossRef] [PubMed]
75. Tseng, C.H.; Tsai, H.J. Sequence analysis of a duck picornavirus isolate indicates that it together with porcine enterovirus type 8 and simian picornavirus type 2 should be assigned to a new picornavirus genus. *Virus Res.* **2007**, *129*, 104–114. [CrossRef]
76. Asnani, M.; Kumar, P.; Hellen, C.U. Widespread distribution and structural diversity of Type IV IRESs in members of *Picornaviridae*. *Virology* **2015**, *478*, 61–74. [CrossRef]
77. Bourke, B.P.; Dusek, R.J.; Ergunay, K.; Linton, Y.M.; Drovetski, S.V. Viral pathogen detection in U.S. game-farm mallard (*Anas platyrhynchos*) flags spillover risk to wild birds. *Front. Vet. Sci.* **2024**, *11*, 1396552. [CrossRef]
78. Babendure, J.R.; Babendure, J.L.; Ding, J.H.; Tsien, R.Y. Control of mammalian translation by mRNA structure near caps. *RNA* **2006**, *12*, 851–861. [CrossRef]
79. Sangar, D.V.; Newton, S.E.; Rowlands, D.J.; Clarke, B.E. All foot and mouth disease virus serotypes initiate protein synthesis at two separate AUGs. *Nucleic Acids Res.* **1987**, *15*, 3305–3315. [CrossRef]
80. Pöyry, T.A.; Jackson, R.J. Mechanisms governing the selection of translation initiation sites on foot-and-mouth disease virus RNA. *J. Virol.* **2011**, *85*, 10178–10188. [CrossRef]
81. Tolf, C.; Gullberg, M.; Johansson, E.S.; Tesh, R.B.; Andersson, B.; Lindberg, A.M. Molecular characterization of a novel Ljungan virus (Parechovirus; *Picornaviridae*) reveals a fourth genotype and indicates ancestral recombination. *J. Gen. Virol.* **2009**, *90*, 843–853. [CrossRef]
82. Boros, Á.; Pankovics, P.; László, Z.; Urbán, P.; Herczeg, R.; Gáspár, G.; Tóth, F.; Reuter, G. The genomic and epidemiological investigations of enteric viruses of domestic caprine (*Capra hircus*) revealed the presence of multiple novel viruses related to known strains of humans and ruminant livestock species. *Microbiol. Spectr.* **2023**, *11*, e02533-23. [CrossRef] [PubMed]
83. Pestova, T.V.; Kolupaeva, V.G. The roles of individual eukaryotic translation initiation factors in ribosomal scanning and initiation codon selection. *Genes Dev.* **2002**, *16*, 2906–2922. [CrossRef] [PubMed]
84. Kozak, M. Downstream secondary structure facilitates recognition of initiator codons by eukaryotic ribosomes. *Proc. Natl. Acad. Sci. USA* **1990**, *87*, 8301–8305. [CrossRef] [PubMed]
85. Joseph, S.; Weiser, B.; Noller, H.F. Mapping the inside of the ribosome with an RNA helical ruler. *Science* **1997**, *278*, 1093–1098. [CrossRef]
86. Wild, K.; Aleksić, M.; Lapouge, K.; Juaire, K.D.; Flemming, D.; Pfeffer, S.; Sinning, I. MetAP-like Ebp1 occupies the human ribosomal tunnel exit and recruits flexible rRNA expansion segments. *Nat. Commun.* **2020**, *11*, 776. [CrossRef]
87. Kraushar, M.L.; Krup, F.; Harnett, D.; Turko, P.; Ambrozkiewicz, M.C.; Sprink, T.; Imami, K.; Günnigmann, M.; Zinnall, U.; Vieira-Vieira, C.H.; et al. Protein synthesis in the developing neocortex at near-atomic resolution reveals Ebp1-mediated neuronal proteostasis at the 60S tunnel exit. *Mol. Cell* **2021**, *81*, 304–322.e16. [CrossRef]
88. Bhaskar, V.; Desogus, J.; Graff-Meyer, A.; Schenk, A.D.; Cavadini, S.; Chao, J.A. Dynamic association of human Ebp1 with the ribosome. *RNA* **2021**, *27*, 411–419. [CrossRef]
89. Pestova, T.V.; Hellen, C.U.; Wimmer, E. A conserved AUG triplet in the 5′ nontranslated region of poliovirus can function as an initiation codon in vitro and in vivo. *Virology* **1994**, *204*, 729–737. [CrossRef]
90. Lulla, V.; Dinan, A.M.; Hosmillo, M.; Chaudhry, Y.; Sherry, L.; Irigoyen, N.; Nayak, K.M.; Stonehouse, N.J.; Zilbauer, M.; Goodfellow, I.; et al. An upstream protein-coding region in enteroviruses modulates virus infection in gut epithelial cells. *Nat. Microbiol.* **2019**, *4*, 280–292. [CrossRef]
91. Kovacs, E.; Tompa, P.; Liliom, K.; Kalmar, L. Dual coding in alternative reading frames correlates with intrinsic protein disorder. *Proc. Natl. Acad. Sci. USA* **2010**, *107*, 5429–5434. [CrossRef] [PubMed]
92. Sorgeloos, F.; Jha, B.K.; Silverman, R.H.; Michiels, T. Evasion of antiviral innate immunity by Theiler's virus L* protein through direct inhibition of RNase, L. *PLoS Pathog.* **2013**, *9*, e1003474. [CrossRef]
93. Drappier, M.; Jha, B.K.; Stone, S.; Elliott, R.; Zhang, R.; Vertommen, D.; Weiss, S.R.; Silverman, R.H.; Michiels, T. A novel mechanism of RNase L inhibition: Theiler's virus L* protein prevents 2-5A from binding to RNase, L. *PLoS Pathog.* **2018**, *14*, e1006989. [CrossRef] [PubMed]
94. Kuyumcu-Martinez, M.; Belliot, G.; Sosnovtsev, S.V.; Chang, K.O.; Green, K.Y.; Lloyd, R.E. Calicivirus 3C-like proteinase inhibits cellular translation by cleavage of poly(A)-binding protein. *J. Virol.* **2004**, *78*, 8172–8182. [CrossRef]
95. Royall, E.; Doyle, N.; Abdul-Wahab, A.; Emmott, E.; Morley, S.J.; Goodfellow, I.; Roberts, L.O.; Locker, N. Murine norovirus 1 (MNV1) replication induces translational control of the host by regulating eIF4E activity during infection. *J. Biol. Chem.* **2015**, *290*, 4748–4758. [CrossRef] [PubMed]
96. Svitkin, Y.V.; Costa-Mattioli, M.; Herdy, B.; Perreault, S.; Sonenberg, N. Stimulation of picornavirus replication by the poly(A) tail in a cell-free extract is largely independent of the poly(A) binding protein (PABP). *RNA* **2007**, *13*, 2330–2340. [CrossRef]
97. Svitkin, Y.V.; Herdy, B.; Costa-Mattioli, M.; Gingras, A.C.; Raught, B.; Sonenberg, N. Eukaryotic translation initiation factor 4E availability controls the switch between cap-dependent and internal ribosomal entry site-mediated translation. *Mol. Cell. Biol.* **2005**, *25*, 10556–10565. [CrossRef]
98. Lin, T.A.; Kong, X.; Haystead, T.A.; Pause, A.; Belsham, G.; Sonenberg, N.; Lawrence, J.C., Jr. PHAS-I as a link between mitogen-activated protein kinase and translation initiation. *Science* **1994**, *266*, 653–656. [CrossRef]

99. Hosmillo, M.; Chaudhry, Y.; Kim, D.S.; Goodfellow, I.; Cho, K.O. Sapovirus translation requires an interaction between VPg and the cap binding protein eIF4E. *J. Virol.* **2014**, *88*, 12213–12221. [CrossRef]
100. Walter, B.L.; Nguyen, J.H.; Ehrenfeld, E.; Semler, B.L. Differential utilization of poly(rC) binding protein 2 in translation directed by picornavirus IRES elements. *RNA* **1999**, *5*, 1570–1585. [CrossRef]
101. Kaminski, A.; Jackson, R.J. The polypyrimidine tract binding protein (PTB) requirement for internal initiation of translation of cardiovirus RNAs is conditional rather than absolute. *RNA* **1998**, *4*, 626–638. [CrossRef]
102. Asnani, M.; Pestova, T.V.; Hellen, C.U. PCBP2 enables the cadicivirus IRES to exploit the function of a conserved GRNA tetraloop to enhance ribosomal initiation complex formation. *Nucleic Acids Res.* **2016**, *44*, 9902–9917. [CrossRef]
103. Blyn, L.B.; Swiderek, K.M.; Richards, O.; Stahl, D.C.; Semler, B.L.; Ehrenfeld, E. Poly(rC) binding protein 2 binds to stem-loop IV of the poliovirus RNA 5′ noncoding region: Identification by automated liquid chromatography-tandem mass spectrometry. *Proc. Natl. Acad. Sci. USA* **1996**, *93*, 11115–11120. [CrossRef] [PubMed]
104. Beckham, S.A.; Matak, M.Y.; Belousoff, M.J.; Venugopal, H.; Shah, N.; Vankadari, N.; Elmlund, H.; Nguyen, J.H.C.; Semler, B.L.; Wilce, M.C.J.; et al. Structure of the PCBP2/stem-loop IV complex underlying translation initiation mediated by the poliovirus type I IRES. *Nucleic Acids Res.* **2020**, *48*, 8006–8021. [CrossRef] [PubMed]
105. Yu, Y.; Abaeva, I.S.; Marintchev, A.; Pestova, T.V.; Hellen, C.U. Common conformational changes induced in type 2 picornavirus IRESs by cognate trans-acting factors. *Nucleic Acids Res.* **2011**, *39*, 4851–4865. [CrossRef] [PubMed]

viruses

MDPI

Review

Emerging Roles of m7G-Cap Hypermethylation and Nuclear Cap-Binding Proteins in Bypassing Suppression of eIF4E-Dependent Translation

Kathleen Boris-Lawrie [1,*], Jessica Liebau [1], Abdullgadir Hayir [1] and Xiao Heng [2,*]

[1] Department of Veterinary and Biomedical Sciences, Institute for Molecular Virology, University of Minnesota, Saint Paul, MN 55108, USA; lieba005@umn.edu (J.L.); hayir003@umn.edu (A.H.)
[2] Department of Biochemistry, University of Missouri, Columbia, MO 65211, USA
* Correspondence: kbl@umn.edu (K.B.-L.); hengx@health.missouri.edu (X.H.)

Abstract: Translation regulation is essential to the survival of hosts. Most translation initiation falls under the control of the mTOR pathway, which regulates protein production from mono-methyl-guanosine (m7G) cap mRNAs. However, mTOR does not regulate all translation; hosts and viruses alike employ alternative pathways, protein factors, and internal ribosome entry sites to bypass mTOR. Trimethylguanosine (TMG)-caps arise from hypermethylation of pre-existing m7G-caps by the enzyme TGS1 and are modifications known for snoRNA, snRNA, and telomerase RNA. New findings originating from HIV-1 research reveal that TMG-caps are present on mRNA and license translation via an mTOR-independent pathway. Research has identified TMG-capping of selenoprotein mRNAs, junD, TGS1, DHX9, and retroviral transcripts. TMG-mediated translation may be a missing piece for understanding protein synthesis in cells with little mTOR activity, including HIV-infected resting T cells and nonproliferating cancer cells. Viruses display a nuanced interface with mTOR and have developed strategies that take advantage of the delicate interplay between these translation pathways. This review covers the current knowledge of the TMG-translation pathway. We discuss the intimate relationship between metabolism and translation and explore how this is exploited by HIV-1 in the context of CD4+ T cells. We postulate that co-opting both translation pathways provides a winning strategy for HIV-1 to dictate the sequential synthesis of its proteins and balance viral production with host cell survival.

Keywords: cap exchange; CBP80/NCBP3; DHX9/RNA helicase A-responsive structure; epigenetic modification; m2,2,7-guanosine cap; trimethylguanosine cap (TMG-cap)

1. Introduction

Historically, HIV-1 (HIV) and other retroviruses have been invaluable models to expose specialized biological processes [1–4]. Recently, the study of HIV and host mRNAs has identified nuclear cap-binding proteins that drive a specialized translation of mRNAs. This research led to the discovery of a novel pathway, licensed by hypermethylated guanosine cap, m2,2,7G-cap (TMG-cap) [5]. In all domains of life, TMG-cap has critical importance in the assembly of catalytic ribonucleoproteins (RNPs), e.g., ribosomes, spliceosomes, and telomerase [6]. TMG-cap arises from pre-existing mono-methyl guanosine (m7G)-caps, which are acted upon by the enzyme trimethylguanosine synthase (TGS1) (Figure 1a). In 2010, TMG-capping of HIV mRNA was discovered by Yedavalli and Jeang [7]. Studies showed that the HIV transcripts that contain a TMG-cap are the Rev/Rev responsive element (RRE)-dependent mRNAs [7]. In HIV, the TMG-cap is critical for effective viral replication. TGS1 downregulation by siRNA diminishes virion production to 10% of control. Even more significant, the progeny virions are poorly infectious on CD4+ T cells. Compared to control virions of the same quantity, their infectivity is reduced by a factor of 1000 [5]. Therefore, TGS1 activity significantly affects HIV production and progeny virion infectivity.

The discovery of TMG-cap in HIV mRNA added to the previous discoveries of TMG-cap in alphaviruses, which are responsible for vector-borne diseases (Figure 1b); the virological significance of the TMG-cap remains to be determined [8,9].

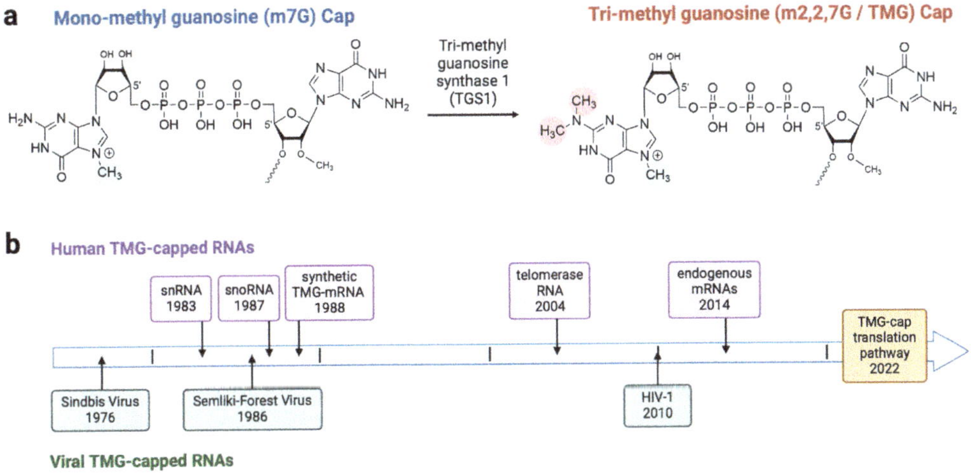

Figure 1. Discovery timeline of TMG-capped RNAs in humans and viruses. (**a**) Chemical structures of mono-methyl guanosine (m7G)- and trimethylguanosine (m2,2,7G, TMG)-caps attached to the transcription start site. Pre-existing m7G-caps are hypermethylated to m2,2,7G-cap by trimethylguanosine synthase 1 (TGS1). Methyl groups: blue, at the 7-position of the m7G-cap; red, at the 2,2 positions of the TMG-cap. (**b**) Timeline of publications. Human RNAs: snRNA in 1983 [10–12] and snoRNA in 1987 [13,14]; in vitro translation of synthetic TMG-mRNA in 1988 [15]; telomerase RNA in 2004 [16]; host mRNAs in 2014 [17]. Viral RNAs: Sindbis in 1976 [8]; Semliki-Forest virus in 1986 [9]; HIV-1 in 2010 [7]. Specialized translation licensed by TMG-cap was characterized in 2022 [5].

A persisting barrier to the complete understanding of HIV post-transcriptional regulation has been the role of the viral complex 5′ untranslated region (UTR) in controlling the biogenesis of the complicated viral proteome [18]. The 5′UTR contains many structural domains that coordinate critical events throughout the viral replication cycle. Thermodynamic equilibrium exists between two mutually exclusive conformations of the 5′UTR, which migrate as a monomer or dimer on native gel electrophoresis [19]. Chemical probing, enzymatic probing, and NMR studies have identified distinct base pairings in the monomeric and dimeric 5′UTR conformations [20]. Recent studies have shown that transcription start site heterogeneity significantly affects the 5′UTR conformation and thus the monomer:dimer equilibrium, positing an additional variable in the complicated life of the multi-functional HIV transcript [20,21].

In solution, ribosomal scanning of HIV mRNA structures poses a significant energetic barrier, so the selection of RNAs to translate remains a longstanding issue. The role of ribosome scanning in HIV was made clear by the insertion of upstream AUG codons (uAUGs) at different locations within the HIV sequences. The uAUGs were severely inhibitory to translation in cells and supportive of ribosome scanning rather than internal ribosome entry or ribosomal shunting [22]. Additionally, there is ample evidence of non-canonical ribosome recruitment.

This article reviews the major host translation pathways adapted by HIV. We discuss the evidence that dual control of translation initiation, virus-induced stress, and the mTOR pathway are consequential throughout viral replication. We conclude that HIV-specialized translation likely influences the survival of HIV-infected CD4+ cells.

2. Overview of TMG-Cap Biogenesis in Human Hosts

2.1. CBC Is a Central Hub for the Maturation of All Pol II Transcripts

RNA polymerase II (Pol II) synthesizes protein-coding transcripts and an abundance of host noncoding RNAs, in addition to retroviral transcripts. The RNAs are processed in a co-transcriptional manner. First, all transcription start sites are modified by the addition of a mono-methylated guanosine (m7G)-cap [23,24]. This guanosine is appended in an atypical 5′ to 5′ linkage and bound by heterodimeric CBP20/CBP80, forming the nuclear cap-binding complex (CBC). The CBC RNA-protein complex tethers the nascent transcript to proteins that carry out splicing and 3′-end processing during the transcription cycle [25,26]. The mature transcript is prepared for CBC replacement by another m7G-cap-binding protein, e.g., TGS1 or eIF4E [27].

2.2. Effects of TMG-Cap on Small Nuclear (snRNA), Small Nucleolar RNAs (snoRNAs), and Telomerase RNA (hTR)

The CBC of uridine-rich small noncoding RNAs exchangees to TGS1, as does human telomerase RNA, hTR. TMG-cap facilitates the nuclear retention of snoRNA, which assembles in nucleolar aggregates and assist in ribosome assembly, e.g., U3 snoRNA [28,29]. CRM1 inactivation by leptomycin B (LMB) was shown to trap TGS1 in a complex with m7G-snoRNA [30]. The authors identified that recombinant TGS1 interacted with CRM1 in a Ran-GTP-dependent manner. They postulated that the direct binding of CRM1 to TGS1 promotes TMG-U3 snoRNAs trafficking to the nucleolus [30].

CBC-bound m7G-snRNA recruits ARS2 and PHAX for nuclear export to the cytoplasm via CRM1/RanGTP, where it interacts with SmB and the survival motor neuron (SMN) complex to replace CBC with TGS1. The resulting hypermethylated cap of snRNA is bound by Snurportin, the nuclear import receptor that facilitates the re-import of TMG-snRNA into the nucleus and trafficking to the nucleolus [31–34].

The noncoding telomerase component-1 RNA (hTR) also acquires a TMG-cap via TGS1 [16,35]. Capping occurs in the nucleus, followed by CRM1-mediated export to the cytoplasm [36]. While not essential for telomerase activity, the TMG-cap of hTR is important for the recruitment of telomerase to telomeres and for the engagement of nuclear Cajal bodies [35]. Failure ofTMG-capping leads to retention of CBC, increased binding of these transcripts to Sm and TERT, and activation of the alternative pathway of telomere maintenance. Also increased telomere elongation is an outcome of diminishing TMG-capping [36]. These emerging studies indicate that TMG-cap is crucial for the regulation of telomere maintenance and telomerase activity.

3. Overview of Major mRNA Translation Pathways

3.1. In Metabolically Active Cells, the m7G-eIF4E Pathway Is the Predominant Translation Initiation Mechanism

CBC exchange to the eIF4E m7G-cap-binding protein has been considered a rate-limiting step in translation control [37,38]. eIF4E activates the recruitment of eIF4G and eIF4A, and together these proteins comprise the tripartite eIF4F complex (Figure 2a). eIF4F joining to a 43S small ribosomal subunit activates the scanning of the 5′UTR for a suitable start codon; the 60S ribosomal subunit joins to form the 80S complex, and polypeptide synthesis begins.

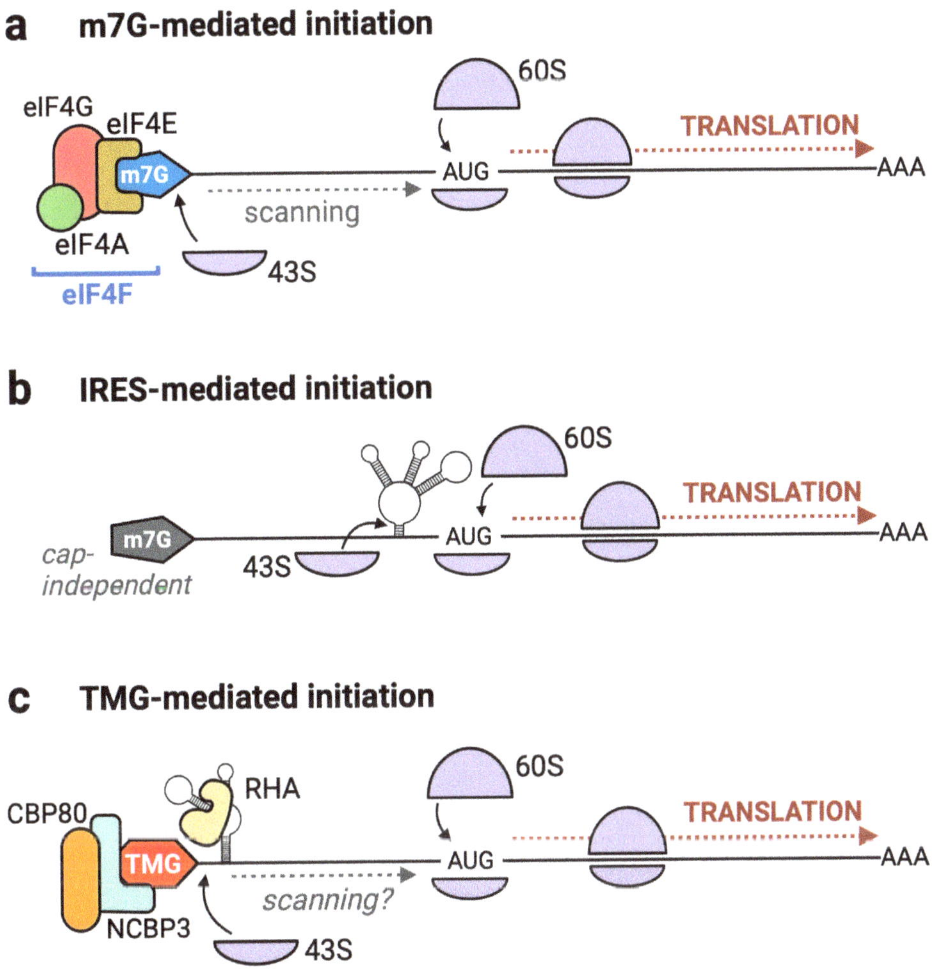

Figure 2. Pathways for host translation initiation. (**a**) eIF4E binds the m7G-cap, which recruits eIF4G and eIF4A to form eIF4F. The 43S ribosome is recruited, scans the mRNA to find a start codon, and then recruits the 60S ribosomal subunit to form the 80S ribosome. (**b**) The internal ribosome entry sequence (IRES) within the 5′UTR experiences direct binding of 43S ribosome, followed by 60S ribosome joining. (**c**) TMG-mediated translation requires RNA helicase A/DHX9 (RHA) recognition of a three-way junction structure within the HIV 5′UTR for retention of the NCBP3/CBP80 heterodimer. Following 43S recruitment, either through ribosome scanning or internal ribosome entry, the 60S subunit joins to form the 80S complex, which initiates protein synthesis.

3.2. Signal Transduction via mTOR Connects Metabolism with Global Translation

Early studies of translation control were focused on nutrient-rich conditions that enable cell growth and proliferation, e.g., the G1/S phase of the cell cycle [38]. But just as important to cell health and resilience is translation during nutrient-depleted conditions.

When resources are scarce, eIF4E binding proteins (4EBPs) bind eIF4E to block eIF4F assembly, repressing translation [39]. When amino acids, tRNA, and energy resources are high, the serine-threonine kinase mTOR phosphorylates 4EBPs to repress their activity, which allows protein translation to occur using the eIF4E pathway [40]. The term "global translation" refers to mTOR-driven protein synthesis, which accounts for ~65% of the total cellular translation [41–45].

mTOR control of eIF4E-mediated translation is critical for the host's response to metabolic or environmental stress. Additionally, mTOR is intimately involved in cell cycle progression. During G1 and S, mTOR activity is high, 4EBP is repressed, and eIF4E translation occurs. G2 and M coincide with mTOR inhibition, 4EBP activation, and eIF4E translation abrogation. 4EBP activity thus serves as a checkpoint for the transition between these two metabolic states [38–40,46].

Logically, alternative pathways to initiate translation seem necessary: how else would cells adapt during environmental stress that inactivates eIF4E translation? How would cells maintain short-lived proteins for mitosis or resume eIF4E translation? In the case of Merkel cell polyomavirus, alternative means are used to hyper-phosphorylate 4EBP, which allows for eIF4E-mediated translation even when mTOR is inactive [47]. Some host mRNAs use alternative m7G-binding proteins to initiate translation. These include the DAP5/p97 isoform of eIF4G [48–52], FXR1a/PARN [53,54], eIF3d [55], and CTIF [56]. Some RNA viruses even antagonize eIF4F to suppress host translation yet maintain viral translation, e.g., picornaviruses. The predominant explanation for the continued translation is internal translation initiation on specific mRNAs.

3.3. *Internal Ribosome Entry in Animal Viruses Overcomes Viral Antagonism of eIF4F*

Internal ribosome entry is the targeted recruitment of the 43S ribosome to a specific region of an RNA known as the internal ribosome entry sequence (IRES) (Figure 2b). Several types of IRES exist. Typically, IRES motifs form complex structures for direct ribosome binding that are independent of mRNA caps. As presented in this Special Issue, elegant biochemical studies have classified the events for IRES translation in viruses [57]. For detailed information, the reader is referred to comprehensive reviews on IRES translation [18,58–60].

For HIV, evidence has been presented in favor of ribosome recruitment within the 5′UTR and gag coding region [18,61,62]. IRES activity in a bicistronic context was shown to be resistant to conditions that specifically inhibit eIF4F-dependent translation. These conditions include the presence of picornaviral 2A, HIV protease, or viral infections, all of which downregulate eIF4F activity [63,64]. IRES activity conveyed by the HIV 5′UTR was shown to be increased in cell cycle-arrested lysates [62,65,66]. Studies identified that HIV IRES activity exhibits cell-type specificity, with four-fold more reporter activity observed in Jurkat T cells compared to the HeLa fibroblast cell line [67].

3.4. *HIV IRES Activity Is Measured in Bicistronic Reporter RNAs*

Gendron et al. studied the HIV 5′UTR residues (+1–369) inclusive of the gag AUG and 30 nt of the gag sequence in transfected Jurkat T cells [68]. HIV 5′UTR residues +104 to 336 have IRES activity in bicistronic reporter constructs. Their investigation compared IRES activity between HIV and the hepatitis C virus (HCV). It was found that the HIV IRES produced only 20% as much translation as the HCV-positive control. HIV IRES activity increased 2-fold in response to the induction of oxidative stress, a condition that is relevant to the stress caused by HIV infection.

Deleting a portion of the IRES (nt +158–167) increased activity 2-fold; the authors speculated this was due to an internal ribosome entry negative element (IRENE) [68]. Notably, this same region was later shown to encompass a three-way junction structure necessary for HIV TMG-capping in immune cells [5]. The A140C structural mutation converts the three-way junction to a coaxial stem-loop structure that eliminates viral protein translation when eIF4E is inactive [69]. This outcome is the opposite of the IRENE phenotype identified in the bicistronic reporter mRNA by Gendron et al. [68], implicating that different molecular interactions operate in the context of the viral RNA and reporter RNA. We suggest studies are warranted to reconcile the different molecular interactions operating in the context of the TMG-capped viral RNA and the 5′UTR within a bicistronic reporter RNA.

4. RNA Helicase A (RHA) and NCBP3 as Novel Actors Within the TMG-Translation Pathway

4.1. RHA Is Necessary for Translation of Select mRNAs

Structural, biochemical, and genetic studies of the HIV 5′UTR revealed that the three-way junction structure (nt +135 to 235) is responsive to RNA helicase A (RHA/DHX9) [69]. Prior studies identified that RHA binds the HIV 5′UTR to facilitate translation and virion infectivity [69–71]. RHA downregulation or the A140C mutation redirects these transcripts to the mTOR-affected eIF4E-m7G-cap translation pathway, as shown in Singh et al. 2022 (Figures 4C and S6; Table S3 of reference [5]). Rev/RRE-dependent HIV RNAs retain RHA binding, whereas fully processed Rev/RRE-independent transcripts do not [5].

The RHA translation mechanism may be conserved across the *Retroviridae*. The 5′UTRs of HIV, human T cell leukemia virus type 1 (HTLV-1), spleen necrosis virus (SNV), feline leukemia virus, Mason-Pfizer monkey virus, and reticuloendotheliosis virus-A (RelA) are interchangeable for translation activity in HIV unspliced reporter RNA [71–75]. The 5′UTR sequences fail to support IRES activity in bicistronic RNAs, but HTLV-1, RelA, and SNV allow for translation during picornavirus-induced eIF4E downregulation [73]. SNV 5′UTR studies were the first to identify that RHA interaction with structural motifs of the retrovirus 5′UTRs bolsters translation [70,72,73,76].

Just as RHA binds the 5′UTR structure in retroviruses, it also associates with the 5′UTR of endogenous RNAs. Pull-down experiments identified that junD mRNA were among those that coprecipitated with RHA [70]. The JUND protein is an AP-1 transcription factor that is translated from unspliced mRNA. It has a complex 5′UTR but lacks IRES activity [77]. The identity of its non-canonical pathway was a longstanding gap in knowledge [78].

4.2. TMG-Cap Facilitates Non-Canonical Translation

In addition to junD, select selenoprotein mRNAs are translated independently of m7G-eIF4E activity, e.g., selT (thioredoxin reductase-like enzyme) and selR (methionine-R-sulfoxide reductase B1) [5,79]. The concept of TMG-cap on selenoprotein mRNAs arose from the work of Allmang and colleagues [17]. m7G-cap selenoprotein transcripts coprecipitate with eIF4E, whereas those with TMG-caps do not [17]. As a group, selenium-containing proteins are best known for their redox activity and anti-inflammatory properties, and their translation is stimulated by oxidative stress [80]. Selenoprotein mRNA is so named for its incorporation of selenocysteine (Sec), the 21st amino acid [81].

These transcripts harbor UGA termination codons that can be recoded to Sec codons. UGA recoding to the Sec codon requires the activity of the Sec insertion sequence (SECIS) in the 3′ UTR and the SECIS-binding protein (SBP2). The steps required for the Sec insertion begin with SBP2/SECIS interaction in the nucleus and CRM1-dependent nuclear export [82].

A later investigation of junD identified that it is also TMG-capped, which licenses its non-canonical translation. Furthermore, junD does not incorporate into RNPs that contain eIF4E; instead, it requires a TMG-cap for translation and associates with RNPs containing NCBP3/CBP80-RHA (Figure 2c) [83]. The TMG-capped selenoprotein mRNAs selT and selR are also associated with these distinct RNPs [5,79].

TGS1 was identified as a host protein that binds the Rev/RRE-dependent HIV RNA, and these transcripts obtain a TMG-cap [5,7,84]. These RNAs retained CBP80 in polysomes, whereas fully processed, Rev/RRE-independent HIV RNA was associated with eIF4E-containing polysomes [5]. RHA downregulation by siRNA eliminates TMG-cap from the viral transcripts, as well as junD. Co-immunoprecipitation assays targeting TMG-capped transcripts have shown RHA binding within the HIV and junD 5′UTR subverts exchange of CBC for eIF4E and is necessary for the formation of NCBP3/CBP80-TMG-mRNPs [70,71,77,83,85].

The biological results support the following process to assemble TMG-mRNPs (Figure 3): (1) CBP20/80 binds m7G-capped mRNA to form CBC; (2) RHA tethers TGS1 near m7G-cap for its hypermethylation to TMG-cap; (3) TGS1's cap-binding pocket cannot accommodate the new TMG-cap; and (4) NCBP3 replaces TGS1 to bind CBP80-associated TMG-cap.

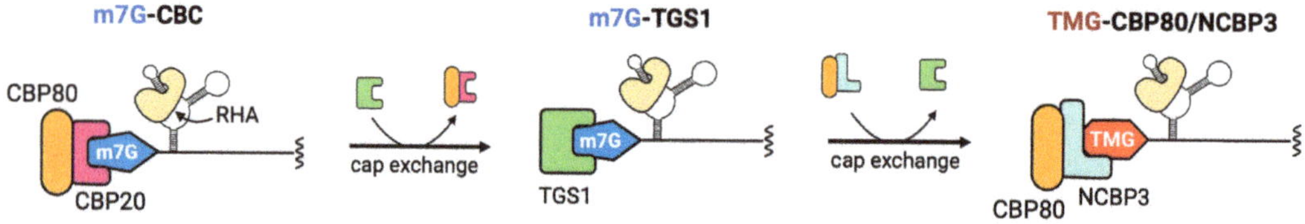

Figure 3. Model for hypermethylation of HIV m7G-cap to m2,2,7G TMG-cap. The m7G-cap-binding complex (CBC) is composed of heterodimeric CBP80 (orange) and CBP20, which directly binds m7G-cap (mauve). RNA helicase A/DHX9 (RHA, tan) recognizes a three-way junction structure within the HIV 5′ untranslated region and facilitates the exchange of CBC to TGS1 (green). The m7G-TGS1 complex is shown with RHA binding the three-way junction for m7G hypermethylation to m2,2,7G (red, TMG-cap). The TGS1 binding pocket does not accommodate the bulkier TMG structure and is released. The TMG-CBP80/NCBP3 complex results from the cap exchange to NCBP3 (blue) and CBP80.

4.3. *Structural Studies Have Helped to Elucidate the Formation of These RNPs*

TMG-cap has poor affinity toward m7G-cap-binding proteins eIF4E and CBP20/CBP80. Their binding of m7G-cap has been characterized as an aromatic sandwich motif that traps the guanosine in π-π stacking interactions [86]. Snurportin1, the snRNP-specific import adaptor protein, is the only known TMG-cap-binding protein [23]. Computational modeling of NCBP3 identified similarities to the cap-binding pocket of Snurportin [87]. Snurportin has been shown to bind the TMG-cap through π-π interactions between an aromate and the RNA's transcription start site. CBP80/NCBP3 binding the TMG-cap is likewise proposed between the NCBP3 aromate and the RNA's transcription start site through larger π-π stacking [87].

Structural studies help to understand the arrangement of NCBP3 in CBC RNPs. A 3.2 Å resolution cryoEM structure of the CBC in complex with the NCBP3 C-term domain (560–620) places the α-helical residues 574–594 of NCBP3 against CBP80 at its MIF4G domain [88]. One hydrogen bond between R105 of CBP20 and D574 of NCBP3 was observed in the CBP20/80-NCBP3 complex, although CBP80 alone can pull down NCBP3 [89]. Together, these data support that the CBP80/CBP20-NCBP3 tripartite complex represents a transition state when the m7G-cap is hypermethylated to TMG-cap and retains heterodimeric CBP80/NCBP3.

5. Adaptive Translation: Dual Initiation Strategies and eIF4E Repression

5.1. HIV-1 mRNAs Are Efficiently Translated When eIF4E Is Inactivated and Translation of Hosts Is Attenuated

In newly infected cells, proviral transcripts are fully processed, which removes the RRE motif. These RNAs, which encode the regulatory proteins Tat and Rev, retain their m7G-cap and are translated by the eIF4E pathway [90]. Tat binds the Transactivation Response (TAR) element to trans-activate proviral transcription. On the other hand, Rev binds the HIV RRE motif and leads to an increase in HIV RNA containing this element. Rev/RRE trans-activates the nuclear export of these RNAs in a CRM1-dependent manner [91]. The Rev/RRE-dependent transcripts template the synthesis of viral accessory proteins that modify virus–host interplay (Vpr, Vif, Vpu) and structural proteins that are components of progeny virions (Gag, Gag-Pol, Env).

Viral protein R (Vpr) activates 4EBP to repress eIF4E activity, which inhibits host protein translation by 65%, as well as the translation of Rev/RRE-independent transcripts, yet HIV virion protein synthesis is maintained [92]. Rev/RRE-dependent transcripts display an unexpected property; rather than obtaining eIF4E, these RNAs retain CBP80 in polysomes and thus use a different translation pathway [92]. This led to the hypothesis that the CBP80-containing polysomes coincide with viral protein synthesis during eIF4E inactivation. By altering the presence of RRE, HIV can utilize dual translation initiation strategies.

HIV translation thus adapts to the host cell environment to balance viral protein expression with cell survival. How HIV contributes to the metabolic reprogramming of hosts has been an active area of research [93–105]. Increased metabolic activity engenders T cells and monocytes that are more susceptible to HIV-1 infection [100,106].

5.2. Translation Machinery of Hosts Is Fundamentally Responsive to Metabolism Shifts

New findings demonstrate that HIV co-opts the interplay of m7G- and TMG-translation pathways to moderate viral protein production in favor of host cell survival. HIV requires host resources and translation mechanisms to synthesize its proteins. Early replication events, including reverse transcription, integration, and provirus transcription [96,107–114], require mTOR activity present in activated T cells.

The transition of Tcells from Effector to Memory to Reactivated states coincides with mTOR inhibition and restoration, respectively, of the eIF4E-m7G translation pathway [115]. Experiments have documented that Tat, Rev, and Nef protein synthesis in CD4+ T cells is curtailed by supplementation with an mTOR inhibitor and restored by the replacement of fresh medium [5,83]. These results indicate that virion protein synthesis can continue as T cells transition to a memory state, even when viral regulatory proteins turnover and transactivation of new proviral transcription is curtailed.

6. Conclusions

To produce its complicated proteome, HIV utilizes both mTOR-dependent and -independent strategies for translation initiation. While IRES is generally considered the alternative to the mTOR pathway of protein synthesis, new evidence indicates that the novel TMG-cap translation pathway is a new explanation for translation that is sustained when eIF4F is inactive.

This article has presented evidence that the TMG-cap pathway is a critical component of biphasic HIV protein synthesis. Since HIV fully processed mRNAs are m7G-capped and licensed for translation by eIF4F mRNPs, regulatory proteins cease translation when eIF4E is inactive. TMG-capping of the Rev/RRE-dependent RNAs licenses assembly of a novel translation mRNP composed of NCBP3/CBP80 and RHA. This specialized translation pathway ensures virion protein translation when eIF4E is inactive. We postulate that the interplay between these translation pathways strategically moderates HIV regulatory protein production to tune infectious virion production to favor host immune cell survival.

The characterization of other components of the NCBP3/CBP80-RHA mRNP that is ongoing will provide clues to a comprehensive understanding of the translation of host TMG-capped mRNAs. Soto-Rifo and colleagues identified that CBP80 interaction with Rev promotes nuclear export and translation of the HIV unspliced RNA [116]. Rev also associates with CTIF, the CBP80/20-dependent translation initiation factor associated with nonsense-mediated decay [117]. The CTIF overexpression inhibits Gag synthesis in a dominant negative manner [116]. We speculate that NCBP3 may take the place of CTIF to bridge Rev-CBP80 and the 43S ribosomal subunit for steady state translation unaffected by mTOR [118]. With the development of ribosome profiling of HIV mRNAs [119–121], the relationship between the dual translation mechanisms licensed by TMG-cap or m7G-cap can now be investigated with great precision.

Some people living with HIV exhibit significantly reduced plasma concentrations of essential amino acids relative to healthy controls ($p < 0.05$), indicative of metabolic stress [122]. Low resource abundance leads to mTOR inhibition, and hosts abrogate eIF4F-dependent translation while specialized translation continues (Figure 4). A new perspective on viral pathogenesis is emerging that HIV-induced metabolism shifts and gene expression remodels can moderate viral production in tune with the immune cell population susceptible to infection [100].

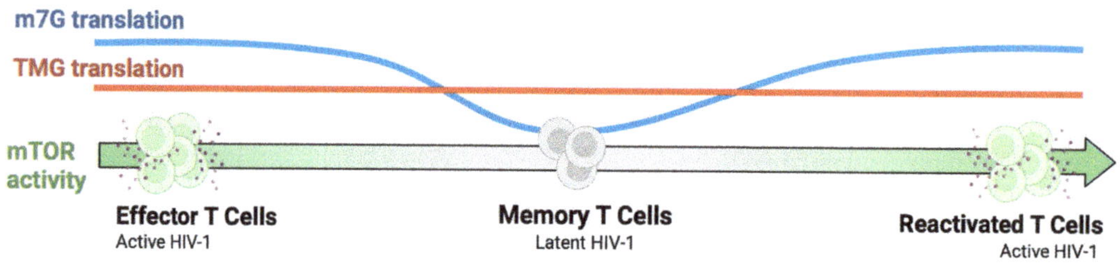

Figure 4. Model for the HIV translation shift across T cell states. mTOR activity supports provirus formation, transcription and processing of viral transcripts. Translation of m7G-capped mRNA (blue) is synchronous with mTOR activity, whereas TMG-capped viral and host mRNAs engage in constitutive translation (red). As metabolic resources decline, mTOR activity wanes (green arrow transition to gray) and Effector T cells progress toward a memory state. Suppressed translation of HIV m7G-capped mRNAs to regulatory proteins diminishes new viral gene expression, diminishing capacity for virion production. Constitutive translation of residual TMG-capped Rev/RRE-dependent viral mRNAs prolongs the generation of virions despite mTOR inhibition, seeding the latent viral reservoir. Restored mTOR activity in reactivated Effector T cells reactivates productive viral infection and dissemination. Gray dots, progeny virions.

Funding: This research was supported by the National Institutes of Health R01AI179590, R21AI167742, and R56AI179590 to K.B.-L. and X.H.

Institutional Review Board Statement: Not applicable.

Informed Consent Statement: Not applicable.

Data Availability Statement: New data were not part of this manuscript.

Conflicts of Interest: The authors declare no conflicts of interest. The funders had no role in the design or in the writing of the manuscript.

References

1. Temin, H.M.; Mizutani, S. RNA-dependent DNA polymerase in virions of Rous sarcoma virus. *Nature* **1970**, *226*, 1211–1213. [CrossRef]
2. Baltimore, D. RNA-dependent DNA polymerase in virions of RNA tumour viruses. *Nature* **1970**, *226*, 1209–1211. [CrossRef] [PubMed]
3. Coffin, J.M.; Hughes, S.H.; Varmus, H.E. *Retroviruses*; Springer: Berlin/Heidelberg, Germany, 1997.
4. Bray, M.; Prasad, A.; Dubay, J.W.; Hunter, E.; Jeang, K.T.; Rekosh, D.; Hammarskjöld, M.L. A small element from the Mason-Pfizer monkey virus genome makes human immunodeficiency virus type 1 expression and replication Rev-independent. *Proc. Natl. Acad. Sci. USA* **1994**, *91*, 1256–1260. [CrossRef]
5. Singh, G.; Seufzer, B.; Song, Z.; Zucko, D.; Heng, X.; Boris-Lawrie, K. HIV-1 hypermethylated guanosine cap licenses specialized translation unaffected by mTOR. *Proc. Natl. Acad. Sci. USA* **2022**, *119*, e2105153118. [CrossRef] [PubMed]
6. Ohtani, M. Plant snRNP Biogenesis: A Perspective from the Nucleolus and Cajal Bodies. *Front. Plant Sci.* **2017**, *8*, 2184. [CrossRef]
7. Yedavalli, V.S.; Jeang, K.T. Trimethylguanosine capping selectively promotes expression of Rev-dependent HIV-1 RNAs. *Proc. Natl. Acad. Sci. USA* **2010**, *107*, 14787–14792. [CrossRef] [PubMed]
8. HsuChen, C.C.; Dubin, D.T. Di-and trimethylated congeners of 7-methylguanine in Sindbis virus mRNA. *Nature* **1976**, *264*, 190–191. [CrossRef]
9. van Duijn, L.P.; Kasperaitis, M.; Ameling, C.; Voorma, H.O. Additional methylation at the N(2)-position of the cap of 26S Semliki Forest virus late mRNA and initiation of translation. *Virus Res.* **1986**, *5*, 61–66. [CrossRef]
10. Bringmann, P.; Reuter, R.; Rinke, J.; Appel, B.; Bald, R.; Lührmann, R. 5′-terminal caps of snRNAs are accessible for reaction with 2,2,7-trimethylguanosine-specific antibody in intact snRNPs. *J. Biol. Chem.* **1983**, *258*, 2745–2747. [CrossRef]
11. Smith, J.H.; Eliceiri, G.L. Antibodies elicited by N2,N2-7-trimethylguanosine react with small nuclear RNAs and ribonucleoproteins. *J. Biol. Chem.* **1983**, *258*, 4636–4640. [CrossRef]
12. Bringmann, P.; Rinke, J.; Appel, B.; Reuter, R.; Lührmann, R. Purification of snRNPs U1, U2, U4, U5 and U6 with 2,2,7-trimethylguanosine-specific antibody and definition of their constituent proteins reacting with anti-Sm and anti-(U1)RNP antisera. *EMBO J.* **1983**, *2*, 1129–1135. [CrossRef] [PubMed]
13. Parker, K.A.; Steitz, J.A. Structural analysis of the human U3 ribonucleoprotein particle reveal a conserved sequence available for base pairing with pre-rRNA. *Mol. Cell. Biol.* **1987**, *7*, 2899–2913.
14. Tyc, K.; Steitz, J.A. U3, U8 and U13 comprise a new class of mammalian snRNPs localized in the cell nucleolus. *EMBO J.* **1989**, *8*, 3113–3119. [CrossRef] [PubMed]
15. Darzynkiewicz, E.; Stepinski, J.; Ekiel, I.; Jin, Y.; Haber, D.; Sijuwade, T.; Tahara, S.M. Beta-globin mRNAs capped with m7G, m2.7(2)G or m2.2.7(3)G differ in intrinsic translation efficiency. *Nucleic Acids Res.* **1988**, *16*, 8953–8962. [CrossRef]
16. Jády, B.E.; Bertrand, E.; Kiss, T. Human telomerase RNA and box H/ACA scaRNAs share a common Cajal body-specific localization signal. *J. Cell Biol.* **2004**, *164*, 647–652. [CrossRef]
17. Wurth, L.; Gribling-Burrer, A.S.; Verheggen, C.; Leichter, M.; Takeuchi, A.; Baudrey, S.; Martin, F.; Krol, A.; Bertrand, E.; Allmang, C. Hypermethylated-capped selenoprotein mRNAs in mammals. *Nucleic Acids Res.* **2014**, *42*, 8663–8677. [CrossRef]
18. Guerrero, S.; Batisse, J.; Libre, C.; Bernacchi, S.; Marquet, R.; Paillart, J.C. HIV-1 replication and the cellular eukaryotic translation apparatus. *Viruses* **2015**, *7*, 199–218. [CrossRef]
19. Berkhout, B.; van Wamel, J.L. The leader of the HIV-1 RNA genome forms a compactly folded tertiary structure. *RNA* **2000**, *6*, 282–295. [CrossRef]
20. Lu, K.; Heng, X.; Garyu, L.; Monti, S.; Garcia, E.L.; Kharytonchyk, S.; Dorjsuren, B.; Kulandaivel, G.; Jones, S.; Hiremath, A.; et al. NMR detection of structures in the HIV-1 5′-leader RNA that regulate genome packaging. *Science* **2011**, *334*, 242–245. [CrossRef]
21. Brown, J.D.; Kharytonchyk, S.; Chaudry, I.; Iyer, A.S.; Carter, H.; Becker, G.; Desai, Y.; Glang, L.; Choi, S.H.; Singh, K.; et al. Structural basis for transcriptional start site control of HIV-1 RNA fate. *Science* **2020**, *368*, 413–417. [CrossRef] [PubMed]
22. Berkhout, B.; Arts, K.; Abbink, T.E. Ribosomal scanning on the 5′-untranslated region of the human immunodeficiency virus RNA genome. *Nucleic Acids Res.* **2011**, *39*, 5232–5244. [CrossRef]
23. Boris-Lawrie, K.; Singh, G.; Osmer, P.S.; Zucko, D.; Staller, S.; Heng, X. Anomalous HIV-1 RNA, How Cap-Methylation Segregates Viral Transcripts by Form and Function. *Viruses* **2022**, *14*, 935. [CrossRef]
24. Kataoka, N. The Nuclear Cap-Binding Complex, a multitasking binding partner of RNA polymerase II transcripts. *J. Biochem.* **2023**, *175*, 9–15. [CrossRef] [PubMed]
25. Gonatopoulos-Pournatzis, T.; Cowling, V.H. Cap-binding complex (CBC). *Biochem. J.* **2014**, *457*, 231–242. [CrossRef]
26. Rambout, X.; Maquat, L.E. The nuclear cap-binding complex as choreographer of gene transcription and pre-mRNA processing. *Genes. Dev.* **2020**, *34*, 1113–1127. [CrossRef]
27. Giacometti, S.; Benbahouche, N.E.H.; Domanski, M.; Robert, M.C.; Meola, N.; Lubas, M.; Bukenborg, J.; Andersen, J.S.; Schulze, W.M.; Verheggen, C.; et al. Mutually Exclusive CBC-Containing Complexes Contribute to RNA Fate. *Cell Rep.* **2017**, *18*, 2635–2650. [CrossRef] [PubMed]

28. Verheggen, C.; Bertrand, E. CRM1 plays a nuclear role in transporting snoRNPs to nucleoli in higher eukaryotes. *Nucleus* **2012**, *3*, 132–137. [CrossRef]
29. Pradet-Balade, B.; Girard, C.; Boulon, S.; Paul, C.; Azzag, K.; Bordonné, R.; Bertrand, E.; Verheggen, C. CRM1 controls the composition of nucleoplasmic pre-snoRNA complexes to licence them for nucleolar transport. *EMBO J.* **2011**, *30*, 2205–2218. [CrossRef]
30. Lemm, I.; Girard, C.; Kuhn, A.N.; Watkins, N.J.; Schneider, M.; Bordonné, R.; Lührmann, R. Ongoing U snRNP biogenesis is required for the integrity of Cajal bodies. *Mol. Biol. Cell* **2006**, *17*, 3221–3231. [CrossRef]
31. Franke, J.; Gehlen, J.; Ehrenhofer-Murray, A.E. Hypermethylation of yeast telomerase RNA by the snRNA and snoRNA methyltransferase Tgs1. *J. Cell Sci.* **2008**, *121 Pt 21*, 3553–3560. [CrossRef]
32. Hausmann, S.; Ramirez, A.; Schneider, S.; Schwer, B.; Shuman, S. Biochemical and genetic analysis of RNA cap guanine-N2 methyltransferases from Giardia lamblia and Schizosaccharomyces pombe. *Nucleic Acids Res.* **2007**, *35*, 1411–1420. [CrossRef] [PubMed]
33. Hausmann, S.; Zheng, S.; Costanzo, M.; Brost, R.L.; Garcin, D.; Boone, C.; Shuman, S.; Schwer, B. Genetic and biochemical analysis of yeast and human cap trimethylguanosine synthase: Functional overlap of 2,2,7-trimethylguanosine caps, small nuclear ribonucleoprotein components, pre-mRNA splicing factors, and RNA decay pathways. *J. Biol. Chem.* **2008**, *283*, 31706–31718. [CrossRef] [PubMed]
34. Buemi, V.; Schillaci, O.; Santorsola, M.; Bonazza, D.; Broccia, P.V.; Zappone, A.; Bottin, C.; Dell'Omo, G.; Kengne, S.; Cacchione, S.; et al. TGS1 mediates 2,2,7-trimethyl guanosine capping of the human telomerase RNA to direct telomerase dependent telomere maintenance. *Nat. Commun.* **2022**, *13*, 2302. [CrossRef]
35. Gallardo, F.; Olivier, C.; Dandjinou, A.T.; Wellinger, R.J.; Chartrand, P. TLC1 RNA nucleo-cytoplasmic trafficking links telomerase biogenesis to its recruitment to telomeres. *EMBO J.* **2008**, *27*, 748–757. [CrossRef] [PubMed]
36. Chen, L.; Roake, C.M.; Galati, A.; Bavasso, F.; Micheli, E.; Saggio, I.; Schoeftner, S.; Cacchione, S.; Gatti, M.; Artandi, S.E.; et al. Loss of Human TGS1 Hypermethylase Promotes Increased Telomerase RNA and Telomere Elongation. *Cell Rep.* **2020**, *30*, 1358–1372.e5. [CrossRef]
37. Mars, J.C.; Ghram, M.; Culjkovic-Kraljacic, B.; Borden, K.L.B. The Cap-Binding Complex CBC and the Eukaryotic Translation Factor eIF4E: Co-Conspirators in Cap-Dependent RNA Maturation and Translation. *Cancers* **2021**, *13*, 6185. [CrossRef]
38. Sonenberg, N.; Hinnebusch, A.G. Regulation of translation initiation in eukaryotes: Mechanisms and biological targets. *Cell* **2009**, *136*, 731–745. [CrossRef]
39. Xie, Y.; Lei, X.; Zhao, G.; Guo, R.; Cui, N. mTOR in programmed cell death and its therapeutic implications. *Cytokine Growth Factor. Rev.* **2023**, *71*, 66–81. [CrossRef]
40. Fritz, S.; Boris-Lawrie, K. The RNPs of eukaryotic translation control. *Res. Trends Cell Mol. Biol.* **2015**, *10*, 105–130.
41. Cinti, A.; Le Sage, V.; Milev, M.P.; Valiente-Echeverría, F.; Crossie, C.; Miron, M.J.; Panté, N.; Olivier, M.; Mouland, A.J. HIV-1 enhances mTORC1 activity and repositions lysosomes to the periphery by co-opting Rag GTPases. *Sci. Rep.* **2017**, *7*, 5515. [CrossRef]
42. Alard, A.; Katsara, O.; Rios-Fuller, T.; Parra, C.; Ozerdem, U.; Ernlund, A.; Schneider, R.J. Breast cancer cell mesenchymal transition and metastasis directed by DAP5/eIF3d-mediated selective mRNA translation. *Cell Rep.* **2023**, *42*, 112646. [CrossRef]
43. Roiuk, M.; Neff, M.; Teleman, A.A. eIF4E-independent translation is largely eIF3d-dependent. *Nat. Commun.* **2024**, *15*, 6692. [CrossRef]
44. Thoreen, C.C.; Chantranupong, L.; Keys, H.R.; Wang, T.; Gray, N.S.; Sabatini, D.M. A unifying model for mTORC1-mediated regulation of mRNA translation. *Nature* **2012**, *485*, 109–113. [CrossRef] [PubMed]
45. Ramírez-Valle, F.; Braunstein, S.; Zavadil, J.; Formenti, S.C.; Schneider, R.J. eIF4GI links nutrient sensing by mTOR to cell proliferation and inhibition of autophagy. *J. Cell Biol.* **2008**, *181*, 293–307. [CrossRef]
46. Pyronnet, S.; Pradayrol, L.; Sonenberg, N. A cell cycle-dependent internal ribosome entry site. *Mol. Cell* **2000**, *5*, 607–616. [CrossRef] [PubMed]
47. Shuda, M.; Velásquez, C.; Cheng, E.; Cordek, D.G.; Kwun, H.J.; Chang, Y.; Moore, P.S. CDK1 substitutes for mTOR kinase to activate mitotic cap-dependent protein translation. *Proc. Natl. Acad. Sci. USA* **2015**, *112*, 5875–5882. [CrossRef]
48. Henis-Korenblit, S.; Strumpf, N.L.; Goldstaub, D.; Kimchi, A. A novel form of DAP5 protein accumulates in apoptotic cells as a result of caspase cleavage and internal ribosome entry site-mediated translation. *Mol. Cell Biol.* **2000**, *20*, 496–506. [CrossRef] [PubMed]
49. Nevins, T.A.; Harder, Z.M.; Korneluk, R.G.; Holcík, M. Distinct regulation of internal ribosome entry site-mediated translation following cellular stress is mediated by apoptotic fragments of eIF4G translation initiation factor family members eIF4GI and p97/DAP5/NAT1. *J. Biol. Chem.* **2003**, *278*, 3572–3579. [CrossRef]
50. Nousch, M.; Reed, V.; Bryson-Richardson, R.J.; Currie, P.D.; Preiss, T. The eIF4G-homolog p97 can activate translation independent of caspase cleavage. *RNA* **2007**, *13*, 374–384. [CrossRef]

51. Yoffe, Y.; David, M.; Kalaora, R.; Povodovski, L.; Friedlander, G.; Feldmesser, E.; Ainbinder, E.; Saada, A.; Bialik, S.; Kimchi, A. Cap-independent translation by DAP5 controls cell fate decisions in human embryonic stem cells. *Genes. Dev.* **2016**, *30*, 1991–2004. [CrossRef]
52. Bukhari, S.I.A.; Truesdell, S.S.; Lee, S.; Kollu, S.; Classon, A.; Boukhali, M.; Jain, E.; Mortensen, R.D.; Yanagiya, A.; Sadreyev, R.I.; et al. A Specialized Mechanism of Translation Mediated by FXR1a-Associated MicroRNP in Cellular Quiescence. *Mol. Cell* **2016**, *61*, 760–773. [CrossRef] [PubMed]
53. Bukhari, S.I.; Vasudevan, S. FXR1a-associated microRNP: A driver of specialized non-canonical translation in quiescent conditions. *RNA Biol.* **2017**, *14*, 137–145. [CrossRef] [PubMed]
54. Jeong, S.J.; Park, S.; Nguyen, L.T.; Hwang, J.; Lee, E.Y.; Giong, H.K.; Lee, J.S.; Yoon, I.; Lee, J.H.; Kim, J.H.; et al. A threonyl-tRNA synthetase-mediated translation initiation machinery. *Nat. Commun.* **2019**, *10*, 1357. [CrossRef]
55. Lee, A.S.; Kranzusch, P.J.; Doudna, J.A.; Cate, J.H. eIF3d is an mRNA cap-binding protein that is required for specialized translation initiation. *Nature* **2016**, *536*, 96–99. [CrossRef] [PubMed]
56. Kim, K.M.; Cho, H.; Choi, K.; Kim, J.; Kim, B.W.; Ko, Y.G.; Jang, S.K.; Kim, Y.K. A new MIF4G domain-containing protein, CTIF, directs nuclear cap-binding protein CBP80/20-dependent translation. *Genes. Dev.* **2009**, *23*, 2033–2045. [CrossRef]
57. Arhab, Y.; Pestova, T.V.; Hellen, C.U.T. Translation of Overlapping Open Reading Frames Promoted by Type 2 IRESs in Avian Calicivirus Genomes. *Viruses* **2024**, *16*, 1413. [CrossRef]
58. Martinez-Salas, E.; Francisco-Velilla, R.; Fernandez-Chamorro, J.; Embarek, A.M. Insights into Structural and Mechanistic Features of Viral IRES Elements. *Front. Microbiol.* **2017**, *8*, 2629. [CrossRef]
59. de Breyne, S.; Soto-Rifo, R.; López-Lastra, M.; Ohlmann, T. Translation initiation is driven by different mechanisms on the HIV-1 and HIV-2 genomic RNAs. *Virus Res.* **2013**, *171*, 366–381. [CrossRef]
60. Shatsky, I.N.; Dmitriev, S.E.; Terenin, I.M.; Andreev, D.E. Cap- and IRES-independent scanning mechanism of translation initiation as an alternative to the concept of cellular IRESs. *Mol. Cells* **2010**, *30*, 285–293. [CrossRef]
61. Buck, C.B.; Shen, X.; Egan, M.A.; Pierson, T.C.; Walker, C.M.; Siliciano, R.F. The human immunodeficiency virus type 1 gag gene encodes an internal ribosome entry site. *J. Virol.* **2001**, *75*, 181–191. [CrossRef]
62. Vallejos, M.; Carvajal, F.; Pino, K.; Navarrete, C.; Ferres, M.; Huidobro-Toro, J.P.; Sargueil, B.; López-Lastra, M. Functional and structural analysis of the internal ribosome entry site present in the mRNA of natural variants of the HIV-1. *PLoS ONE* **2012**, *7*, e35031. [CrossRef]
63. Ventoso, I.; Blanco, R.; Perales, C.; Carrasco, L. HIV-1 protease cleaves eukaryotic initiation factor 4G and inhibits cap-dependent translation. *Proc. Natl. Acad. Sci. USA* **2001**, *98*, 12966–12971. [CrossRef]
64. Amorim, R.; Costa, S.M.; Cavaleiro, N.P.; da Silva, E.E.; da Costa, L.J. HIV-1 transcripts use IRES-initiation under conditions where Cap-dependent translation is restricted by poliovirus 2A protease. *PLoS ONE* **2014**, *9*, e88619. [CrossRef] [PubMed]
65. Russell, R.S.; Roldan, A.; Detorio, M.; Hu, J.; Wainberg, M.A.; Liang, C. Effects of a single amino acid substitution within the p2 region of human immunodeficiency virus type 1 on packaging of spliced viral RNA. *J. Virol.* **2003**, *77*, 12986–12995. [CrossRef] [PubMed]
66. Vallejos, M.; Deforges, J.; Plank, T.D.; Letelier, A.; Ramdohr, P.; Abraham, C.G.; Valiente-Echeverría, F.; Kieft, J.S.; Sargueil, B.; López-Lastra, M. Activity of the human immunodeficiency virus type 1 cell cycle-dependent internal ribosomal entry site is modulated by IRES trans-acting factors. *Nucleic Acids Res.* **2011**, *39*, 6186–6200. [CrossRef]
67. Plank, T.D.; Whitehurst, J.T.; Kieft, J.S. Cell type specificity and structural determinants of IRES activity from the 5′ leaders of different HIV-1 transcripts. *Nucleic Acids Res.* **2013**, *41*, 6698–6714. [CrossRef]
68. Gendron, K.; Ferbeyre, G.; Heveker, N.; Brakier-Gingras, L. The activity of the HIV-1 IRES is stimulated by oxidative stress and controlled by a negative regulatory element. *Nucleic Acids Res.* **2011**, *39*, 902–912. [CrossRef] [PubMed]
69. Song, Z.; Gremminger, T.; Singh, G.; Cheng, Y.; Li, J.; Qiu, L.; Ji, J.; Lange, M.J.; Zuo, X.; Chen, S.J.; et al. The three-way junction structure of the HIV-1 PBS-segment binds host enzyme important for viral infectivity. *Nucleic Acids Res.* **2021**, *49*, 5925–5942. [CrossRef]
70. Hartman, T.R.; Qian, S.; Bolinger, C.; Fernandez, S.; Schoenberg, D.R.; Boris-Lawrie, K. RNA helicase A is necessary for translation of selected messenger RNAs. *Nat. Struct. Mol. Biol.* **2006**, *13*, 509–516. [CrossRef]
71. Bolinger, C.; Sharma, A.; Singh, D.; Yu, L.; Boris-Lawrie, K. RNA helicase A modulates translation of HIV-1 and infectivity of progeny virions. *Nucleic Acids Res.* **2010**, *38*, 1686–1696. [CrossRef]
72. Roberts, T.M.; Boris-Lawrie, K. Primary sequence and secondary structure motifs in spleen necrosis virus RU5 confer translational utilization of unspliced human immunodeficiency virus type 1 reporter RNA. *J. Virol.* **2003**, *77*, 11973–11984. [CrossRef] [PubMed]
73. Bolinger, C.; Yilmaz, A.; Hartman, T.R.; Kovacic, M.B.; Fernandez, S.; Ye, J.; Forget, M.; Green, P.L.; Boris-Lawrie, K. RNA helicase A interacts with divergent lymphotropic retroviruses and promotes translation of human T-cell leukemia virus type 1. *Nucleic Acids Res.* **2007**, *35*, 2629–2642. [CrossRef] [PubMed]
74. Hull, S.; Boris-Lawrie, K. RU5 of Mason-Pfizer monkey virus 5′ long terminal repeat enhances cytoplasmic expression of human immunodeficiency virus type 1 gag-pol and nonviral reporter RNA. *J. Virol.* **2002**, *76*, 10211–10218. [CrossRef]

75. Butsch, M.; Hull, S.; Wang, Y.; Roberts, T.M.; Boris-Lawrie, K. The 5′ RNA terminus of spleen necrosis virus contains a novel posttranscriptional control element that facilitates human immunodeficiency virus Rev/RRE-independent Gag production. *J. Virol.* **1999**, *73*, 4847–4855. [CrossRef]
76. Dangel, A.W.; Hull, S.; Roberts, T.M.; Boris-Lawrie, K. Nuclear interactions are necessary for translational enhancement by spleen necrosis virus RU5. *J. Virol.* **2002**, *76*, 3292–3300. [CrossRef] [PubMed]
77. Short, J.D.; Pfarr, C.M. Translational regulation of the JunD messenger RNA. *J. Biol. Chem.* **2002**, *277*, 32697–32705. [CrossRef]
78. Hernandez, J.M.; Floyd, D.H.; Weilbaecher, K.N.; Green, P.L.; Boris-Lawrie, K. Multiple facets of junD gene expression are atypical among AP-1 family members. *Oncogene* **2008**, *27*, 4757–4767. [CrossRef]
79. Zucko, D.; Boris-Lawrie, K. Blocking tri-methylguanosine synthase 1 (TGS1) stops anchorage-independent growth of canine sarcomas. *Cancer Gene Ther.* **2023**, *30*, 1274–1284. [CrossRef]
80. Touat-Hamici, Z.; Legrain, Y.; Bulteau, A.L.; Chavatte, L. Selective up-regulation of human selenoproteins in response to oxidative stress. *J. Biol. Chem.* **2014**, *289*, 14750–14761. [CrossRef]
81. Vindry, C.; Ohlmann, T.; Chavatte, L. Translation regulation of mammalian selenoproteins. *Biochim. Biophys. Acta Gen. Subj.* **2018**, *1862*, 2480–2492. [CrossRef]
82. Copeland, P.R.; Ohlmann, T.; Chavatte, L. A novel RNA binding protein, SBP2, is required for the translation of mammalian selenoprotein mRNAs. *EMBO J.* **2000**, *19*, 306–314. [CrossRef] [PubMed]
83. Singh, G.; Fritz, S.E.; Seufzer, B.; Boris-Lawrie, K. The mRNA encoding the JUND tumor suppressor detains nuclear RNA-binding proteins to assemble polysomes that are unaffected by mTOR. *J. Biol. Chem.* **2020**, *295*, 7763–7773. [CrossRef]
84. Enünlü, I.; Pápai, G.; Cserpán, I.; Udvardy, A.; Jeang, K.T.; Boros, I. Different isoforms of PRIP-interacting protein with methyltransferase domain/trimethylguanosine synthase localizes to the cytoplasm and nucleus. *Biochem. Biophys. Res. Commun.* **2003**, *309*, 44–51. [CrossRef]
85. Schreiber, K.H.; Ortiz, D.; Academia, E.C.; Anies, A.C.; Liao, C.Y.; Kennedy, B.K. Rapamycin-mediated mTORC2 inhibition is determined by the relative expression of FK506-binding proteins. *Aging Cell* **2015**, *14*, 265–273. [CrossRef] [PubMed]
86. Worch, R.; Niedzwiecka, A.; Stepinski, J.; Mazza, C.; Jankowska-Anyszka, M.; Darzynkiewicz, E.; Cusack, S.; Stolarski, R. Specificity of recognition of mRNA 5′ cap by human nuclear cap-binding complex. *RNA* **2005**, *11*, 1355–1363. [CrossRef]
87. Gebhardt, A.; Habjan, M.; Benda, C.; Meiler, A.; Haas, D.A.; Hein, M.Y.; Mann, A.; Mann, M.; Habermann, B.; Pichlmair, A. mRNA export through an additional cap-binding complex consisting of NCBP1 and NCBP3. *Nat. Commun.* **2015**, *6*, 8192. [CrossRef] [PubMed]
88. Dubiez, E.; Pellegrini, E.; Finderup Brask, M.; Garland, W.; Foucher, A.E.; Huard, K.; Heick Jensen, T.; Cusack, S.; Kadlec, J. Structural basis for competitive binding of productive and degradative co-transcriptional effectors to the nuclear cap-binding complex. *Cell Rep.* **2024**, *43*, 113639. [CrossRef]
89. Schulze, W.M.; Stein, F.; Rettel, M.; Nanao, M.; Cusack, S. Structural analysis of human ARS2 as a platform for co-transcriptional RNA sorting. *Nat. Commun.* **2018**, *9*, 1701. [CrossRef]
90. Karn, J.; Stoltzfus, C.M. Transcriptional and posttranscriptional regulation of HIV-1 gene expression. *Cold Spring Harb. Perspect. Med.* **2012**, *2*, a006916. [CrossRef]
91. Pollard, V.W.; Malim, M.H. The HIV-1 Rev protein. *Annu. Rev. Microbiol.* **1998**, *52*, 491–532. [CrossRef]
92. Sharma, A.; Yilmaz, A.; Marsh, K.; Cochrane, A.; Boris-Lawrie, K. Thriving under stress: Selective translation of HIV-1 structural protein mRNA during Vpr-mediated impairment of eIF4E translation activity. *PLoS Pathog.* **2012**, *8*, e1002612. [CrossRef] [PubMed]
93. Hegedus, A.; Williamson, M.K.; Huthoff, H. HIV-1 pathogenicity and virion production are dependent on the metabolic phenotype of activated CD4+ T cells. *Retrovirology* **2014**, *11*, 98. [CrossRef]
94. Arjona, S.P.; Allen, C.N.S.; Santerre, M.; Gross, S.; Soboloff, J.; Booze, R.; Sawaya, B.E. Disruption of Mitochondrial-associated ER membranes by HIV-1 tat protein contributes to premature brain aging. *CNS Neurosci. Ther.* **2023**, *29*, 365–377. [CrossRef] [PubMed]
95. Huang, C.Y.; Chiang, S.F.; Lin, T.Y.; Chiou, S.H.; Chow, K.C. HIV-1 Vpr triggers mitochondrial destruction by impairing Mfn2-mediated ER-mitochondria interaction. *PLoS ONE* **2012**, *7*, e33657. [CrossRef]
96. Crater, J.M.; Nixon, D.F.; O'Brien, R.L.F. HIV-1 replication and latency are balanced by mTOR-driven cell metabolism. *Front. Cell Infect. Microbiol.* **2022**, *12*, 1068436. [CrossRef] [PubMed]
97. Hollenbaugh, J.A.; Munger, J.; Kim, B. Metabolite profiles of human immunodeficiency virus infected CD4+ T cells and macrophages using LC-MS/MS analysis. *Virology* **2011**, *415*, 153–159. [CrossRef]
98. Lahouassa, H.; Daddacha, W.; Hofmann, H.; Ayinde, D.; Logue, E.C.; Dragin, L.; Bloch, N.; Maudet, C.; Bertrand, M.; Gramberg, T. SAMHD1 restricts the replication of human immunodeficiency virus type 1 by depleting the intracellular pool of deoxynucleoside triphosphates. *Nat. Immunol.* **2012**, *13*, 223–228. [CrossRef]
99. Rahman, A.N.; Liu, J.; Mujib, S.; Kidane, S.; Ali, A.; Szep, S.; Han, C.; Bonner, P.; Parsons, M.; Benko, E.; et al. Elevated glycolysis imparts functional ability to $CD8^+$ T cells in HIV infection. *Life Sci. Alliance* **2021**, *4*, e202101081. [CrossRef]

100. Palmer, C.S.; Cherry, C.L.; Sada-Ovalle, I.; Singh, A.; Crowe, S.M. Glucose Metabolism in T Cells and Monocytes: New Perspectives in HIV Pathogenesis. *EBioMedicine* **2016**, *6*, 31–41. [CrossRef]
101. Loisel-Meyer, S.; Swainson, L.; Craveiro, M.; Oburoglu, L.; Mongellaz, C.; Costa, C.; Martinez, M.; Cosset, F.L.; Battini, J.L.; Herzenberg, L.A.; et al. Glut1-mediated glucose transport regulates HIV infection. *Proc. Natl. Acad. Sci. USA* **2012**, *109*, 2549–2554. [CrossRef]
102. Wester, C.W.; Okezie, O.A.; Thomas, A.M.; Bussmann, H.; Moyo, S.; Muzenda, T.; Makhema, J.; van Widenfelt, E.; Musonda, R.; Novitsky, V.; et al. Higher-than-expected rates of lactic acidosis among highly active antiretroviral therapy-treated women in Botswana: Preliminary results from a large randomized clinical trial. *J. Acquir. Immune Defic. Syndr.* **2007**, *46*, 318–322. [CrossRef]
103. Mikaeloff, F.; Gelpi, M.; Escos, A.; Knudsen, A.D.; Høgh, J.; Benfield, T.; de Magalhães, J.P.; Nielsen, S.D.; Neogi, U. Transcriptomics age acceleration in prolonged treated HIV infection. *Aging Cell* **2023**, *22*, e13951. [CrossRef] [PubMed]
104. Tripathy, M.K.; Mitra, D. Differential modulation of mitochondrial OXPHOS system during HIV-1 induced T-cell apoptosis: Up regulation of Complex-IV subunit COX-II and its possible implications. *Apoptosis* **2010**, *15*, 28–40. [CrossRef]
105. Kallianpur, K.J.; Walker, M.; Gerschenson, M.; Shikuma, C.M.; Gangcuangco, L.M.A.; Kohorn, L.; Libutti, D.E.; Nir, T.M.; Jahanshad, N.; Thompson, P.M.; et al. Systemic Mitochondrial Oxidative Phosphorylation Protein Levels Correlate with Neuroimaging Measures in Chronically HIV-Infected Individuals. *AIDS Res. Hum. Retroviruses* **2020**, *36*, 83–91. [CrossRef]
106. Valle-Casuso, J.C.; Angin, M.; Volant, S.; Passaes, C.; Monceaux, V.; Mikhailova, A.; Bourdic, K.; Avettand-Fenoel, V.; Boufassa, F.; Sitbon, M.; et al. Cellular Metabolism Is a Major Determinant of HIV-1 Reservoir Seeding in CD4. *Cell Metab.* **2019**, *29*, 611–626.e5. [CrossRef] [PubMed]
107. Besnard, E.; Hakre, S.; Kampmann, M.; Lim, H.W.; Hosmane, N.N.; Martin, A.; Bassik, M.C.; Verschueren, E.; Battivelli, E.; Chan, J.; et al. The mTOR Complex Controls HIV Latency. *Cell Host Microbe* **2016**, *20*, 785–797. [CrossRef] [PubMed]
108. Dias, J.; Cattin, A.; Bendoumou, M.; Dutilleul, A.; Lodge, R.; Goulet, J.P.; Fert, A.; Raymond Marchand, L.; Wiche Salinas, T.R.; Ngassaki Yoka, C.-D.; et al. Retinoic acid enhances HIV-1 reverse transcription and transcription in macrophages via mTOR-modulated mechanisms. *Cell Rep.* **2024**, *43*, 114414. [CrossRef]
109. Taylor, H.E.; Calantone, N.; Lichon, D.; Hudson, H.; Clerc, I.; Campbell, E.M.; D'Aquila, R.T. mTOR Overcomes Multiple Metabolic Restrictions to Enable HIV-1 Reverse Transcription and Intracellular Transport. *Cell Rep.* **2020**, *31*, 107810. [CrossRef]
110. Henrich, T.J.; Bosch, R.J.; Godfrey, C.; Mar, H.; Nair, A.; Keefer, M.; Fichtenbaum, C.; Moisi, D.; Clagett, B.; Buck, A.M.; et al. Sirolimus reduces T cell cycling, immune checkpoint marker expression, and HIV-1 DNA in people with HIV. *Cell Rep. Med.* **2024**, *5*, 101745. [CrossRef]
111. Campbell, G.R.; Bruckman, R.S.; Herns, S.D.; Joshi, S.; Durden, D.L.; Spector, S.A. Induction of autophagy by PI3K/MTOR and PI3K/MTOR/BRD4 inhibitors suppresses HIV-1 replication. *J. Biol. Chem.* **2018**, *293*, 5808–5820. [CrossRef]
112. van Montfort, T.; van der Sluis, R.; Darcis, G.; Beaty, D.; Groen, K.; Pasternak, A.O.; Pollakis, G.; Vink, M.; Westerhout, E.M.; Hamdi, M.; et al. Dendritic cells potently purge latent HIV-1 beyond TCR-stimulation, activating the PI3K-Akt-mTOR pathway. *EBioMedicine* **2019**, *42*, 97–108. [CrossRef] [PubMed]
113. Jin, S.; Liao, Q.; Chen, J.; Zhang, L.; He, Q.; Zhu, H.; Zhang, X.; Xu, J. TSC1 and DEPDC5 regulate HIV-1 latency through the mTOR signaling pathway. *Emerg. Microbes Infect.* **2018**, *7*, 138. [CrossRef] [PubMed]
114. Martin, A.R.; Pollack, R.A.; Capoferri, A.; Ambinder, R.F.; Durand, C.M.; Siliciano, R.F. Rapamycin-mediated mTOR inhibition uncouples HIV-1 latency reversal from cytokine-associated toxicity. *J. Clin. Investig.* **2017**, *127*, 651–656. [CrossRef] [PubMed]
115. Panwar, V.; Singh, A.; Bhatt, M.; Tonk, R.K.; Azizov, S.; Raza, A.S.; Sengupta, S.; Kumar, D.; Garg, M. Multifaceted role of mTOR (mammalian target of rapamycin) signaling pathway in human health and disease. *Signal Transduct. Target. Ther.* **2023**, *8*, 375. [CrossRef]
116. Toro-Ascuy, D.; Rojas-Araya, B.; García-de-Gracia, F.; Rojas-Fuentes, C.; Pereira-Montecinos, C.; Gaete-Argel, A.; Valiente-Echeverría, F.; Ohlmann, T.; Soto-Rifo, R. A Rev-CBP80-eIF4AI complex drives Gag synthesis from the HIV-1 unspliced mRNA. *Nucleic Acids Res.* **2018**, *46*, 11539–11552. [CrossRef]
117. García-de-Gracia, F.; Gaete-Argel, A.; Riquelme-Barrios, S.; Pereira-Montecinos, C.; Rojas-Araya, B.; Aguilera, P.; Oyarzún-Arrau, A.; Rojas-Fuentes, C.; Acevedo, M.L.; Chnaiderman, J.; et al. CBP80/20-dependent translation initiation factor (CTIF) inhibits HIV-1 Gag synthesis by targeting the function of the viral protein Rev. *RNA Biol.* **2021**, *18*, 745–758. [CrossRef]
118. Choe, J.; Oh, N.; Park, S.; Lee, Y.K.; Song, O.K.; Locker, N.; Chi, S.G.; Kim, Y.K. Translation initiation on mRNAs bound by nuclear cap-binding protein complex CBP80/20 requires interaction between CBP80/20-dependent translation initiation factor and eukaryotic translation initiation factor 3g. *J. Biol. Chem.* **2012**, *287*, 18500–18509. [CrossRef]
119. Bertrand, L.; Nelde, A.; Ramirez, B.C.; Hatin, I.; Arbes, H.; François, P.; Demais, S.; Labaronne, E.; Decimo, D.; Guiguettaz, L.; et al. Unveiling conserved HIV-1 open reading frames encoding T cell antigens using ribosome profiling. *Nat. Commun.* **2025**, *16*, 1707. [CrossRef]
120. Labaronne, E.; Décimo, D.; Bertrand, L.; Guiguettaz, L.; Thibault, J.M.; Sohier, S.; Cluet, D.; Vivet-Boudou, V.; Chaves Valadão, A.L.; Dahoui, A.; et al. Non-AUG HIV-1 uORF translation elicits specific T cell immune response and regulates viral transcript expression. *Nat. Commun.* **2025**, *16*, 1706. [CrossRef]

121. Kibe, A.; Buck, S.; Gribling-Burrer, A.S.; Gilmer, O.; Bohn, P.; Koch, T.; Mireisz, C.N.M.; Schlosser, A.; Erhard, F.; Smyth, R.P.; et al. The translational landscape of HIV-1 infected cells reveals key gene regulatory principles. *Nat. Struct. Mol. Biol.* **2025**. [CrossRef]
122. Laurichesse, H.; Tauveron, I.; Gourdon, F.; Cormerais, L.; Champredon, C.; Charrier, S.; Rochon, C.; Lamain, S.; Bayle, G.; Laveran, H.; et al. Threonine and methionine are limiting amino acids for protein synthesis in patients with AIDS. *J. Nutr.* **1998**, *128*, 1342–1348. [CrossRef] [PubMed]

MDPI AG
Grosspeteranlage 5
4052 Basel
Switzerland
Tel.. +41 61 683 77 34

Viruses Editorial Office
E-mail: viruses@mdpi.com
www.mdpi.com/journal/viruses

www.ingramcontent.com/pod-product-compliance
Lightning Source LLC
Chambersburg PA
CBHW061452210926
51850CB00008B/1693
* 9 7 8 3 7 2 5 8 4 9 6 9 7 *